NONLINEAR VALUATION AND NON-GAUSSIAN RISKS IN FINANCE

What happens to risk as the economic horizon goes to zero and risk is seen as an exposure to a change in state that may occur instantaneously at any time? All activities that have been undertaken statically at a fixed finite horizon can now be reconsidered dynamically at a zero time horizon, with arrival rates at the core of the modeling.

This book, aimed at practitioners and researchers in financial risk, delivers the theoretical framework and various applications of the newly established dynamic conic finance theory. The result is a nonlinear non-Gaussian valuation framework for risk management in finance. Risk-free assets disappear and low-risk portfolios must pay for their risk reduction with negative expected returns. Hedges may be constructed to enhance value by exploiting risk interactions. Dynamic trading mechanisms are synthesized by machine learning algorithms. Optimal exposures are designed for option positioning simultaneously across all strikes and maturities.

DILIP B. MADAN is Professor Emeritus at the Robert H. Smith School of Business at the University of Maryland. He has been a consultant to Morgan Stanley since 1996 and to Norges Bank Investment Management since 2012. He is a founding member and past president of the Bachelier Finance Society. He was a Humboldt Awardee in 2006, was named Quant of the Year in 2008, and was inducted into the University of Maryland's Circle of Discovery in 2014. He is the cocreator of the variance gamma model (1990, 1998) and of conic finance. He coauthored, with Wim Schoutens, *Applied Conic Finance* (Cambridge University Press, 2016).

WIM SCHOUTENS is a professor at KU Leuven, Belgium. He has extensive practical experience of model implementation and is well known for his consulting work to the banking industry and other institutions. He served as an expert witness for the General Court of the European Union, Luxembourg, and has worked as an expert for the International Monetary Fund and for the European Commission. In 2012, he was awarded the John von Neumann Visiting Professorship of the Technical University of Munich. He has authored several books on financial mathematics and is a regular lecturer to the financial industry. Finally, he is a member of the Belgium CPI commission.

NONLINEAR VALUATION AND NON-GAUSSIAN RISKS IN FINANCE

DILIP B. MADAN
Robert H. Smith School of Business, University of Maryland

WIM SCHOUTENS
KU Leuven

CAMBRIDGE
UNIVERSITY PRESS

CAMBRIDGE
UNIVERSITY PRESS

University Printing House, Cambridge CB2 8BS, United Kingdom

One Liberty Plaza, 20th Floor, New York, NY 10006, USA

477 Williamstown Road, Port Melbourne, VIC 3207, Australia

314–321, 3rd Floor, Plot 3, Splendor Forum, Jasola District Centre, New Delhi – 110025, India

103 Penang Road, #05–06/07, Visioncrest Commercial, Singapore 238467

Cambridge University Press is part of the University of Cambridge.

It furthers the University's mission by disseminating knowledge in the pursuit of education, learning, and research at the highest international levels of excellence.

www.cambridge.org
Information on this title: www.cambridge.org/9781316518090
DOI: 10.1017/9781108993876

First published 2022

A catalogue record for this publication is available from the British Library.

ISBN 978-1-316-51809-0 Hardback

To our perennial supporters:
to Vimla, Meena, Maneka, Sherif, Shivali, Sabrina, and Maia
—Dilip
to Ethel, Jente, and Maitzanne
—Wim

Contents

	Preface	*page* xi
	Acknowledgments	xiii
1	**Introduction**	1
2	**Univariate Risk Representation Using Arrival Rates**	5
	2.1 Pure Jump Finite Variation Probability Models	7
	2.2 Probability Densities and Arrival Rates	11
	2.3 The Complex Exponential Variation	16
	2.4 Evaluating Event Arrival Rates	20
	2.5 Variation Outcomes	22
	2.6 Drift, Volatility, Risk Dimensions, and Their Compensation	26
3	**Estimation of Univariate Arrival Rates from Time Series Data**	30
	3.1 Complex Exponential Variations and Data	30
	3.2 Digital Moment Estimation	31
	3.3 Variance Gamma, Bilateral Gamma, and Bilateral Double Gamma Estimation Results	33
	3.4 Assessing Parameter Contributions	35
4	**Estimation of Univariate Arrival Rates from Option Surface Data**	39
	4.1 Depreferencing Option Prices	39
	4.2 Estimation Results	43
5	**Multivariate Arrival Rates Associated with Prespecified Univariate Arrival Rates**	48
	5.1 Multivariate Model for Bilateral Gamma Marginals	49
	5.2 The Role of Dependency Parameters in the Multivariate Bilateral Gamma Model	52
	5.3 Multivariate Bilateral Gamma Lévy Copulas	53
	5.4 Multivariate Model for Bilateral Double Gamma Marginals	55
	5.5 Simulated Count of Multivariate Event Arrival Rates	56
6	**The Measure-Distorted Valuation As a Financial Objective**	59
	6.1 Linear Valuation Issues	61

	6.2	Modeling Risk Acceptability	63
	6.3	Nonlinear Conservative Valuation	65
	6.4	Risk Reward Decompositions of Value	66
	6.5	Remarks on Modigliani–Miller Considerations	67
	6.6	Probability Distortions	67
	6.7	Measure Distortions Proper	73
	6.8	Dual Formulation of Measure Distortions	78
	6.9	Explicit Representation of Dual Distortions Φ, $\widetilde{\Phi}$.	82
	6.10	Generic Considerations in the Maximization of Market Valuations	84
7	**Representing Market Realities**		85
	7.1	Risk Charges and the Measure Distortion Parameters	86
	7.2	Measure Distortions and Option Prices	87
	7.3	Measure-Distorted Value-Maximizing Hedges for a Short Gamma Target	90
	7.4	Measure Distortions Implied by Hedges for a Long Gamma Target	95
8	**Measure-Distorted Value-Maximizing Hedges in Practice**		98
	8.1	Hedging Overview	99
	8.2	The Hedge-Implementing Enterprise	100
	8.3	Summarizing Option Surfaces Using Gaussian Process Regression	101
	8.4	Selecting the Hedging Arrival Rates	104
	8.5	Approximating Variation Exposures	105
	8.6	Measure Distortion Parameters	106
	8.7	Backtest Hedging Results for Multiple Strangles on SPX	108
9	**Conic Hedging Contributions and Comparisons**		110
	9.1	Univariate Exposure Hedging Study	112
	9.2	Distorted Least Squares	113
	9.3	Example Illustrating Distorted Least-Squares Hedges	116
	9.4	Incorporating Weightings	118
	9.5	Measure-Distorted Value Maximization	119
	9.6	Greek Hedging	120
	9.7	Theta Issues in Exposure Design	120
	9.8	Incorporating Spreads	122
	9.9	No Spread Access and Theta Considerations	124
10	**Designing Optimal Univariate Exposures**		126
	10.1	Exposure Design Objectives	127
	10.2	Exposure Design Constraints	128
	10.3	Exposure Design Problem	128
	10.4	Lagrangean Analysis of the Design Problem	129
	10.5	Discretization and Solution	130
	10.6	Details Related to Lévy Measure Singularities at Zero	131
	10.7	Sample Optimal Exposure Designs	131
	10.8	Further Details about Some Particular Cases	132

11 Multivariate Static Hedge Designs Using Measure-Distorted Valuations 135
11.1 A Two-Dimensional Example 136
11.2 A 10-Dimensional Example 142

12 Static Portfolio Allocation Theory for Measure-Distorted Valuations 150
12.1 Measure Integrals by Simulation 152
12.2 Dual Formulation of Portfolio Problem 152
12.3 Approximation by Probability Distortion 154
12.4 Implementation of Portfolio Allocation Problems 154
12.5 Mean Risk Charge Efficient Frontiers 156
12.6 Sensitivity of Required Returns to Choice of Points on Frontiers 163
12.7 Conic Alpha Construction Based on Arrival Rates 164
12.8 Fixed Income Asset Efficient Exposure Frontiers 165

13 Dynamic Valuation via Nonlinear Martingales and Associated Backward Stochastic Partial Integro-Differential Equations 171
13.1 Backward Stochastic Partial Integro-Differential Equations and Valuations 173
13.2 Nonlinear Valuations and BSPIDE 175
13.3 Spatially Inhomogeneous Bilateral Gamma 176
13.4 Dynamic Implementation of Hedging Problems 179

14 Dynamic Portfolio Theory 184
14.1 The Dynamic Law of Motion 185
14.2 Relativity Dynamics 187
14.3 The Full Sample 188
14.4 Portfolio Construction 188
14.5 Stationary Exposure Valuation 191
14.6 Stationary Value and Policy Results 192
14.7 Building Neural Net Policy Functions and Simulating Trades 192

15 Enterprise Valuation Using Infinite and Finite Horizon Valuation of Terminal Liquidation 195
15.1 Bilateral Gamma Enterprise Returns 197
15.2 Prudential Capital for Bilateral Gamma Returns 201
15.3 Regulatory Risk Capital for Enterprises with Bilateral Gamma Returns 209
15.4 Calibration of Measure-Distortion Parameters 210
15.5 Results for Equity Enterprises 216
15.6 Results for Treasury Bond Investments 216
15.7 Results for Hedge Fund Enterprises 217
15.8 Short Position Capital Requirements 220
15.9 Equity versus Leveraged Equity 220

16 Economic Acceptability 223
16.1 Interplay between Equity Markets and Regulators 224
16.2 Candidate Physical Laws of Motion 225
16.3 Adapted Measure Distortions 226
16.4 Equity and Regulatory Capital Constructions 228
16.5 Financial Sector Capital during and after the Financial Crisis 230

17 Trading Markovian Models 235
17.1 Return Dependence on States 237
17.2 Markovian State Dynamics 239
17.3 Formulation and Solution of Market Value Maximization 240
17.4 Results on Policy Functions for 10 Stocks 242
17.5 Results for Sector ETFs and SPY 243

18 Market-Implied Measure-Distortion Parameters 245
18.1 Designing the Time Series Estimation of Measure-Distortion Parameters 245
18.2 Estimation Results 247
18.3 Distribution of Measure-Distorted Valuations for Equity Underliers 247
18.4 Structure of Measure-Distorted Valuation-Level Curves 249
18.5 Valuation Frontiers 250
18.6 Acceptability Indices 251
18.7 Acceptability-Level Curves 252
18.8 Equilibrium Return Distributions 253
18.9 Empirical Construction of Return Distribution Equilibria and Their Properties 254

References 257
Index 265

Preface

Financial valuation as an expectation with respect to a probability both ignores the naturally present uncertainty in the probability and makes value a linear function of the risk. Nonlinear valuation instead is based on a diverse set of probabilities and two nonlinear valuations naturally arise. The lower valuation takes the infimum of expectations over the set of probabilities while the upper valuation takes the supremum. The result is a concave conservative valuation for assets that may be maximized and a convex conservative valuation for liabilities that may be minimized. The classical linear valuation, by virtue of its linearity, is problematic as an optimization objective. The resulting new financial objectives can be applied to all aspects of financial decision-making. Madan and Schoutens (2016) investigated and reported on applying these new nonlinear objectives for financial analysis to many aspects of finance.

The classical view of risk as multiple future outcomes at some horizon described by their probabilities is here revised. Risk is here viewed as an exposure to instantaneous changes in states with the resulting time horizon being sent to zero. A form of continuity can be modeled by allowing for infinite small or minuscule changes in state. The changes are described by the time frequency of their arrival rates with the aggregate arrival across all possible changes potentially infinite. As a result there is not a probability for the changes in states, but just a measure given by the arrival rate. Normalization to a probability is not possible; the aggregate arrival rate is infinite.

Risk is described by the arrival rate measure. Expectations are replaced by variations that are integrals of state changes with respect to the arrival rate measure. Nonlinear valuations allow for sets of alternate measures incorporating natural uncertainties in the measure. The lower and upper valuations are then infima and suprema or variations over the sets of measures. Probability distortions encountered in Madan and Schoutens (2016) are replaced by measure distortions introduced in Madan et al. (2017a).

The objectives of financial analysis are then reformulated with risk described by multidimensional arrival rate measures and the nonlinear valuations of risk are based on measure-distorted variations. The first five chapters present this implementable reformulation of risk and its value. The rest of the book presents applications that include multidimensional hedging, portfolio allocation, and derivative positioning. All activities may be undertaken statically at the level of the instant, which is the limit of letting the horizon go to zero, or dynamically by solving nonlinear partial integro-differential equations backward from a finite or infinite terminal date. In the latter case the arrival rates may be adapted to evolving filtrations of information. The difficulty with dynamic modeling is the general inability to commit future actions to be consistent with plans generated earlier. Hence only the initial decision of a dynamic plan may see implementation. Be that as it may, both formulations are considered. The dynamic formulation here works with a probability at a terminal date or horizon. A generalization to an infinite measure at the terminal date that cannot be normalized to a probability is an interesting subject for subsequent research and beyond the scope of research accomplished to date.

All implementation requires the estimation of multidimensional arrival rates from data, and the first chapters present the details for undertaking this work. Upon completion one has the ability to estimate the count of loss and gain arrival rates of positions in financial instruments. The instruments considered are assets and liabilities undertaken relative to a numeraire accounting for time value considerations. All positions are always risky.

Acknowledgments

This book is the work of the authors, but without the support of our colleagues, PhD students, coauthors, people we met at seminars and conferences, and referees of the papers we published, it would not have been possible.

We would like to thank explicitly Peter Carr, José Manuel Corcuera, Freddy Delbaen, Jan De Spiegeleer, Ernst Eberlein, Robert Elliott, Pankaj Khandelwal, Ajay Khanna, Peter Leoni, Shige Peng, Martijn Pistorius, Sofie Reyners, Yazid Sharaiha, Mitja Stadje, Juan Jorge Vicente Alvarez, King Wang, and Marc Yor.

Our gratitude and respect also go to the team of Cambridge University Press for the enthusiasm and professionalism with which they have embraced this book project.

1
Introduction

Risk is often defined by the probabilities of possible outcomes, be they the tossing of coins, the rolling of dice, or the prices of assets at some future date. Uncertainty exists as the possible outcomes are many and the actual outcome is not known. The risk may be valued statistically at its expected value or in a market at the current price to be paid or received for acquiring or delivering a unit of currency on the resolution of the risk. The market value is also understood to be a discounted expected value under altered probabilities that reflect prices of events as opposed to their real probabilities. By construction the value of a risk is hence a linear function on the space of risks with the value of a combination being equal to an equivalent combination of values. As a consequence value maximization is not possible as nonconstant linear functions have no maximal values. Optimization becomes possible only after introducing constraints that limit the set of possibilities.

This view of risk, as probabilities for multiple outcomes, and its valuation, as an expectation, is reformulated here, with important consequences for the practice of financial decision-making and analysis. Risk is viewed in a substantially different way with the principles employed in valuing risk also suitably revised.

Risk is seen as an exposure to a change in state that may occur instantaneously at any time. The sets of alternate current states, like possible outcomes, are many. But the probabilities of outcomes are replaced by the arrival rates of possible state changes. However, unlike probabilities that sum to or integrate to unity, aggregate arrival rates may be infinite. Indeed, in general, there could be many small and insignificant changes in state. The total measure across all possibilities could be infinite and the normalization of counts to unity is lost. Risk is then defined by an infinite measure on possible changes of state with all substantive state changes having a finite measure, but insignificant or small changes may occur infinitely often. The aggregate effect of the infinite number of small changes is, however, finite.

Apart from the reformulation of risk, we follow Peng (2019) in recognizing that the valuation of risk should not be based on a single probability or a single measure. Such an approach ignores our natural uncertainty about the probability or measure. In valuation, we allow for alternate probabilities or measures being relevant. The resulting relevant valuations then turn out to be infima and suprema of expectations over the set of all alternates, be they probabilities or measures. There are then naturally two valuations, one lower associated with the infima, the other upper associated with the suprema. The lower is concave and may be maximized while the upper is convex and may be minimized. Value optimization is thereby restored to some extent.

In this regard it may be noted that the maximization of market value has been a rallying cry for decision-making in capitalist economies. Milton Friedman argues for it in a classic article in the *New York Times*, September 13, 1970, entitled "The Social Responsibility of Business Is to Increase Its Profits." Yet another, more detailed advocacy is presented in Jensen (2002). However, it has also long been recognized in economics and finance that value maximization is problematic given its linearity. The linearity was observed, for example, as a consequence of the law of one price, that is, using a single probability, and the absence of arbitrage in Ross (1978). The value of a portfolio is then the sum of the component values and independent of how they are combined. Similarly a position hedged at zero cost has the same value no matter the hedge. Under linearity, value maximization just does not work, and many authors resort to personalized utility-based valuations. These are controversial as markets may disregard the particular utility being employed with serious consequences for the enterprise. A way out is to build a market-based nonlinear valuation function. This is precisely what the Peng (2019) theory of nonlinear expectations or martingales delivers. The purpose of this book, partially, is to bring these ideas to finance in a practical and implementable way.

An earlier work (Madan and Schoutens, 2016) dealt with nonlinear valuations associated with alternate probabilities of outcomes realized at some time horizon. The purpose here is to develop the results, with applications, in the context of infinite arrival rate measures for instantaneous exposures at a zero time horizon.

An important financial implication of a zero time horizon, encountered in Chapter 12, is the absence of any sense of a positive risk-free rate at a zero time horizon. Recognizing furthermore the perpetual presence of risk exposures, however assets are held, leads to an empirically observed negative return on exposures with a low risk level. The reference rate in an instantaneous exposure economy can become a negative rate. As a consequence required returns on many equity assets turn out to be negative. The assertion of a positive required return on equity turns out to be just an accidental or unintended consequence of assuming a static model with a horizon coupled with a positive risk-free rate over this horizon. Participants in actual economies exposed to substantial instantaneous or zero horizon risks have no such artificial horizon effects. Negative required returns for some equity assets are completely consistent with zero horizon economic risks.

Many financial situations are managed by a partial hedging of risks. Hedging targets the conditional expectation of the risk to hedge given the outcomes on the hedging assets. But for a liability, one would like bias positions to be a little over the liability while for an asset delivering on the hedge, one may want to be a little under the asset value. In the setting of hedging by least-squares minimization, the structure of upper and lower valuations leads to the concept of distorted least squares that accomplish parametrically these value-sensitive upper and under hedges. These hedging improvements are encountered in Chapters 9 and 11 in the univariate and multivariate contexts. The direct maximization of lower valuations and the minimization of upper valuations for hedging purposes are also studied in Chapters 7 and 8.

Aside from hedging considerations, risk exposures are occasionally designed in limited ways in portfolio theory and more aggressively in options markets, providing access to a richer class of exposures. Chapters 11 and 12 take up the question of portfolio theory. For option exposures, the question has been addressed by applying utility theory in a static, single-period context using nonlinear valuations in Madan and Schoutens (2016). Chapter 10 takes up the problem of constructing optimal arrival rates made possible by positions in a portfolio of options.

Instantaneous exposure design is an important part of financial management, but on occasion one has to take account of longer-term objectives in a dynamic setting. Nonlinear valuation in a dynamic setting takes us to the developing literature on backward stochastic partial integro-differential equations, addressed here in a Markovian setting. The dynamic hedging problem for nonlinear value optimization is addressed in Chapter 13 with the portfolio problem taken up in Chapter 14. For both problems, one needs to usefully represent value and asset allocations as functions of state variables in possibly high dimensions. Applications are made of numerous machine learning methods to synthesize the functions involved. For the value maximization, we use Gaussian process regression (GPR) while for portfolio allocations feedforward neural nets are used. A wealth accumulation comparison is made between a myopic investor and a long-term investor in a Markovian context. It is observed that the long-term investor delivers the superior performance.

A special case of critical importance in finance is the valuation of equity itself. The lower valuation of equity is the capital the equity markets safely provide economic enterprises, and it must be sufficient to cover the risks being taken. The latter are monitored and formulated by financial regulatory bodies in modern capitalist economies and the question arises as to whether equity markets are providing sufficient capital. The lower valuation is a prudent valuation and as such fits into the prudential framework. If so, one has an economically acceptable enterprise. These issues are formulated and empirically analyzed in Chapters 15 and 16.

An important trading strategy pursued in the financial markets is termed "pairs trading." Often this involves trading one stock against another, related stock. Higher-dimensional forms of such trading strategies are formulated and solved for a nonlinear dynamic valuation objective in Chapter 17. The methods are empirically illustrated in up to 10 dimensions.

Nonlinear valuation objectives are formulated from the perspective of determining a lower market value of a risk and not a personalized value. It is not about what you think something is worth but what you would get for it in the market. Parameters in formulating nonlinear valuation objectives must therefore be derived from market data. This issue is taken up in Chapter 18. It is followed up in Elliot et al. (2020).

Chapters 7 onward deliver the many financial applications of nonlinear valuation applied to finance. The construction of practical and implementable nonlinear valuation objective functions is the subject matter of Chapter 6. A prerequisite for this is the description of risk to which nonlinear valuation is applied. The application requires the full description of the risk with all parameters estimated. Chapters 2 through 5 accomplish these tasks in both a univariate and multivariate context from both the backward-looking time series data and the forward-looking data of option markets.

2

Univariate Risk Representation Using Arrival Rates

The starting point of our analysis is the representation of risk exposure resulting from taking a position in a single financial security with current value S dollars. A typical example can be a long position in a stock, bond, or other asset. The risk one is exposed to is that the value can suddenly change to a level S'. For assets with positive values, one focuses attention on the logarithm of relative values or

$$x = \ln\left(\frac{S'}{S}\right). \tag{2.1}$$

In principle, as zero is an admissible jump size, the possible values for x range across all the nonzero real numbers.

The basic premise on values adopted in this work is that when they move, they do so instantaneously. For example, the value could be at level S just before time t and then jump to S' at time t. The resulting value process is what is called in the literature a pure jump process with the difference of values between two time points being the sum of all the jumps in the interim. In principle one can count the number of times that a jump of size x occurs over a period. This count is the realized number of such jump arrivals in the period. It is a random entity and associated with it is the expected number of such arrivals per unit time at time t, that is, the jump arrival rate function denoted by $k_t(x)$. We often work with a context in which the arrival rates do not depend on t; we then write $k(x)$.

Risk is represented by the arrival rate function at any time. Critical to the management of risk is an understanding of the underlying arrival rates for value change and their properties. It is useful therefore to connect such a risk representation with more classical ones. Classically, one considers S' to be the value at a later time $t + H$ with S the value at the current time t. The logarithm of the relative values on division by the horizon is then the continuously compounded return x_H over the interim

$$x_H = \frac{1}{H}\ln\left(\frac{S'}{S}\right). \tag{2.2}$$

The possible values for x_H range across the real numbers and risk is described by specifying the probability density $p_{t,t+H}(x)$ for the outcomes x. The risk representation is specific to the three entities, t, H, and x.

In sum, the analysis in this book replaces the classical risk representation in terms of the probability density $p_{t,t+H}(x)$ at a horizon H by a representation with the arrival rate function $k_t(x)$ for the instantaneous exposure x at time t. The relationship between the arrival rate functions and the probability densities for pure jump processes is given by

$$k_t(x) = \lim_{H \to 0} \frac{p_{t,t+H}(x)}{H}.$$

Models for probability are complicated by the dependence on the horizon H with analytical formulations often inaccessible. The arrival rate functions, on the other hand, are simpler expressions that are analytical in many cases.

The connection between arrival rate functions and probability densities at horizons can also be made via characteristic functions. In this case $k_t(x) = \lim_{H \to 0}(p_{t,t+H}(x)/H)$. The characteristic function is then related to the probability density by

$$\begin{aligned}
\phi_{t,t+H}(u) &= E\left[\exp\left(iu\ln(S'/S)\right)\right] \\
&= E\left[\exp\left(\int_t^{t+H}\int_{-\infty}^{\infty}\left(e^{iux}-1\right)k_s(x)dxds\right)\right] \\
&= \int_{-\infty}^{\infty} e^{iux} p_{t,t+H}(x)dx.
\end{aligned}$$

One may also observe that for compound Poisson processes of the form

$$X(H) = \sum_{j=1}^{N(H)} X_j,$$

where $N(H)$ is a Poisson process with arrival rate λ and jumps X_j independently and identically distributed with density $g(x)$, it is the case that

$$\begin{aligned}
\phi_{X(H)}(u) &= E\left[\exp\left(iuX(H)\right)\right] \\
&= E\left[\exp\left(iu\sum_{j=1}^{N(H)} X_j\right)\right] \\
&= \sum_{n=0}^{\infty}\frac{(\lambda H)^n e^{-\lambda H}}{n!}E\left[\exp\left(iu\sum_{j=1}^{n} X_j\right)\right] \\
&= \sum_{n=0}^{\infty}\frac{(\lambda H)^n e^{-\lambda H}}{n!}E[\exp\left(iuX_j\right)]^n \\
&= \sum_{n=0}^{\infty}\frac{(\lambda H)^n e^{-\lambda H}}{n!}\left(\int_{-\infty}^{\infty} e^{iux}g(x)dx\right)^n \\
&= \exp\left(\lambda H\left(\int_{-\infty}^{\infty} e^{iux}g(x)dx - 1\right)\right) \\
&= \exp\left(\int_{-\infty}^{\infty}(e^{iux}-1)\lambda Hg(x)dx\right).
\end{aligned}$$

Recognizing that as λ is the arrival rate of a jump while $g(x)$ is the probability that the jump has size x, then $\lambda g(x) = k(x)$ is the arrival rate of a jump of size x. For a compound Poisson process, we then write that

$$\phi_{X(H)}(u) = \exp\left(H\int_{-\infty}^{\infty}(e^{iux}-1)k(x)dx\right).$$

One may fix attention to the case $H = 1$ and write $p(x) = p_1(x)$ and then note that

$$\phi(u) = \int_{-\infty}^{\infty} e^{iux}p(x)dx, \tag{2.3}$$

$$\phi(u) = \exp\left(\int_{-\infty}^{\infty}(e^{iux}-1)k(x)dx\right). \tag{2.4}$$

The equations (2.3) and (2.4) hold not only for compound Poisson processes but also more generally. For example, they hold for all pure jump processes of independent increments and finite variation. These are the processes that may be split into the difference of the sum of their positive and negative jumps taken separately. They may also be extended to certain Markov processes with time- and space-dependent arrival rate functions. These relations lead us to consider probability models for the logarithm of relative values and their associated arrival rate functions. The next section introduces numerous examples.

2.1 Pure Jump Finite Variation Probability Models

There is an extensive literature on pure jump finite variation probability models. Examples include the normal inverse Gaussian model (Barndorff-Nielsen, 1998), the Variance Gamma model (Madan and Seneta, 1990; Madan et al., 1998), the CGMY model (Carr et al., 2002), and the generalized hyperbolic model (Eberlein and Keller, 1995; Eberlein, 2001). For other examples, see Schoutens (2003) and Cont and Tankov (2004). The specific models we work with here are related to the criterion of entropy maximization.

2.1.1 The Entropy Maximization Principle

The principle, enunciated by Jaynes (1957), states that one should select the distribution that leaves the largest remaining uncertainty (i.e., the maximum entropy) consistent with the constraints of what is known. Consider a discrete random variable X with outcomes x_i, $i = 1, \ldots, M$. Assume there is a large number N of observed outcomes, that is, where x_i occurs n_i times with

$$\sum_{i=1}^{M} n_i = N.$$

The number of ways in which the N outcomes may occur is given by the multinomial coefficient

$$\frac{N!}{n_1!\cdots n_M!}.$$

If we assume that in reality forces work toward maximizing the number of different configurations (i.e., the most uncertainty), one has to maximize the multinomial coefficient, or equivalently, its logarithm. Applying Stirling's approximation for the factorial function, we find that maximizing the logarithm of the multinomial coefficient is equivalent to maximizing

$$N \ln N - N - \sum_{i=1}^{M} (n_i \ln n_i - n_i) = -N \sum_{i=1}^{M} p_i \ln p_i, \qquad p_i = \frac{n_i}{N}. \tag{2.5}$$

Forces in reality are then seen as maximizing the entropy, which is given by equation (2.5). Hence, if one has to select a distribution p_i, $i = 1, \ldots, M$, the maximum entropy principle prescribes to choose the p_i to maximize equation (2.5).

For continuous densities $p(x)$ the entropy is given by

$$I = -\int_{-\infty}^{\infty} p(x) \ln p(x) dx, \tag{2.6}$$

and the maximum entropy distribution $p^*(x)$ is obtained by maximizing I of equation (2.6).

2.1.2 The Gaussian Case

Suppose that the mean rate μ and variance rate σ^2 are known at the macro level.

$$\text{Max}_{p(x)\geq 0} - \int p(x) \ln p(x) dx,$$

$$\text{subject to:} \int p(x) dx = 1, \quad \int x p(x) dx = \mu, \quad \int x^2 p(x) dx - \mu^2 = \sigma^2.$$

The Lagrangean function $\mathcal{L}$ of this optimization problem is given by

$$\begin{aligned} \mathcal{L} = & - \int p(x) \ln p(x) dx - \lambda \left(\int p(x) dx - 1 \right) \\ & - \theta \left(\int x p(x) dx - \mu \right) - \eta \left(\int x^2 p(x) dx - \mu^2 - \sigma^2 \right). \end{aligned}$$

The first-order conditions yield

$$-1 - \ln p(x) - \lambda - \theta x - \eta x^2 = 0.$$

The solution subject to the constraints is the normal density

$$p(x) = \frac{\exp\left(-\frac{(x-\mu)^2}{2\sigma^2}\right)}{\sigma\sqrt{2\pi}},$$

the classical model for risk analysis. The two parameters of mean μ and variance σ^2 describe all possible probabilities that one may wish to evaluate. The normal distribution is therefore the maximum entropy distribution for a random variable with known mean and variance.

2.1.3 The Gamma Case

Consider now a symmetric distribution with a zero mean over a small time step h where the variance is random with a known mean μ. The Lagrangean function for entropy maximization of the density of the variance, $p(y)$ for $y > 0$ a positive variance is

$$\mathcal{L} = -\int_0^\infty p(y)\ln p(y)dy - \lambda\left(\int_0^\infty p(y)dy - 1\right) - \theta\left(\int_0^\infty yp(y)dy - \mu\right).$$

The first-order condition yields

$$-1 - \ln p(y) - \lambda - \theta y = 0,$$

and the maximum entropy distribution is exponential with mean μ. As there is a real number of small time steps in any large time step, the density for the variance is a sum of exponentials or a gamma density.

2.1.4 The Symmetric Variance Gamma

When the variance of a normal distribution, for example, to model the stock return, is itself modeled by a gamma density, the model that results for the density of the return over the large time step t is then the symmetric variance gamma (svg) model of Madan and Seneta (1990). A lot is known about this density. First, its characteristic function is given by

$$\phi_t^{svg}(u) = \left(\frac{1}{1+\frac{\sigma^2\nu u^2}{2}}\right)^{\frac{t}{\nu}}.$$

The associated process is easily seen to be the difference of two independent gamma processes on noting that

$$1+\frac{\sigma^2\nu u^2}{2} = \left(1-\frac{i\sigma\sqrt{\nu}u}{\sqrt{2}}\right)^{\frac{t}{\nu}}\left(1+\frac{i\sigma\sqrt{\nu}u}{\sqrt{2}}\right)^{\frac{t}{\nu}}.$$

The two gamma processes for the upward and downward motion for the stock are here parametrically identical, that is, they have the same scale and speed or the same mean and variance rates. The two gamma processes are pure jump, increasing, infinitely divisible processes. For the svg model, the arrival rates for jumps of size $\pm x$ are given by the Lévy density

$$k_{svg}(\pm x) = \frac{1}{\nu}\frac{\exp\left(-\frac{\sqrt{2}}{\sigma\sqrt{\nu}}x\right)}{x}, \quad x > 0.$$

Given that the svg model provides us with our first example of an arrival rate function, some remarks on its properties are in order. First we note that this arrival rate function is not a probability as its integral over the real line is in fact infinite. It provides us with a measure, but not a probability measure. Second, the measure of all sets A bounded away from zero with $|x| > \varepsilon > 0$ for all $x \in A$ is finite in that

$$\int_A k_{svg}(x)dx < \infty.$$

One may recognize this property as the assertion that interesting things happen finitely often. Equivalently, nothing happens all the time as all infinite sets encompass a neighborhood of zero. All arrival rate functions we consider have comparable properties.

Even if aggregate arrival rates are finite, an important distinction between arrival rates as opposed to probabilities is that the former is not a normalized entity while the latter is normalized to unity. The normalizing constant is an additional entity that must be determined to work with probability. Arrival rates are therefore also simpler objects as they are akin to unnormalized probabilities. By not normalizing one is able to work with just the frequency with which the known outcomes are occurring.

2.1.5 The Variance Gamma

The svg of Madan and Seneta (1990) was generalized to the VG in Madan et al. (1998). The VG model is formed on evaluating a Brownian motion with drift μ and variance σ^2 at a random time given by a gamma process with unit mean rate and variance rate ν. The resulting characteristic function is

$$\phi_t^{\mathrm{VG}}(u) = \left(\frac{1}{1 - i\theta\nu u + \frac{\sigma^2\nu}{2}u^2} \right)^{\frac{t}{\nu}}.$$

Equivalently, the process can be written as the difference of two independent gamma processes with different scale and the same speed or rate parameter. With scales b_p and b_n for the positive and negative motion, the characteristic function for the difference of the two gamma processes is

$$\phi_t^{\mathrm{VG}}(u) = \left(\frac{1}{1 - ib_p u} \right)^{\frac{t}{\nu}} \left(\frac{1}{1 + ib_n u} \right)^{\frac{t}{\nu}} = \left(\frac{1}{1 - i\left(b_p - b_n\right)u + b_p b_n u^2} \right)^{\frac{t}{\nu}}.$$

The two representations are identical on defining

$$\theta = \frac{b_p - b_n}{\nu}, \qquad \sigma^2 = \frac{2}{\nu} b_p b_n.$$

The VG arrival rate function is given by

$$k_{\mathrm{VG}}(x) = \frac{1}{\nu}\left(\frac{\exp(-x/b_p)}{x} \mathbf{1}_{x>0} + \frac{\exp(-|x|/b_n)}{|x|} \mathbf{1}_{x<0} \right) \tag{2.7}$$

and the rates of exponential decay in arrival rates on the two sides are different. Generally in estimates the up scale b_p is smaller than the down scale b_n.

2.1.6 The Bilateral Gamma

Generalizing further, Küchler and Tappe (2008) proposed a bilateral gamma (BG) model that differentiated both speed and scale on the two sides. One could consider an entropy-maximizing approach to the up and down moves leading to separate gamma processes for the two sides. With speed parameters c_p and c_n representing the rate of the positive and negative motions, the BG characteristic function is

$$\phi_t^{\mathrm{BG}}(u) = \left(\frac{1}{1 - ib_p u}\right)^{c_p t} \left(\frac{1}{1 + ib_n u}\right)^{c_n t},$$

and the process representation is

$$X^{\mathrm{BG}}(t) = b_p \gamma_p(c_p t) - b_n \gamma_n(c_n t)$$

where γ_p, γ_n are independent standard gamma processes with unit mean and variance rates. The BG arrival rate function is given by

$$k_{\mathrm{BG}}(x) = c_p \frac{\exp(-x/b_p)}{x} \mathbf{1}_{x>0} + c_n \frac{\exp(-|x|/b_n)}{|x|} \mathbf{1}_{x<0}.$$

Recently Madan and Wang (2017) and Madan et al. (2017b) have shown that data on financial returns across many asset classes support a differentiation of the speeds. For equity assets, upside speeds dominated their downside counterparts. The BG model allows us to capture the phrase that markets take the escalator up and the elevator down.

2.1.7 The Bilateral Double Gamma

If one now randomizes the speed parameters on the two sides to have positive maximal entropy gamma densities in their own right, one obtains the bilateral double gamma (BGG) model introduced in Madan et al. (2018). The characteristic function is now given as

$$\phi_t^{\mathrm{bgg}}(u) = \left(\frac{1}{1 + \beta_p \ln(1 - iub_p)}\right)^{\eta_p} \times \left(\frac{1}{1 + \beta_n \ln(1 + iub_n)}\right)^{\eta_n}$$

with six parameters b_p, β_p, η_p and b_n, β_n, η_n.

The BGG arrival rate functions on the positive or negative side are numerically obtained and derived in Madan et al. (2018) as

$$k^{\mathrm{bgg}}(x) = \mathbf{1}_{x<0} \int_0^\infty \eta_n \frac{e^{-y/\beta_n}}{y} \frac{x^{y-1} e^{-x/b_n}}{b_n^y \Gamma(y)} dy + \mathbf{1}_{x>0} \int_0^\infty \eta_p \frac{e^{-y/\beta_p}}{y} \frac{x^{y-1} e^{-x/b_p}}{b_p^y \Gamma(y)} dy.$$

The BGG is on either side a gamma process time changed by another independent gamma process. This process can be continued, but our applications stop at the double gamma. The literature provides many other examples of arrival rate functions generated by subordinating Brownian motion with drift and volatility using other increasing processes for the time change. The reader is referred to Schoutens (2003) and Cont and Tankov (2004) for such a wider list. The applications presented here build on the BG and the BGG.

2.2 Probability Densities and Arrival Rates

One of the advantages of focusing decision theory directly on arrival rates is that in many cases these are known analytically while probability densities may often be accessed indirectly at specific horizons using Fourier inversion of characteristic functions. The general relationship between the characteristic function $\phi(u)$ and the arrival rate function $k(x)$ was observed in equation (2.4). From this relationship one may often derive the explicit arrival rate function from analytical expressions for the characteristic function. We illustrate such a derivation in the case of the gamma density.

For the gamma density with scale $b > 0$ and speed $c > 0$, the characteristic function is

$$\phi^g(u) = \left(\frac{1}{1 - iub}\right)^c. \tag{2.8}$$

The derivative of the log characteristic function is

$$\frac{ic}{\frac{1}{b} - iu} = i\int_0^\infty e^{iux} ce^{-x/b} dx = i\int_0^\infty e^{iux} xk(x)dx,$$

where the expression on the left follows from differentiating equation (2.8) and the expression on the right from differentiating (2.4). On noting that as

$$\frac{c}{\frac{1}{b} - iu} = \int_0^\infty e^{iux} ce^{-x/b} dx$$

one finds the arrival rate function

$$k(x) = c\frac{\exp(-x/b)}{x}.$$

2.2.1 The Variance Gamma

For the VG model, discussed in §2.1.5, both the arrival rate function and the characteristic function are simple, but the density is explicit only in terms of the modified Bessel functions $K_\nu(x)$,

$$K_\nu(x) = \frac{1}{2}\left(\frac{z}{2}\right)^\nu \int_0^\infty \exp\left(-\left(t + \frac{z^2}{4t}\right)\right) t^{-\nu-1} dt.$$

The closed form, useful for estimation, was presented in Carr and Madan (2014) and is given by

$$f_{vg}(x) = \frac{(GM)^C}{2^{C-1}\Gamma(C)\sqrt{2\pi}\left(\frac{G+M}{2}\right)^{C-1/2}} \times \exp\left(\frac{G-M}{2}x\right)|x|^{C-1/2} \times K_{C-1/2}\left(\frac{G+M}{2}|x|\right), \tag{2.9}$$

$$M = 1/b_p; \; G = 1/b_n.$$

2.2.2 The Bilateral Gamma

For the BG, the density may be expressed in terms of Whittaker functions $W_{\lambda,\mu}$,

$$W_{\lambda,\mu}(x) = \frac{x^\lambda e^{-x/2}}{\Gamma(\mu - \lambda + \frac{1}{2})} \int_0^\infty t^{\mu-\lambda-1/2} e^{-t} \left(1 + \frac{t}{x}\right)^{\mu+\lambda-1/2} dt, \; x > 0$$

for $\mu - \lambda > -1/2$. The density presented in Küchler and Tappe (2008) is given by

$$f_{bg}(x) = \frac{M^{c_p} G^{c_n}}{(M+G)^{\frac{c_p+c_n}{2}} \Gamma(c_p)} x^{\frac{c_p+c_n}{2}-1} e^{-\frac{x}{2}(M-G)} \times W_{\frac{c_p-c_n}{2}, \frac{c_p+c_n-1}{2}}(x(G+M)),$$
$$\text{for } x > 0 \; M = 1/b_p; \; G = 1/b_n,$$

where for negative x we evaluate at $|x|$ with the roles of the up and down processes flipped.

At the moment there isn't an explicit form for the density of the BGG. It is clear that the densities are more complex and involved than the associated arrival rate functions. This will be even more the case when we come to the analysis of multivariate problems as is called for in portfolio theory and the hedging of multidimensional exposures. We close with some other examples of characteristic functions of processes and associated arrival rate functions.

2.2.3 The Normal Inverse Gaussian Model

The normal inverse Gaussian model (Barndorff-Nielsen, 1998) is obtained on time-changing Brownian motion with drift by T_t^ν the first passage time of and independent Brownian motion with drift ν to the level t. The Laplace transform of T_t^ν is

$$E\left[\exp\left(-\lambda T_t^\nu\right)\right] = \exp\left(-t\left(\sqrt{2\lambda + \nu^2} - \nu\right)\right).$$

The normal inverse Gaussian (NIG) process is defined as

$$X_{\text{NIG}}(t) = \theta T_t^\nu + \sigma W(T_t^\nu), \tag{2.10}$$

for an independent Brownian motion $W(t)$. The characteristic function is obtained by conditioning on the time change as

$$E\left[\exp\left(iuX_{\text{NIG}}(t)\right)\right] = \exp\left(-\delta t\left(\sqrt{\alpha^2 - (\beta + iu)^2} - \sqrt{\alpha^2 - \beta^2}\right)\right),$$

where

$$\alpha^2 = \frac{\nu^2}{\sigma^2} + \frac{\theta^2}{\sigma^4}, \qquad \beta = \frac{\theta}{\sigma^2}, \qquad \delta = \sigma.$$

In terms of the parameterization α, β, δ,

$$X_{\text{NIG}}(t) = \beta\delta^2 T_t^{\delta\sqrt{\alpha^2-\beta^2}} + \delta W\left(T_t^{\delta\sqrt{\alpha^2-\beta^2}}\right).$$

The process is infinitely divisible with Lévy density

$$k_{\text{NIG}}(x) = \sqrt{\frac{2}{\pi}}\delta\alpha^2 \frac{e^{\beta x} K_1(|x|)}{|x|}$$

where $K_a(x)$ is the modified Bessel function of order a.

2.2.4 The Generalized Hyperbolic Model

The time change for the generalized hyperbolic model (Eberlein and Keller, 1995; Eberlein, 2001) is given by the generalized inverse Gaussian (GIG) process that has a unit time density given by

$$f_{\mathrm{GIG}}(g) = \left(\frac{\gamma}{\delta}\right)^{\lambda} \frac{1}{2K_{\lambda}(\delta\gamma)} x^{\lambda-1} \exp\left(-\frac{1}{2}\left(\frac{\delta^2}{x} + \gamma^2 x\right)\right).$$

Special cases are the inverse Gaussian distribution occurring at $\lambda = -0.5$ and the gamma distribution occurring at $\delta = 0$. The GIG distribution is infinitely divisible, giving rise to a Lévy process, $T_{\mathrm{GIG}}(t)$, but the distribution is generally not closed under convolution and hence the GIG density only arises at unit time increments. The characteristic function for the time change is given by

$$E\left[\exp\left(iuT_{\mathrm{GIG}}(t)\right)\right] = \phi_{\mathrm{GIG}}(u,t) = \left(\left(\frac{\gamma^2}{\gamma^2 - 2iu}\right) \frac{K_{\lambda}\left(\delta\sqrt{\gamma^2 - 2iu}\right)}{K_{\lambda}(\delta\gamma)}\right)^{t}.$$

The generalized hyperbolic (GH) Lévy process is constructed as

$$X_{\mathrm{GH}}(t) = \beta T_{\mathrm{GIG}(\lambda,\delta,\sqrt{\alpha^2-\beta^2})}(t) + W\left(T_{\mathrm{GIG}(\lambda,\delta,\sqrt{\alpha^2-\beta^2})}(t)\right).$$

The characteristic function for the generalized hyperbolic process is given by

$$E\left[\exp\left(iuX_{\mathrm{GH}}(t)\right)\right] = \phi_{\mathrm{GH}}(u,t) = \left(\left(\frac{\alpha^2 - \beta^2}{\alpha^2 - (\beta + iu)^2}\right)^{\frac{\lambda}{2}} \frac{K_{\lambda}\left(\delta\sqrt{\alpha^2 - (\beta + iu)^2}\right)}{K_{\lambda}\left(\delta\sqrt{\alpha^2 - \beta^2}\right)}\right)^{t}.$$

The Lévy measure of the generalized hyperbolic process is given by

$$k_{\mathrm{GH}}(x) = \frac{e^{\beta x}}{|x|}\left[\int_0^{\infty} \frac{e^{-|x|\sqrt{2y+\alpha^2}}}{\pi^2 y\left[J^2_{|\lambda|}(\delta\sqrt{2y}) + Y^2_{|\lambda|}(\delta\sqrt{2y})\right]} dy + \lambda^{+} e^{-\alpha|x|}\right],$$

where J_a, Y_a are the Bessel J and Y functions.

2.2.5 The CGMY Model

Apart from time changes of Brownian motion other Lévy processes have been formulated by generalizing Lévy measures. Carr et al. (2002) proposed the CGMY model as a generalization of the VG model by generalizing the Lévy measure (2.7) to

$$k_{\mathrm{CGMY}}(x) = \mathbf{1}_{x<0}\frac{e^{-G|x|}}{|x|^{1+Y}} + \mathbf{1}_{x>0}\frac{e^{-Mx}}{x^{1+Y}}.$$

This gives rise directly to the Lévy process $X_{\mathrm{CGMY}}(t)$ that traverses the cases of finite activity for $Y < 0$, infinite activity when $Y \geq 0$, and infinite variation for $Y \geq 1$. The activity is finite or infinite depending on whether the process has finitely or infinitely many jumps in any interval of time. When the coefficients C, Y are differentiated on the positive and negative sides, the resulting process, termed the KoBoL process, was studied in Boyarchenko and Levendorskii (2000).

The characteristic function for the CGMY process is given by

$$E\left[iuX_{\mathrm{CGMY}}(t)\right] = \exp\left(Ct\Gamma(-Y)\left((M-iu)^Y - M^Y + (G+iu)^Y - G^Y\right)\right).$$

Subsequently it was shown in Madan and Yor (2008) that the CGMY process is also a Brownian motion with drift, time changed by an independent increasing Lévy process or subordinator $T_{\mathrm{CGMY}}(t)$

$$X_{\mathrm{CGMY}}(t) = \left(\frac{G-M}{2}\right)T_{\mathrm{CGMY}}(t) + W\left(T_{\mathrm{CGMY}}(t)\right).$$

The Laplace transform of the time change $T_{\mathrm{CGMY}}(t)$ is

$$\begin{aligned} E[e^{-\lambda Y(t)}] &= \exp\left(tC\Gamma(-Y)\left[2r^Y\cos(\eta Y) - M^Y - G^Y\right]\right), \qquad (2.11)\\ r &= \sqrt{2\lambda + GM},\\ \eta &= \arctan\left(\frac{\sqrt{2\lambda - \left(\frac{G-M}{2}\right)^2}}{\left(\frac{G+M}{2}\right)}\right). \end{aligned}$$

The Lévy measure for the subordinator is a shaved stable $Y/2$ subordinator and is given by

$$\begin{aligned} \nu(dy) &= \frac{K}{y^{1+\frac{Y}{2}}}f(y)dy,\\ f(y) &= e^{-\frac{(B^2-A^2)y}{2}}E\left[e^{-\frac{B^2 y}{2}\frac{\gamma_{Y/2}}{\gamma_{1/2}}}\right],\\ B &= \frac{G+M}{2}, \qquad K = \left[\frac{C\Gamma\left(\frac{Y}{4}\right)\Gamma\left(1-\frac{Y}{4}\right)}{2\Gamma(1+\frac{Y}{2})}\right] \qquad (2.12) \end{aligned}$$

where the shaving function is $f(y)$.

2.2.6 The Meixner Process

Like the CGMY model, the Meixner process was introduced by Schoutens and Teugels (1998) and Pitman and Yor (2003) directly in terms of its Lévy measure:

$$k_{\mathrm{MXNR}}(x) = \delta\frac{\exp\left(\frac{b}{a}x\right)}{x\sinh\left(\frac{\pi x}{a}\right)}.$$

The resulting process, $X_{\mathrm{MXNR}}(t)$, is a pure jump process of infinite variation and its characteristic function is

$$E\left[\exp(iuX_{\mathrm{MXNR}}(t)\right] = \phi_{\mathrm{MXNR}}(u,t) = \left(\frac{\cos(b/2)}{\cosh((au-ib)/2)}\right)^{2\delta t}.$$

It was shown in Madan and Yor (2008) that this process is also a Brownian motion with drift time changed by an independent subordinator $T_{\mathrm{MXNR}}(t)$. Specifically,

$$X_{\mathrm{MXNR}}(t) = \frac{b}{a}T_{\mathrm{MXNR}}(t) + W\left(T_{\mathrm{MXNR}}(t)\right).$$

The Lévy measure for the subordinator is a shaved stable 1/2, and explicitly we shave by $g(x)$ and

$$\begin{aligned}
k_{\mathrm{MXNR}}(x) &= \frac{\delta a}{\sqrt{2\pi x^3}}g(x),\\
g(x) &= \exp\left(-\frac{A^2x}{2}\right)\sum_{-\infty}^{\infty}(-1)^n e^{-n^2\pi^2/(2C^2x)},\\
A &= \frac{b}{a},\\
C &= \frac{\pi}{a}.
\end{aligned}$$

2.3 The Complex Exponential Variation

Characteristic functions synthesize probability densities with Fourier inversion providing a path back to the density from the characteristic function. However, the complex exponential e^{iux} may not always be integrated with respect to an arrival rate function as it is near unity for x near zero. For arrival rates that integrate to infinity in an interval around zero, the complex exponential is then bounded away from zero on a set of infinite measure and is then not integrable with respect to this measure. Aggregate arrival rates or integrals of arrival rates are infinite for the VG and BG models, as well as all other limit laws described by self-decomposable random variables (Sato, 1999). The structure of these laws is discussed in greater detail later. In general, integration with respect to arrival rates requires the integrand to be near zero for x near zero. We therefore define the complex exponential variation by

$$\chi(u) = \int_{-\infty}^{\infty}\left(e^{iux}-1\right)k(x)dx. \tag{2.13}$$

2.3.1 The Complex Exponential Variation and Tail Measures

Define the signed tail measure associated with an arrival rate function by

$$K(x) = -\int_{-\infty}^{x}k(y)dy\mathbf{1}_{x<0} + \int_{x}^{\infty}k(y)dy\mathbf{1}_{x>0}. \tag{2.14}$$

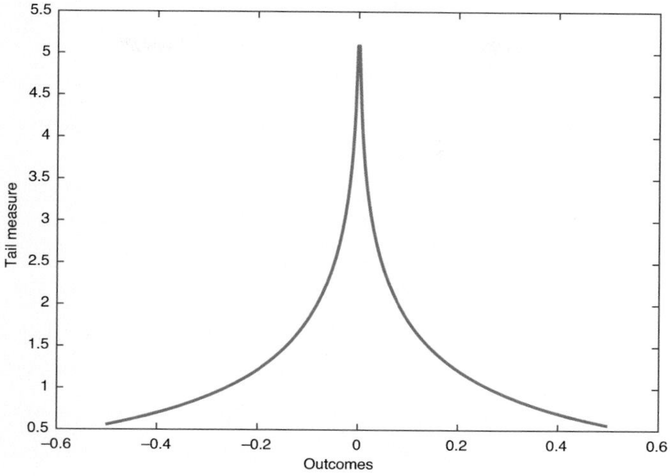

Figure 2.1 Absolute value of the inverse Fourier transform of the bilateral gamma complex exponential variation when all parameters are unity. The transform is negative for negative x and the absolute value is displayed.

For negative jumps x we take the negative of the lower tail measure while for positive x we take the upper tail measure.

One may integrate by parts the complex exponential variation (2.13) to obtain

$$\chi(u) = iu \int_{-\infty}^{0} e^{iux} K(x)dx + iu \int_{0}^{\infty} e^{iux} K(x)dx.$$

Hence we obtain the Fourier transform of the signed tail arrival function

$$\frac{\chi(u)}{iu} = \int_{-\infty}^{\infty} e^{iux} K(x)dx. \tag{2.15}$$

The inverse Fourier transform of $\chi(u)/iu$ is the signed tail arrival rate function (2.14). Thus for arrival rate functions the synthesizing object is the complex exponential variation that may be used on Fourier inversion to access the tail measure function. Figure 2.1 presents the absolute value, $|K(x)|$, of the inverse Fourier transform of $\chi(u)/(iu)$ for the BG case when all parameters are unity. The transform itself is negative for negative x and the absolute value is displayed.

Arrival rates may be directly accessed from option prices at a given maturity using procedures presented in Madan and Wang (2020a). However, they may in general reflect both signed arrival rates and signed probabilities.

2.3.2 Multidimensional Exponential Variations

For a multivariate arrival rate function $k(x)$, for $x \in \mathbb{R}^d$ one may define for $u = (u_1, \ldots, u_d)'$ the multidimensional exponential variation

$$\chi(u) = \int_{\mathbb{R}^d} \left(e^{iu'x} - 1\right) k(x)dx.$$

Signed tail measures may be defined by

$$K(x) = \prod_i \operatorname{sign}(x_i) \int_{A(x)} k(y)dy, \tag{2.16}$$
$$A(x) = \left\{y \in \mathbb{R}^d \mid \operatorname{sign}(x_i)y_i \geq \operatorname{sign}(x_i)x_i\right\}.$$

One may then observe analogously to the univariate case that

$$\frac{\chi(u)}{i^d \prod_j u_j} = \int_{\mathbb{R}^d} e^{iu'x} K(x)dx.$$

It follows that

$$K(x) = \frac{1}{(2\pi)^{d/2}} \int_{\mathbb{R}^d} e^{-iu'x} \frac{\chi(u)}{i^d \prod_j u_j} du.$$

2.3.3 Arrival Rate Properties

The characteristic functions created by complex exponential variations are those for infinitely divisible random variables associated with pure jump processes of independent and identically distributed increments. They are also referred to as Lévy processes. Readers may refer to Sato (1999) for a study of such processes and the characterization of their characteristic functions in the form (2.13) when the process is pure jump, of finite variation, and with no drift term and no continuous component. The arrival rate function $k(x)$ is also referred to as the Lévy density and in general must satisfy the integrability condition

$$\int_{|x|<a} x^2 k(x)dx < \infty, \text{ for } a > 0,$$

in a neighborhood of zero.

Certain process properties are associated with the structure of the arrival rate function $k(x)$. First, we note that if

$$\lambda = \int_{-\infty}^{\infty} k(x)dx < \infty,$$

the distribution of the random variable at unit time is that of a compound Poisson process. Let $N(t)$ be a Poisson process with arrival rate λ at which jumps occur with jump sizes X_i drawn from the distribution with density $k(x)/\lambda$. The unit time random variable X is then given as

$$X = \sum_{i=1}^{N(1)} X_i.$$

Most of the processes appearing in this work are not compound Poisson and are what are called infinite activity processes, for which

$$\int_{-\infty}^{\infty} k(x)dx = \infty.$$

These processes have infinitely many jumps in any time interval, but there are only finitely many jumps above any given absolute size. The infinite activity is associated with arbitrarily small sizes of moves. Though outcomes at specific times may be simulated from the densities accessed analytically or by Fourier inversion of characteristic functions, exact path simulation is difficult if not impossible given the infinity of small jumps necessary for exactness. They may only be approximated. Simulation at discrete time steps is possible using densities accessed by Fourier inversion of characteristic functions, for example.

When it is the case that

$$\int_{-\infty}^{\infty} |x|k(x)dx < \infty, \tag{2.17}$$

the process is said to be of finite variation and can be written as the difference of two independent increasing processes. We may write

$$X(t) = X_p(t) - X_n(t)$$

for increasing processes $X_p(t)$, $X_n(t)$. The VG, BG, and BGG are all examples of finite variation processes.

If on the other hand it is the case that

$$\int_{-\infty}^{\infty} |x|k(x)dx = \infty, \tag{2.18}$$

the process is said to be of infinite variation and it cannot be split into two processes for the upward and downward motions. Brownian motion with continuous sample paths is a classic example of a process of infinite variation; examples of pure jump processes with infinite variation are the CGMY process (with $Y > 1$, Carr et al., 2002). Apart from simulation using densities accessed by Fourier inversion, they may also be simulated as the mean shifted difference of two signed processes with only positive jumps but with a large compensating negative drift. The final mean shift corrects for the two negative drift compensations.

2.3.4 Limit Laws and Self-Decomposability

The standard normal distribution is a classic example of a limit law. It is an example of the limit of taking averages of a large number N of independent random variables that are centered and scaled to unit variance and then scaled by $\sqrt{N}$. However, the normal distribution is not the only limit law. All the other limit laws were characterized by Lévy (1937) and Khintchine (1938) as the class L_0 of self-decomposable random variables.

A random variable X is said to be self-decomposable if for every constant c, with $0 < c < 1$, an independent random variable exists denoted by $X^{(c)}$ such that X has the same distribution as that of $cX + X^{(c)}$. The equality in distribution is expressed as follows:

$$X \overset{(d)}{=} cX + X^{(c)}.$$

The self-decomposable laws are infinitely divisible with a special structure to their arrival rate functions $k(x)$. Sato (1999) shows that X is self-decomposable if and only if $|x|k(x)$ is a decreasing function of $|x|$. For the VG and BG, this decreasing function of the absolute jump size is just the negative exponential function $\exp(-a|x|)$. Given the division by $|x|$ in $k(x)$ for self-decomposable laws they are all processes of infinite activity and compound Poisson processes are hence not limit laws. Most of the processes encountered in this work are self-decomposable laws with the BG a particularly simple example of such a process.

2.4 Evaluating Event Arrival Rates

Apart from tail measures and arrival rates of tail events many risk management situations call for the evaluation of the arrival rates of events represented by other subsets A of the set of moves x. For example, when the payoff $w(x)$ to a straddle is less than $-a$, where $w(x) = -[(K_1 - e^x)^+ + (e^x - K_2)^+]$. We then wish to determine for A bounded away from zero, the arrival rate $K(A)$ of the event A given by

$$K(A) = \int_A k(x)dx < \infty.$$

When A is the union of different intervals, one may use differences of tail arrival rates evaluated at the endpoints to determine the arrival rate of each interval and then aggregate these values. An alternative to such direct evaluation, useful in particular for high-dimensional integrals, is to convert the integral to an expectation computation that may be accomplished by Monte Carlo simulations. For this purpose one chooses a density $p(x)$ on the real line that one may simulate from and then write

$$K(A) = \int_A k(x)dx = \int_A \frac{k(x)}{p(x)} p(x)dx = \mathrm{E}^p\left[\mathbf{1}_A \frac{k(x)}{p(x)}\right] \underset{\text{def}}{=} \mathrm{E}^p\left[\mathbf{1}_A \lambda(x)\right].$$

It is useful, in this regard, to ensure that the measure change function $\lambda(x)$ has a finite squared expectation under the density p on the set A. Otherwise, the Monte Carlo simulations may not converge with the sample sizes. It is useful then not to take k and p so far apart that the ratio diverges to infinity and fails to have a finite squared expectation. There is actually an interest in keeping the squared expectation small by choosing p carefully.

2.4.1 Simulation Densities for Gamma Arrival Rates

Consider first the case of the gamma distribution with arrival rate function

$$k_g(x) = c\frac{e^{-x/b}}{x}, \quad x > 0.$$

One may consider for p a gamma density with shape parameter $\gamma > 0$ and inverse scale parameter $a > 0$ with density

$$p(x) = \frac{a^\gamma x^{\gamma-1} e^{-ax}}{\Gamma(\gamma)}, \quad x > 0.$$

In this case we obtain

$$\lambda(x) = \frac{c\Gamma(\gamma)}{a^\gamma} x^{-\gamma} e^{-\left(\frac{1}{b}-a\right)x}.$$

The expectation of the square on set A is

$$\begin{aligned}\left(\frac{c\Gamma(\gamma)}{a^\gamma}\right)^2 &\int_A x^{-2\gamma} e^{-2(\frac{1}{b}-a)x} \frac{a^\gamma x^{\gamma-1} e^{-ax}}{\Gamma(\gamma)} dx \\ &= \left(\frac{c\Gamma(\gamma)}{a^\gamma}\right)^2 \frac{a^\gamma}{\Gamma(\gamma)} \int_A x^{-\gamma-1} e^{-(\frac{2}{b}-a)x} dx \\ &= \frac{c^2\Gamma(\gamma)}{a^\gamma} \left(\frac{2}{b} - a\right)^\gamma \int_A y^{-\gamma-1} e^{-y} dy.\end{aligned}$$

The choice of $a = 1/b$ leads to

$$c^2\Gamma(\gamma) \int_A y^{-\gamma-1} e^{-y} dy.$$

It is advantageous to take γ positive but not too large so that the squared expectation is finite.

2.4.2 Example: Simulating Arrival Rates in the Bilateral Gamma Model

We consider the BG model with parameter setting $b_p = 0.01$, $c_p = 0.5$, $b_n = 0.02$, and $c_n = 0.4$ and demonstrate how arrival rates of tail events can be computed. The arrival rate function $k_{\mathrm{BG}}(x)$ of the BG model consists of two parts, one for positive x and one for negative x. Hence the arrival rate of subsets of events A can be partitioned as follows:

$$K(A) = \mathrm{E}^{p_{\mathrm{up}}} \left[\mathbf{1}_A \mathbf{1}_{x>0} \lambda_{\mathrm{up}}(x)\right] + \mathrm{E}^{p_{\mathrm{dn}}} \left[\mathbf{1}_A \mathbf{1}_{x<0} \lambda_{\mathrm{dn}}(x)\right].$$

We choose gamma simulation densities as specified in the previous section.

$$\begin{aligned}p_{\mathrm{up}}(x) &= \frac{x^{\gamma-1} e^{-x/b_p}}{b_p^\gamma \Gamma(\gamma)}, \quad x > 0, \\ p_{\mathrm{dn}}(x) &= \frac{x^{\gamma-1} e^{-x/b_n}}{b_n^\gamma \Gamma(\gamma)}, \quad x > 0, \\ \gamma &= 0.075.\end{aligned}$$

We then have

$$\lambda_{\mathrm{up}}(x) = b_p^\gamma c_p \Gamma(\gamma) x^{-\gamma}, \qquad \lambda_{\mathrm{dn}}(-x) = b_n^\gamma c_n \Gamma(\gamma) |x|^{-\gamma}.$$

We then approximate

$$K(x < -0.05) = b_n^\gamma c_n \Gamma(\gamma) \frac{1}{N} \sum_{i=1}^N \mathbf{1}_{(x_i<-.05)} (-x_i)^\gamma,$$

where $-x_i$ is sampled from the gamma density $p_{\mathrm{dn}}(x)$.

Table 2.1 presents the tail measures by direct integration and simulation for sample sizes of 10,000, 100,000, and 1,000,000.

Table 2.1 *Simulated bilateral gamma tail measures. The bilateral gamma parameters were* $b_p = 0.01$, $c_p = 0.5$, $b_n = 0.02$, *and* $c_n = 0.4$. *The value of* γ *the shape parameter for simulated gamma density was* 0.075.

		Sample Size		
Tail Events	Exact	10,000	100,000	1,000,000
$x < -0.05$	0.009966	0.006611	0.010262	0.009904
$x < -0.025$	0.058565	0.054022	0.057314	0.059201
$x < -0.01$	0.223909	0.228952	0.224065	0.225729
$x < -0.005$	0.417713	0.445700	0.410627	0.418764
$x > 0.005$	0.279887	0.306185	0.273544	0.280678
$x > 0.01$	0.109692	0.099253	0.109339	0.109684
$x > 0.025$	0.012457	0.008821	0.012689	0.012835
$x > 0.05$	0.000574	0.000561	0.000618	0.000551

2.5 Variation Outcomes

When working with probabilities, one defines the class of random variables as those real-valued functions on the event space that are integrable with respect to a probability measure. The space of random variables includes all constant functions as they are always integrable to any power. The situation is different when working with arrival rate functions that typically integrate to infinity, as we have already observed. Nonzero constant functions are no longer integrable with respect to the arrival rate function. Hence they are not part of the objects to be studied. The space of integrable and square-integrable functions $c(x)$ satisfying

$$\int_{-\infty}^{\infty} |c(x)|k(x)dx < \infty, \tag{2.19}$$

$$\int_{-\infty}^{\infty} c(x)^2 k(x)dx < \infty \tag{2.20}$$

we term the space of variation outcomes. The nomenclature variation is motivated on noting that $c(0)$ must be zero and $c(x)$ measures the variation in the object of interest when the underlying is exposed to a shock or jump of size x. Variation outcomes of finite variation satisfy condition (2.19), and those satisfying condition (2.20) are referred to as variation outcomes of finite quadratic variation. When $c(x) = e^{iux} - 1$, we have the complex exponential variation as already encountered.

2.5.1 Variation Spaces

More generally the triple $(\Omega, \mathcal{F}, \nu)$, for an abstract event space Ω, a $\sigma - field$ of subsets $\mathcal{F}$, and a positive measure ν defined on the sets in $\mathcal{F}$ with $\nu(\Omega) = \infty$ may be referred to as a variation space. The collection of subsets $A \in \mathcal{F}$ for which $\nu(A) < \infty$ are the interesting sets representing the occurrence of interesting outcomes. On sets of infinite measure the only occurrence allowed by integrability is the zero or almost zero outcome.

For a variation space, the variation of a variation outcome $x(\omega)$ of finite variation is

$$V(x) = \int_\Omega x(\omega)\nu(d\omega).$$

Each variation outcome of finite variation has associated with it a tail arrival rate function defined by

$$A(y) = \begin{cases} \nu\left\{\omega | x(\omega) > y\right\} & y > 0, \\ \nu\left\{\omega | x(\omega) < y\right\} & y < 0 \end{cases}$$

and one may observe that

$$V(x) = -\int_0^\infty A(y)dy + \int_0^\infty A(y)dy. \tag{2.21}$$

The expression (2.21) evaluates variation using a Choquet type of integral of the tail arrival function.

One may also define the complex exponential variation by

$$\phi_x(u) = \int_\Omega \left(e^{iux(\omega)} - 1\right)\nu(d\omega).$$

This exponential variation may also be written as

$$\phi_x(u) = -\int_{-\infty}^\infty \left(e^{iuy} - 1\right)\text{sign}(y)dA(y) = iu\left(\int_{-\infty}^\infty e^{iuy}\left(-A(y)\mathbf{1}_{y<0} + A(y)\mathbf{1}_{y>0}\right)dy\right),$$

as already observed. The absolute value of the Fourier inverse of $\phi_x(u)/iu$ delivers the tail measure function.

More generally in the multivariate d-dimensional case, if we define multivariate complex exponential variation by

$$\phi_x(u) = \int_\Omega \left(e^{iu'x(\omega)} - 1\right)\nu(d\omega),$$

and the multivariate tail arrival function is defined by

$$A(y) = \prod_j \text{sign}(y_j)\nu\left(\left\{\omega | \prod_j x_j(\omega)\text{sign}(y_j) > \prod_j \text{sign}(y_j)y_j\right\}\right),$$

then, as observed earlier, we have that

$$\frac{\phi_x(u)}{i^n \prod_j u_j} = \int_{-\infty}^\infty \cdots \int_{-\infty}^\infty e^{iu'y}A(y)dy_1 \cdots dy_n.$$

2.5.2 Variation Space Example

Let $\Omega = \mathbb{R}$ with measure ν being just the Lebesgue measure. A simple element of finite variation is

$$c(x) = -\exp(-|x|)\mathbf{1}_{x<0} + \exp(-x)\mathbf{1}_{x>0},$$

Table 2.2 *Prices and implied volatilities for options. Presented are prices and implied volatilities for AAPL options for fixed delta levels on August 25, 2017. The spot price was 160.02.*

	Puts			Calls		
Delta	Strike	Price	IV	Strike	Price	IV
0.2	146	3.1844	0.2803	183	1.4109	0.2424
0.3	152	4.6833	0.2700	174	2.9693	0.2424
0.4	157	6.4781	0.2616	167	5.0152	0.2467
0.5	162	8.8057	0.2533	162	7.4593	0.2533

with variation $V(X) = \int_{-\infty}^{\infty} c(x)dx$, equal to zero and absolute variation, $\int_{-\infty}^{\infty} |c(x)|dx$, equal to 2.

Consider the arrival rate function $A(y)$. For $y > 0$,

$$A(y) = \nu\left(\{x \mid c(x) > y\}\right).$$

Note that

$$c(x) > y \quad \text{iff} \quad x > 0 \quad \text{and} \quad e^{-x} > y \quad \text{iff} \quad x < -\ln(y), \quad \text{for} 0 < y < 1,$$

and hence

$$A(y) = -\ln(y),\ 0 < y < 1.$$

Similarly for negative y,

$$A(y) = -\ln(|y|),\ \ -1 < y < 0.$$

The complex exponential variation is then

$$\phi_X(u) = iu\left(\int_{-1}^{0} e^{iuy}\ln(|y|)dy - \int_{0}^{1} e^{iuy}\ln(y)dy\right).$$

2.5.3 Real-World Variation Space Example

For the underlier with ticker AAPL for August 25, 2017, consider a short position in 20, 30, 40, and 50 delta strangles of maturity three months. The spot price was 160.02, and the exact strikes are presented in Table 2.2 along with the option prices and implied volatilities.

Consider variations in the logarithm of relative values of AAPL between −20% and +20%.One may value the short position in strangles at the implied volatility to determine the change in the value of the position. This results in the current variation outcome on the option book. Figure 2.2 presents the realized variation outcome as a function of the jump size. Such variation outcomes are regularly constructed in many financial institutions at the level of a variety of trading desks with respect to numerous underlying asset risks.

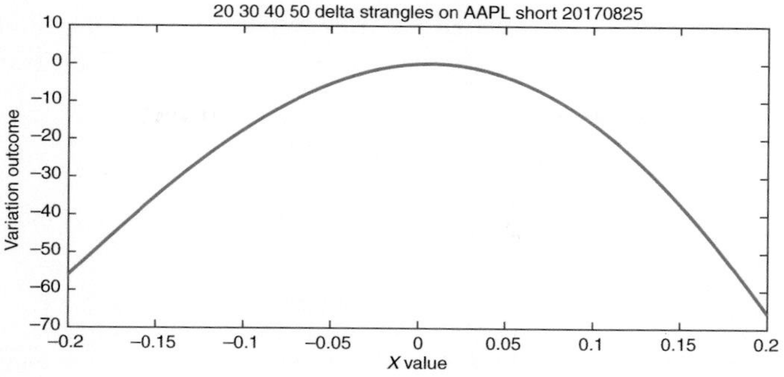

Figure 2.2 Variation outcome for short 20, 30, 40, and 50 delta strangles on AAPL as evaluated on August 25, 2017. The changes in value were computed using implied volatilities as reported in Table 2.2.

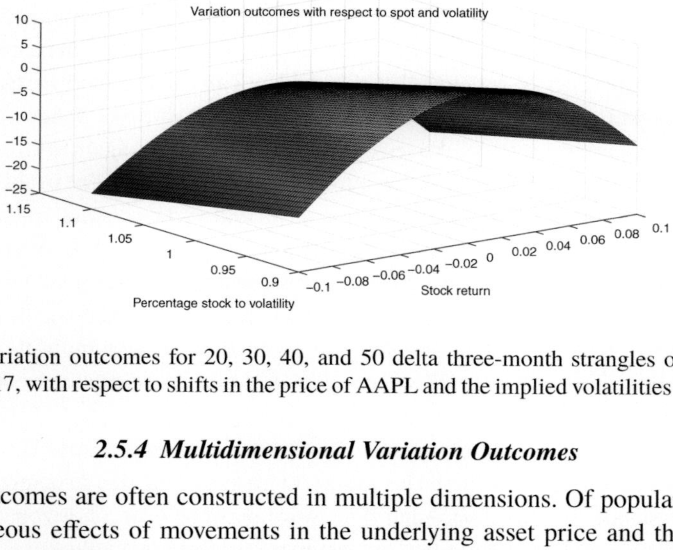

Figure 2.3 Variation outcomes for 20, 30, 40, and 50 delta three-month strangles on AAPL as of August 25, 2017, with respect to shifts in the price of AAPL and the implied volatilities on the options.

2.5.4 Multidimensional Variation Outcomes

Variation outcomes are often constructed in multiple dimensions. Of popular concern are the simultaneous effects of movements in the underlying asset price and the volatilities. Figure 2.3 presents the simultaneous variation outcomes for movements in the spot and implied volatilities for short positions on the 20, 30, 40, and 50 delta three-month strangles on AAPL on August 25, 2017.

2.5.5 Conditional Variation Measures

Let $(\Omega, \mathcal{F}, \mathcal{F}_t, \nu)$ be a filtered space with a variation measure ν satisfying $\nu(\Omega) = \infty$. Further suppose $x(\omega)$ has finite variation given by $V(x)$. We are now interested in defining the conditional variation of x conditional on $\mathcal{F}_t$. For this purpose, we define a measure on $\mathcal{F}_t$ by

$$\theta(A) = \int_\Omega x(\omega)\mathbf{1}_A \nu(d\omega), \; A \in \mathcal{F}_t, \nu(A) < \infty,$$

and observe that it is absolutely continuous with respect to ν. By the Radon–Nikodym theorem (Royden and Fitzpatrick, 2010) for general measures, an $\mathcal{F}_t$ measurable function Y exists such that

$$\frac{\theta(A)}{V(x)} = \int_\Omega Y\mathbf{1}_A \nu(d\omega),$$

and the function $V(x)Y$ satisfies

$$\int_\Omega V(x)Y\mathbf{1}_A \nu(d\omega) = \int_\Omega x\mathbf{1}_A \nu(d\omega) \text{ for } A \in \mathcal{F}_t, \nu(A) < \infty.$$

Hence $V(x)Y$ is the $\mathcal{F}_t$ conditional variation of x. A full dynamic treatment of variation measures is a subject for future research. The dynamic treatment in this book works with a probability measure at the terminal date.

2.6 Drift, Volatility, Risk Dimensions, and Their Compensation

The normal distribution reduces risk discussions to two entities, the mean and the variance. The reduction is a consequence of the fact that these two entities define the entire probability distribution. In a process of independently and identically distributed increments we get the linearity of both the mean and variance with respect to time. When risk is instead represented by the arrival rates of moves of different sizes, then, again with independently and identically distributed increments, the number of moves of each size in any time interval is a Poisson distributed random variable with a mean and variance equal to the arrival rate. All aspects of the outcome over time are a consequence of the structure of arrival rates. In particular, for the pure jump case, we do not have a linear drift as designed by a linear trend term. The trend would induce a continuous component that is not made up of unanticipated jumps. Nor do we have a constant variance rate as engineered by a multiple of a standard Brownian motion. The latter would once again introduce a continuous component and violate the pure jump feature of evolution.

We only allow for structural arrival rates of moves of different sizes. The structure describes how arrival rates vary with the jump size. Once specified, this structure induces both a mean and variance for the upward and downward motions over a horizon. Furthermore, there will then be a decomposition into at least two drifts and two volatilities, one each for the processes of upward and downward moves. For the BG, there is a one–one relationship between the parameters and these entities. The BG risk has four dimensions. This moves to six for the BGG model. In all generality, there will be infinitely many parameters for the description of the upward and downward motions. These may be related to variation moments embedded in arrival rate functions.

2.6.1 Variations As Risk Dimensions

The risk dimensions when working with probability at some horizon may be mapped from the parameters to the moments of the distribution. When the risk is characterized in terms of arrival rates, the focus shifts to power variations. These are given on the two sides as follows. Define by $v^p(m)$ and $v^n(m)$ the mth power variation on the positive and negative sides as

$$v_m^p = \left(\int_0^\infty x^m k(x)dx\right)^{\frac{1}{m}}, \qquad v_m^n = \left(\int_{-\infty}^0 |x|^m k(x)dx\right)^{\frac{1}{m}},$$

when they are finite. For the BG and the BGG, they are finite for all m. If one recognizes that in reality there are always just a finite set of jumps that can occur, then they are also finite for all m. Since it is difficult to identify a particular finite set of relevant jump sizes, one allows for analytical convenience the entire real line. Departure from finite power variations for all powers may become necessary for the effective and analytically tractable description of some data, but by and large we entertain structures with finite power variations for all powers. In terms of mapping these entities to the parameters, one employs the first few levels of m, the first two for the BG and the first three for the BGG. The explicit mappings for these two models are provided in the following sections.

2.6.2 Bilateral Gamma Risk Dimensions

The mapping from parameters to variations for the BG is:

$$v_1^p = b_p c_p \quad v_2^p = b_p\sqrt{c_p} \quad v_1^n = b_n c_n \quad v_2^n = b_n\sqrt{c_n}.$$

2.6.3 Bilateral Double Gamma Risk Dimensions

For the BGG, one employs three variations on each side as follows:

$$\begin{aligned}
v_1^p &= b_p\beta_p\eta_p \\
v_2^p &= b_p\sqrt{\eta_p}\sqrt{\beta_p(1+\beta_p)} \\
v_3^p &= b_p\eta_p^{1/3}(2\beta_p + 3\beta_p^2 + 2\beta_p^3)^{1/3} \\
v_1^n &= b_n\beta_n\eta_n \\
v_2^n &= b_n\sqrt{\eta_n}\sqrt{\beta_n(1+\beta_n)} \\
v_3^n &= b_n\eta_n^{1/3}(2\beta_n + 3\beta_n^2 + 2\beta_n^3)^{1/3}.
\end{aligned}$$

2.6.4 Risk Compensation

When risk is described by a simple distribution like the normal distribution, it is fully characterized by the two parameters of the mean and the variance. Somewhat more generally the distribution may not be specified, but one supposes a finite mean and variance and attention is focused on these two moments. The first is viewed as a reward and the second as a measure of the risk. It is then anticipated that markets will not allow return distributions to occur with very large means coupled with a very low variance. The distribution would become too attractive, leading to actions that eliminate it from a continued existence. The market may allow higher means if they are coupled with higher variances. The trade-off between the two is seen as a risk reward compensation, or a reward that is commensurate with the risk.

For more complex risk descriptions requiring multiparameter risk characterizations, the questions of risk compensation arise once again as markets will not allow all possibilities to eventuate. Compensations, however, may be described by how movements in some parameters are accompanied by movements in others to be economically viable. The trade-offs may be described in terms of value-neutral movements. This section illustrates risk compensation issues in richer risk characterization environments.

Risk exposures have to be valued. They may be valued collectively as part of a larger portfolio, but they may also be valued individually in isolation. Under linear valuation the value of the package is just the sum of component valuations, but more generally we anticipate that for nonlinear conservative valuations as occur in two-price economies, the value of the package may be higher. Before taking on the issues of packages and portfolios, one may consider just the valuation in isolation. Given the risk is defined by the parameters of its arrival rate functions, its value in isolation is a function of these parameters. As variations are entities measured in the same units, one may consider value as a function of the variations. Let V be the value of the risk exposure in isolation; then one anticipates an ability to construct the function

$$V = \Phi\left(v_1^p, v_1^n, v_2^p, v_2^n, \ldots\right). \tag{2.22}$$

In such a rich context for risk characterization, it is natural to associate reward with v_1^p the positive variation. The negative variation is a risk rather than a reward. Even for short positions, the positive variation of the short position would constitute the reward. The classical mean variance trade-off is then replaced by the trade-off permitted or expected in markets between the other power variations and v_1^p. Here we study risk compensation locally along curves of value invariance by asking what movements in other power variations require what changes in v_1^p so as to leave the value V of equation (2.22) unchanged. The relevant trade-offs may be estimated from market data at any time.

Given a structure of arrival rates that remains constant over a short period, then one has a process of independent and identically distributed increments over a short period. The result is a distribution of returns to an invested dollar for which the current value is unity. Now if the first upward variation is raised, then one anticipates a higher value and adjustments on risk will have to occur to get back to a value of unity.

If, however, value is seen as a classically constructed application of financial valuation (Harrison and Kreps, 1979), then we have the formula

$$V = E\left[ZR\right] \tag{2.23}$$

for a pricing kernel $Z \geq 0$ and a single period asset return R. The presence of time discounting and a horizon-specific interest rate of r yields that

$$E[Z] = \frac{1}{1+r}.$$

The value of the return V is unity, for it is accessed for a dollar by construction, and hence we write

$$1 = \frac{E[R]}{1+r} + \text{Cov}\left(Z - \frac{1}{1+r}, R\right),$$

which yields the classical compensation equation

$$E[R] - r \approx -\text{Cov}\left(Z - \frac{1}{1+r}, R\right).$$

However, in terms of risk components, the return distribution is completely given by the arrival rate function and the associated power variations. The classical value compensations among risk components cannot be determined as a function of the distribution and/or its parameters because the computation of value by equation (2.23) requires the joint law of Z, R and more exactly the effects on covariations between Z and R. The marginal law of R is insufficient for this exercise as is implicit in equation (2.22).

One may get toward a valuation like that of equation (2.22) via the use of copulas. Given the selection of Z and its marginal law $L(z)$ and a marginal law for the return R, $F(R)$, the choice of a copula function $C(u_1, u_2)$ that yields joint distribution is given by

$$G(Z, R) = C(L(Z), F(R)).$$

The marginal distribution of R given Z is

$$L(R|Z = z) = \frac{C(L(z), F(r))}{C(L(z), 1)}.$$

The conditional expectation of R given Z is

$$-\int_{-\infty}^{0} \frac{C(L(z), F(r))}{C(L(z), 1)} dr + \int_{0}^{\infty} \left(1 - \frac{C(L(z), F(r))}{C(L(z), 1)}\right) dr,$$

and the value of the claim on R is

$$V = \int_{0}^{\infty} \left[-\int_{-\infty}^{0} \frac{C(L(z), F(r))}{C(L(z), 1)} dr + \int_{0}^{\infty} \left(1 - \frac{C(L(z), F(r))}{C(L(z), 1)}\right) dr\right] dL(z). \tag{2.24}$$

Equation (2.24) may then be employed to analyze the implicit trade-offs between higher- and first-power variations and the two first-power variations implicit in a classical valuation of risk. A popular candidate for the copula could be a Gaussian copula with a fixed correlation parameter. The valuation equation (2.24) may also be employed to ascertain the compensating effects on first-power variations of changes in the first- and higher-power variations of the systematic risk encapsulated by Z. Additionally one may consider the effects of shifts in the copula defining the joint return.

3

Estimation of Univariate Arrival Rates from Time Series Data

Arrival rate functions for the moves in the logarithm of the price of a single asset may, quite generally, be related to probability densities of returns. The connection between arrival rates and densities is accomplished via the complex exponential variation. Assuming a certain model for the arrival rates, one can employ the corresponding density to determine tail probabilities. On the other hand, from time series data on returns at a specific horizon, one can estimate the observed tail probabilities based on the observed frequency of comparable tail events. A least-squares matching of model and observed tail probabilities then gives an estimate for the parameters of the arrival rate function. Such an estimation technique was proposed and investigated by Madan (2015a), who demonstrated that in cases when it is known that the data do not come from the model, the use of bounded moments like probabilities of tail events can yield better results than maximum likelihood estimation.

Arrival rate functions may be enriched with functional forms employing a greater number of parameters. A question that arises is whether the additional parameters are needed and what the additional flexibility delivers. One may inquire into the directions in which tail probabilities may move with respect to a richer model and the ability of the poorer model to explain these movements. We then need to assess the possibility of such moves in practice.

3.1 Complex Exponential Variations and Data

Let the complex exponential variation be given by equation (2.13). We have seen in equations (2.3) and (2.4) that the exponential of the complex exponential variation is the characteristic function of the density for the random variable, say X, that represents the sum of all the shocks associated with the arrival rate function $k(x)$. More precisely, the complex exponential variation is the Fourier transform or characteristic function of such a probability density. Fourier inversion of the characteristic function delivers the density and the associated tail probabilities. Parameters in analytically specified arrival rate functions may then be estimated by least-squares matching of model tail probabilities with their observed counterparts. The tail probability is a digital moment evaluating the expected or observed number of outcomes above/below an outcome x that is respectively positive/negative. Such a digital moment matching estimation was studied in Madan (2015a), where it was favorably compared with maximum likelihood in contexts when the data were known not to come from the model. The boundedness of the digital moment rendered it less susceptible to outliers that naturally arise when the data are not from the model.

Maximum likelihood is also known to be an example of the generalized method of moments as presented in Hansen (1982). The associated moments, however, can be unbounded and adversely impacted in estimation by the presence of outliers. Outliers can be present with greater frequency when the data do not come from the model. There are of course, other bounded moments that could be employed, like the trigonometric transforms associated with empirical characteristic functions, as proposed in Feuerverger and McDunnough (1981). However, they can also be highly correlated, as noted in Madan and Seneta (1987). Furthermore, they lack the visual structure present in the probabilities of tail events.

Denote the tail probability by $T(x)$. For $x > 0$, we define $T(x)$ as the probability of a return exceeding x, while for $x < 0$ it is the probability of a return below x. Tail probabilities may be considered digital moments that evaluate the expected or observed number of outcomes above/below a positive/negative outcome x. Fourier inversion of the characteristic function of X then delivers the density function of this random variable and hence the associated tail probabilities. A particularly effective tool for this inversion is the fast Fourier transform. Parameters in analytically specified arrival rate functions may then be estimated by least-squares matching of the model tail probabilities with their observed counterparts.

3.2 Digital Moment Estimation

Given a data sample of N observations on asset returns x_i, $i = 1, \ldots, N$ at some specific horizon, be it a day, week, or month, we estimate the parameters of an arrival rate function. It is supposed for this purpose that the data are drawn from a distribution with a characteristic function associated with the arrival rate function by the complex exponential variation (2.13). Two issues to address are the construction of digital moments or tail probabilities and the weighting scheme employed in least-squares computations.

3.2.1 Constructing Digital Moments or Tail Probabilities

The reference to tail probabilities as digital moments comes from the literature on option pricing where a digital option has the payoff of unity if the asset is above or below a level. The former is a digital call and the latter a digital put. They are also the derivatives of call and put option payoffs with respect to the underlying asset price. Attention is typically focused on out-of-the-money options or calls for strikes above the spot and puts for strikes below the spot. With respect to returns, one considers call payoffs at positive returns and put payoffs at negative returns. Hence we are interested in the probabilities of being above positive return levels and the probabilities of being below negative return levels. The associated tail probabilities are digital call and put option prices constructed here under the physical measure. They may be constructed from the empirical cumulative distribution function implied by the data sample.

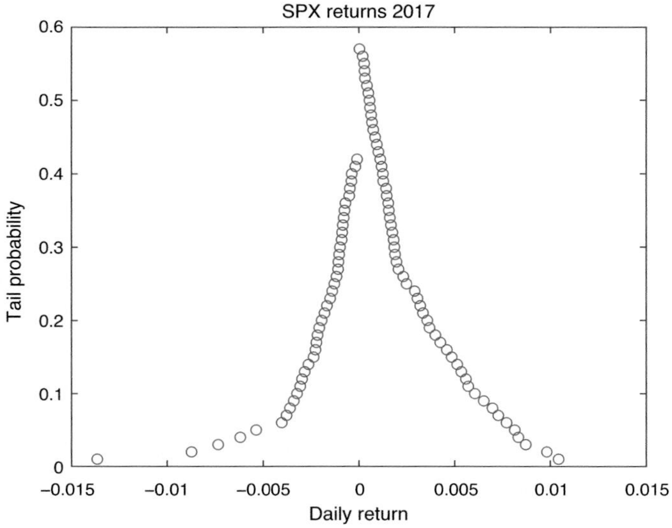

Figure 3.1 Tail probabilities or out-of-the-money digital moments on S&P 500 daily returns for the year 2017

For the empirical distribution function, the data are sorted in increasing order $x_{(1)} < x_{(2)} < \cdots < x_{(N)}$ with the cumulative probability level of i/N assigned to the point $x_{(i)}$. Assuming $N > 100$, the percentile points $y_{(j)}$ for probability $j/100$ may be obtained by interpolating the empirical cumulative distribution function at the points $j/100$ to construct points $y_{(j)}$ for $j = 1, \ldots, 99$. The tail probability associated with $y_{(j)} < 0$ is then $w_j = j/100$ and for $y_{(j)} > 0$ the tail probability is $w_j = 1 - j/100$.

Figure 3.1 presents the tail probabilities for daily returns on the S&P 500 index for the year 2017.

3.2.2 Anderson–Darling Weighting

The probability model may be used to compute the model tail probability for outcome $y_{(j)}$ that we denote by $\widehat{w}_j$. The parameters of the arrival rate function can now be estimated by matching the empirical tail probabilities w_j with the model tail probabilities $\widehat{w}_j$. One could do this by minimizing the least-squares objective

$$\sum_{j=1}^{N} \left(w_j - \widehat{w}_j\right)^2 .$$

However, with a view to giving tail events, with lower probabilities, a greater weight, we adopt the Anderson and Darling (1952) weighting scheme and minimize

$$\sum_{j=1}^{N} \frac{\left(w_j - \widehat{w}_j\right)^2}{w_j(1 - w_j)}.$$

3.2.3 Data Centering Issues

It was reported in Madan (2018a) that the process of centering the data and estimating the centered model on the centered data introduces noise and can lead to a poorer estimation of even the mean. Further, Madan (2017b) demonstrated an increased efficiency in estimating mean returns by estimating the uncentered model on the uncentered data and deriving the mean analytically from the estimated parameters. A critical issue is the typical estimation strategy of demeaning or centering the data and then estimating the centered model on the centered data. However, demeaning by evaluating the sample mean is particularly noisy for log price relatives as it is just the difference between the last and first item divided by the number of time points. The centered data are then perturbed by this noise and its quality could be rendered questionable for even the purpose of finding the mean. We present evidence here that one may obtain far superior estimates of even the mean by estimating the uncentered model on the uncentered data.

Often it is also advocated that one may even work with normalized data that are both centered and scaled to unit variance. The latter activity, though less noisy than centering, employs the maximum likelihood estimate for the normal distribution to estimate the variance on data that may be far from normally distributed. As a result it is a source of noisy perturbation. Critical to the use of uncentered and unnormalized data is the standing assumption that no mean shifts are occurring and that any mean that arises is only a consequence of the asymmetry in the arrival rates of positive and negative returns. There are no trend terms in the data. There is also no dominating or major Gaussian component justifying the variance calculation. There may in fact be considerable asymmetries between the positive and negative small moves to invalidate the classical variance computation. These asymmetries have been further documented for financial data in Madan and Wang (2017) and Madan et al. (2017b).

For the VG model, for instance, the mean of the log price relative is simply the parameter θ. Here we present the results of such an exercise for 207 stocks with returns from daily data over the years 2014 through 2017. Table 3.1 presents the percentiles for the two mean estimates in basis points. The uncentered estimates are more widely consistent with financial priors of a positive return on equity. They are also less volatile in terms of the interquartile ranges. We continue to recommend the estimation of the uncentered model on the uncentered data.

3.3 Variance Gamma, Bilateral Gamma, and Bilateral Double Gamma Estimation Results

3.3.1 S&P 500 Index

For returns on the S&P 500 index, over the four years 2014 through 2017, the VG, BG and BGG models were estimated by digital moment estimation using uncentered data and Anderson–Darling weighting. Figure 3.2 presents a plot of the logarithm of the tail

Table 3.1 *Percentiles for mean returns across 207 stocks in basis points. The mean returns are computed using uncentered and centered daily data for the years 2014 through 2017.*

Percentile	Uncentered	Centered
1	−878.80	−432.05
5	−17.75	−52.75
10	3.75	−37.76
25	9.24	−24.48
50	12.80	−15.96
75	16.20	−9.27
90	19.73	−3.17
95	21.23	−1.34
99	26.09	10.63

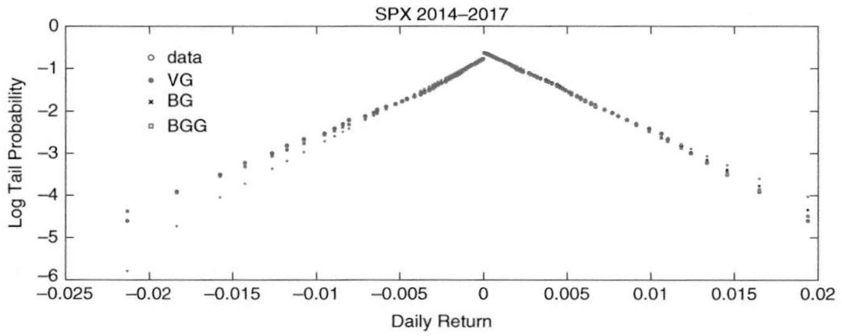

Figure 3.2 Actual and predicted logarithm of tail probabilities on S&P 500 index daily returns over the period 2014 through 2017. One may observe the improvement provided by the bilateral gamma over the variance gamma in the tails. The bilateral double gamma provides a further improvement over the bilateral gamma in the upper tail.

probabilities as observed in the data and the predicted log probability from the three models. The BGG model displays a significant improvement over the other models in the tails.

The estimated parameter values for the VG were

$$\sigma = 0.0073397, \quad \nu = 1.039578, \quad \theta = 0.001041.$$

For the BG, the result was

$$b_p = 0.005109, \quad c_p = 1.086345, \quad b_n = 0.007062, \quad c_n = 0.597112.$$

The BGG estimated parameters were

$$\begin{aligned} b_p &= 0.003573, & \beta_p &= 0.411850, & \eta_p &= 3.670280, \\ b_n &= 0.006778, & \beta_n &= 0.084859, & \eta_n &= 7.146159. \end{aligned}$$

We estimate the parameters of the three models for 207 stocks over the period 2014–2017. For the purpose of presenting representative parameter values for these models, the

Table 3.2 *Representative variance gamma parameter sets quantized from estimates for 207 stocks. Also shown are the proportion of points represented by each set in percentages.*

σ	ν	θ	prop.(%)
0.0028	6.8381	−0.0649	3.86
0.0139	0.5039	0.0012	10.63
0.0137	0.6173	0.0013	18.36
0.0135	0.7161	0.0013	18.36
0.0132	0.3936	0.0013	5.79
0.0142	0.7952	0.0014	6.76
0.0194	1.2867	0.0005	7.73
0.0143	0.4445	0.0014	8.69
0.0127	0.5576	0.0013	13.52
0.0108	0.2982	0.0012	6.28

207 parameter sets were quantized into 10 sets. Quantization algorithms form clusters of points using a distance minimization technique and then represent each cluster by its centroid along with providing the proportion of points in each cluster. Tables 3.2, 3.3, and 3.4 present the quantized points along with the proportion of the 207 points they represent. For the estimation of BG and BGG the starting values were taken from the VG estimation. The VG estimation employed the closed form for the density provided in equation (2.9). Fourier inversion was used for BG and BGG.

3.4 Assessing Parameter Contributions

Given two models predicting tail probabilities $\widehat{w}$, we consider how these tail probabilities change in response to a change in a parameter. More specifically we consider what variations are possible in a richer model that cannot be delivered by the poorer model.

3.4.1 Eigen Analysis of Responses

For two models with parameter sets θ, η, with the second being the richer model, one may evaluate the matrix of derivatives

$$\frac{\partial w}{\partial \theta}, \quad \frac{\partial w}{\partial \eta}$$

of predicted tail probabilities where we write w for $\widehat{w}$. The changes in probability for each model may be approximated locally by

$$\Delta w = \frac{\partial w}{\partial \theta}\Delta\theta, \qquad \Delta w = \frac{\partial w}{\partial \eta}\Delta\eta.$$

Suppose the actual change in probabilities is given by the second model or by

$$\frac{\partial w}{\partial \eta}\Delta\eta.$$

Table 3.3 *Representative bilateral gamma parameter sets quantized from estimates for 207 stocks. Also shown are the proportion of points represented by each set in percentages.*

b_p	c_p	b_n	c_n	prop.(%)
0.0076	2.6314	0.0086	2.1495	5.31
0.0091	1.6177	0.0089	1.5194	13.04
0.0039	3.3226	0.0065	1.8377	3.86
0.0054	2.2552	0.0077	1.4211	11.11
0.0062	1.8998	0.0076	1.3735	21.25
0.0043	3.8472	0.0043	3.4777	1.45
0.0084	2.1029	0.0088	1.8454	5.31
0.0044	2.7363	0.0064	1.7053	4.83
2.0724	1.0752	0.0149	0.8862	17.39
0.0070	1.5514	0.0085	1.1405	16.43

Table 3.4 *Representative bilateral double gamma parameter sets quantized from estimates for 207 stocks. Also shown are the proportion of points represented by each set in percentages.*

b_p	β_p	η_p	b_n	β_n	η_n	$prop.$(%)
0.0033	2.4997	2.9052	0.0019	5.3582	1.7577	11.11
0.0027	1.6489	2.7277	0.0067	0.2917	5.2874	22.71
0.0080	8.2569	4.1681	0.0054	1.2087	9.7749	4.34
0.0022	1.7051	3.4368	0.0058	0.1526	10.438	11.11
0.0021	1.0414	5.3169	$6.9e-5$	99.703	1.4266	1.45
0.0022	3.2354	2.1149	0.0068	0.4567	4.0701	15.45
9.2879	0.0245	573.28	0.0125	0.0072	807.67	2.42
0.0016	4.8964	3.1149	0.0049	24.2143	1.7651	3.86
0.0045	2.0770	2.2635	0.0053	1.6519	1.8751	13.53
0.0019	2.3299	2.5629	0.0064	0.2083	6.9321	14.01

The first model tries to explain them with a change in its parameters, satisfying

$$\frac{\partial w}{\partial \eta}\Delta\eta = \frac{\partial w}{\partial \theta}\Delta\theta + \varepsilon.$$

The least-squares solution for $\Delta\theta$ is

$$\widehat{\Delta\theta} = \left[\left(\frac{\partial w}{\partial \theta}\right)'\left(\frac{\partial w}{\partial \theta}\right)\right]^{-1}\left[\left(\frac{\partial w}{\partial \theta}\right)'\left(\frac{\partial w}{\partial \eta}\right)\right]\Delta\eta.$$

The predicted probability changes are

$$\widehat{\Delta w} = \frac{\partial w}{\partial \theta}\widehat{\Delta\theta}\frac{\partial w}{\partial \theta}\left[\left(\frac{\partial w}{\partial \theta}\right)'\left(\frac{\partial w}{\partial \theta}\right)\right]^{-1}\left[\left(\frac{\partial w}{\partial \theta}\right)'\left(\frac{\partial w}{\partial \eta}\right)\right]\Delta\eta.$$

The prediction errors are then

$$\left[\frac{\partial w}{\partial \eta} - \frac{\partial w}{\partial \theta}\left[\left(\frac{\partial w}{\partial \theta}\right)'\left(\frac{\partial w}{\partial \theta}\right)\right]^{-1}\left[\left(\frac{\partial w}{\partial \theta}\right)'\left(\frac{\partial w}{\partial \eta}\right)\right]\right]\Delta\eta.$$

The squared errors are

$$\Delta\eta' Q \Delta\eta,$$

where

$$\begin{aligned} Q &= A'A \\ A &= \left[\frac{\partial w}{\partial \eta} - \frac{\partial w}{\partial \theta}\left[\left(\frac{\partial w}{\partial \theta}\right)'\left(\frac{\partial w}{\partial \theta}\right)\right]^{-1}\left[\left(\frac{\partial w}{\partial \theta}\right)'\left(\frac{\partial w}{\partial \eta}\right)\right]\right]. \end{aligned}$$

We may write

$$A = \left[I - \frac{\partial w}{\partial \theta}\left[\left(\frac{\partial w}{\partial \theta}\right)'\left(\frac{\partial w}{\partial \theta}\right)\right]^{-1}\left(\frac{\partial w}{\partial \theta}\right)'\right]\left(\frac{\partial w}{\partial \eta}\right).$$

It follows from the idempotent structure of the first matrix that

$$Q = \left(\frac{\partial w}{\partial \eta}\right)'\left[I - \frac{\partial w}{\partial \theta}\left[\left(\frac{\partial w}{\partial \theta}\right)'\left(\frac{\partial w}{\partial \theta}\right)\right]^{-1}\left(\frac{\partial w}{\partial \theta}\right)'\right]\left(\frac{\partial w}{\partial \eta}\right).$$

The eigenvalues and eigenvectors of Q inform us of how the second model affects probabilities in ways the first model cannot explain.

3.4.2 Analysis of Bilateral Gamma over Variance Gamma and Bilateral Double Gamma over Bilateral Gamma

For the VG, BG, and BGG parameter estimates reported in §3.3.1 for the S&P 500 index, the matrices Q for BG over VG and BGG over BG were constructed. The first is four by four and the latter is six by six. The eigenvalues for the first Q matrix were evaluated to be

$$330.58 \qquad 63.19 \qquad 0.034 \qquad 0.0004.$$

Because the first two eigenvalues are significantly larger, we consider the corresponding first two eigenvectors, which are

b_p	0.9951	0.0971
c_p	−0.0065	0.0041
b_n	−0.0970	0.9953
c_n	0.0183	−0.0005.

Hence we see that the effect of BG variations in b_p, and b_n on tail probabilities is not easily explained by the VG model. We know that the BG differs from the VG by accomodating differences in c_p and c_n, but this difference may account for a differentiated response to movements in b_p, b_n that are differentiated from what the VG can deliver.

For the Q matrix for the BGG over the BG, the eigenvalues are

$$29.5659 \quad 1.0269 \quad 0.0038 \quad 9e\text{-}5 \quad 2e\text{-}7 \quad 5e\text{-}10.$$

The corresponding first two eigenvalues are

$$\begin{array}{lrr} b_p & 0.9968 & -0.0790 \\ \beta_p & 0.0087 & 0.0067 \\ \eta_p & 0.0007 & -0.0017 \\ b_n & -0.0775 & -0.9755 \\ \beta_n & -0.0154 & -0.2053 \\ \eta_n & -0.0002 & -0.0026. \end{array}$$

The movement generated by b_p and a combination of b_n and β_n in the BGG are not well explained by the BG model.

4

Estimation of Univariate Arrival Rates from Option Surface Data

Time series data are both backward looking and sparse, providing just one new piece of information per unit time. Option prices on the other hand provide us with numerous pieces of information at each time point from the extensive quotations on prices of options with many different strikes for any maturity of interest. It is therefore a matter of considerable interest to devise strategies capable of extracting information on the physical density at a given maturity from the prices of options. The data on option prices have grown considerably from a few hundred options on an underlier some 10 years ago to more than 2,000 options currently traded on various underliers. For example, $2,695$ options were traded on February 8, 2018, for the SPY as the underlying asset. They were spread across 28 maturities ranging between a few days to two years. The level of moneyness relative to the forward used to be between plus or minus 30% of the forward but has now stretched to around 80% on the down side.

4.1 Depreferencing Option Prices

Option price data are data on price that directly reflect a risk-neutral density $g(x)$ at the option maturity for the price $g(x)dx$ of a return to maturity of xdx. Simultaneously, a physical density $f(x)$ exists for the realized return to maturity of x. The two densities are not identical. The relationship between them incorporates general considerations of preference related to the underlying risk exposures.

A classical derivation of the impact of preferences on the relationship between the risk-neutral and physical densities follows Huang and Litzenberger (1988). Suppose for this purpose the existence of a representative agent. Suppose the agent has a utility function of wealth $u(W)$ and an optimal forward random wealth position W at the option maturity of T. The optimality of W implies that an offer of taking a positions of t units in the delivery of $c(x)$ at T when purchased now at its forward, time T, market price w, is rejected for all functions c. The expected utility, $V(t)$, from doing this is given by

$$V(t) = E\left[u(W + tc(x) - tw)\right].$$

As $V(t)$ is maximized at $t = 0$, setting $V'(0) = 0$ yields the equation

$$E\left[u'(W)\,(c(x) - w)\right] = 0.$$

It follows that

$$w = \frac{E\left[u'(W)c(x)\right]}{E\left[u'(W)\right]}$$

or that

$$g(x) = \frac{E\left[u'(W)|x\right]}{E\left[u'(W)\right]}. \tag{4.1}$$

The denominator of equation (4.1) is just a normalizing constant while the numerator here is the expectation of marginal utility $u'(W)$, conditioned on the level of the return x. The critical element is the modeling of the numerator in equation (4.1). In modeling a market the relevant marginal utility conditional on the size of the move x may be taken to be a weighted average of personalized marginal utilities of market participants active in the particular option market. Some of these participants will be long in the asset in question while others will be short. Accounting for both sides of the market, we model the local premium or behavior of marginal utility conditional on the post-jump price of the asset as a function of the move of size x to be proportional to

$$e^{-\eta x} + ae^{\zeta x}. \tag{4.2}$$

Here η is the premium rate for market drops extracted from the concerns of long positions. Similarly, ζ is the premium rate for short positions seeking protection. The parameter a is a measure of the weight on the short side. This is a basic measure change candidate reflecting two-sided premia considerations and three parameters. We understand from prior research (Bakshi et al., 2010; Madan, 2016a) that the measure change incorporating preferences is typically U-shaped and this structure is captured by the measure change (4.2).

The representation (4.2) is quite classical and a parametric or nonparametric form has been employed by many authors and more recently by Ross (2015) and Borovička et al. (2016) in the discussions on the recovery theorem. Bakshi et al. (2018) go on to test the restrictions imposed by the recovery theorem. The information for recovery at a theoretical level makes Markovian assumptions and requires information on option prices in all Markovian states. We focus on what may be done locally with the information in just one state, the current one. Other approaches to employing option prices to recover the physical distribution, either market subjective or the real one, include Barkhagen et al. (2016) and Cuesdeanu and Jackwerth (2018). The former requires the selection of a numeraire asset allowing option pricing under the physical measure while the latter addresses some of the puzzles related to the nonmonotonicity of the pricing kernel. We wish to test whether a rich enough modeling of current preferences allows for a realistic and testable extraction of the current physical density.

With regard to variations in the current physical density $f(x)$, the return over a horizon of interest may be thought of as

$$f(x) = \int l(x \mid y)m(y)dy,$$

where $l(x \mid y)$ is the conditional density of x given auxiliary variables y and $m(y)$ is the marginal density of y. Examples of auxiliary variables may be found in Flannery and Protopapadakis (2002). By the mean value theorem we may write

$$f(x) = l(x \mid y^*(x)),$$

and this density may move around substantially as the market focuses attention on different relevant auxiliary variables.

Suppose f and g are viewed as successfully captured by the four- or six-parameter BG or double gamma class of models. From the analysis of time series data and the estimation of risk-neutral distributions at numerous maturities separately such a view may be easily justified. However, generally one also expects that parameters in g differ from their physical counterparts. As a consequence a proposed measure change with two or three parameters lacks sufficient flexibility to accommodate all the required changes in these parameters.

The preference modeling may be further enriched by supposing that risk premia on the long and short side, η and ζ, have distributions of their own. As these risk premia are positive, maximal entropy considerations suggest the use of gamma distribution for the risk premia. But η, ζ are then unbounded and the expectations of $e^{-\eta x}$ and $e^{\zeta x}$ for x BG distributed may then not be finite. These considerations motivate the use of distributions for η, ζ that take values on a finite set. We propose the strategy of employing as distributions for these risk premia the use of Gauss–Laguerre quadrature approximations to the gamma distribution. Under such an approximation the possible outcomes for η and ζ are finite. With two parameters for the distribution of η and two for the distribution of ζ, the preference model has five parameters and perhaps there is then enough flexibility in the modeling of preferences to keep the candidate physical density correct or unperturbed. It is critical to provide sufficient flexibility in the preference modeling to protect f itself from being perturbed away from its status as a candidate physical density. The quality of f as a candidate physical density may be assessed by evaluating the associated empirical distribution function for the realized forward returns at the appropriate maturity and measuring its distance from a uniform distribution. Such an evaluation was employed in Madan and Schoutens (2020) for long horizon returns.

4.1.1 Depreferencing Details

Suppose the candidate physical density $f(x)$ is in a parametric class with characteristic function $\phi(u)$. For the base preference perturbed measure change (4.2) the preference perturbed unnormalized density h is

$$(e^{-\eta x} + ae^{\zeta x})f(x).$$

Its characteristic function on normalization is given by

$$\frac{\phi(u + i\eta) + a\phi(u - i\zeta)}{\phi(i\eta) + a\phi(-i\zeta)}.$$

Going further, we introduce distributions of risk aversion for long positions and a similar distribution of risk aversion for short positions with both distributions being independent and in the class of gamma distributions. The densities for the risk premia are then

$$g_{\mathrm{L}}(\eta) = \frac{c_{\mathrm{L}}^{\gamma_{\mathrm{L}}}}{\Gamma(\gamma_{\mathrm{L}})}\eta^{\gamma_{\mathrm{L}}-1}e^{-c_{\mathrm{L}}\eta}, \qquad g_{\mathrm{S}}(\zeta) = \frac{c_{\mathrm{S}}^{\gamma_{\mathrm{S}}}}{\Gamma(\gamma_{\mathrm{S}})}\zeta^{\gamma_{\mathrm{S}}-1}e^{-c_{\mathrm{S}}\zeta}.$$

The use of a gamma density for the risk aversions may once again be motivated by recognizing that risk aversion is random with finite means for the level and logarithm of risk aversion and then the maximal entropy density is the gamma density. The unnormalized characteristic function of h at maturity t is then as follows:

$$\int_0^\infty \frac{c_{\mathrm{L}}^{\gamma_{\mathrm{L}}}}{\Gamma(\gamma_{\mathrm{L}})}\eta^{\gamma_{\mathrm{L}}-1}e^{-c_{\mathrm{L}}\eta}\phi\,(u+i\eta)\,d\eta + \int_0^\infty \frac{c_{\mathrm{S}}^{\gamma_{\mathrm{S}}}}{\Gamma(\gamma_{\mathrm{S}})}\zeta^{\gamma_{\mathrm{S}}-1}e^{-c_{\mathrm{S}}\zeta}\phi(u-i\zeta)d\zeta,$$

with the normalization factor organized by requiring the characteristic function evaluated at zero to be unity. The normalization factor is then

$$\int_0^\infty \frac{c_{\mathrm{L}}^{\gamma_{\mathrm{L}}}}{\Gamma(\gamma_{\mathrm{L}})}\eta^{\gamma_{\mathrm{L}}-1}e^{-c_{\mathrm{L}}\eta}\phi\,(i\eta)\,d\eta + \int_0^\infty \frac{c_{\mathrm{S}}^{\gamma_{\mathrm{S}}}}{\Gamma(\gamma_{\mathrm{S}})}\zeta^{\gamma_{\mathrm{S}}-1}e^{-c_{\mathrm{S}}\zeta}\phi(-i\zeta)d\zeta.$$

The integrals may be converted to summations using Gauss–Laguerre quadrature.

For the unnormalized long side we write

$$\int_0^\infty \frac{c_{\mathrm{L}}^{\gamma_{\mathrm{L}}}}{\Gamma(\gamma_{\mathrm{L}})}\eta^{\gamma_{\mathrm{L}}-1}e^{-c_{\mathrm{L}}\eta}\phi\,(u+i\eta)\,d\eta = \frac{1}{\Gamma(\gamma_{\mathrm{L}})}\int_0^\infty w^{\gamma_{\mathrm{L}}-1}\phi\left(u+\frac{iw}{c_{\mathrm{L}}}\right)e^{-w}dw.$$

We use a weight function generator for the gamma density available as downloadable matlab code `gaulagwa.m` that delivers weights and abscissas for Gauss–Laguerre quadrature for the weight function $x^a\exp(-x)$:

$$[\chi(\gamma_{\mathrm{L}}), x(\gamma_{\mathrm{L}})] = \texttt{gaulagwa}(\gamma_{\mathrm{L}}, N),$$

where N is the number of points in the quadrature approximation. In application we used a value of 20 for the number of points in the quadrature approximation. We then write the approximation

$$\frac{1}{\Gamma(\gamma_{\mathrm{L}})}\int_0^\infty w^{\gamma_{\mathrm{L}}-1}\phi\left(u+\frac{iw}{c_{\mathrm{L}}}\right)e^{-w}dw \approx \frac{1}{\Gamma(\gamma_{\mathrm{L}})}\sum_{j=1}^{N}\chi_j(\gamma_{\mathrm{L}}-1)\phi\left(u+\frac{ix_j(\gamma_{\mathrm{L}}-1)}{c_{\mathrm{L}}}\right).$$

The characteristic function is then

$$\frac{\sum_j \frac{\chi_j(\gamma_{\mathrm{L}}-1)}{\Gamma(\gamma_{\mathrm{L}})}\phi\left(u+\frac{ix_j(\gamma_{\mathrm{L}}-1)}{c_{\mathrm{L}}}\right) + a\sum_j \frac{\chi_j(\gamma_{\mathrm{S}}-1)}{\Gamma(\gamma_{\mathrm{S}})}\phi\left(u-\frac{ix_j(\gamma_{\mathrm{S}}-1)}{c_{\mathrm{S}}}\right)}{\sum_j \frac{\chi_j(\gamma_{\mathrm{L}}-1)}{\Gamma(\gamma_{\mathrm{L}})}\phi\left(\frac{ix_j(\gamma_{\mathrm{L}}-1)}{c_{\mathrm{L}}}\right) + a\sum_j \frac{\chi_j(\gamma_{\mathrm{S}}-1)}{\Gamma(\gamma_{\mathrm{S}})}\phi\left(-\frac{ix_j(\gamma_{\mathrm{S}}-1)}{c_{\mathrm{S}}}\right)}$$

with parameters

$$c_{\mathrm{L}},\ \gamma_{\mathrm{L}},\ c_{\mathrm{S}},\ \gamma_{\mathrm{S}},\ a,\ b_p,\ c_p,\ b_n,\ c_n,$$

with five preference parameters and four parameters for the implied BG physical density with parameters b_p, c_p, b_n, c_n. We refer to this risk-neutral model as the double gamma tilted bilateral gamma model. Similarly, when the physical model is BGG the risk-neutral model has 11 parameters given by c_{L}, γ_{L}, c_{S}, γ_{S}, a, b_p, β_p, η_p, b_n, β_n, and η_n. We refer to this model as the double gamma tilted bilateral double gamma.

4.1.2 Evaluating the Quality of Physicality

An implied physical density may be extracted from option prices using the calibration and estimation of preference parameters and the associated BG or double gamma parameters. We now wish to ascertain whether we are actually estimating a physical density. For this purpose suppose we estimate a physical density for say a monthly maturity on a particular day t for a particular underlying asset i. Let r_{it} be the actual one month ahead return on asset i on day t. For $h = 21$ days we determine

$$r_{it} = \ln\left(\frac{S_{i,t+h}}{S_{i,t}}\right),$$

where $S_{i,t}$ is the closing price of asset i on day t.

Restricting the discussion to the BG model, let the BG parameters estimated from one month maturity options on day t for asset i be $b_{p,it}, c_{p,it}$ $b_{n,it}$ and $c_{n,it}$. Denote by $F_{\mathrm{BG}}(x, b_p, c_p, b_n, c_n)$ the BG distribution function evaluated at x for the parameter values b_p, c_p, b_n, and c_n. Define

$$u_{it} = F_{\mathrm{BG}}\left(r_{it}, b_{p,it}, c_{p,it}, b_{n,it}, c_{n,it}\right).$$

The variables u_{it} are the estimated distribution functions evaluated for each asset and date at the immediately following actual realized forward monthly return. For a successful extraction of physical laws the data u_{it} should be uniformly distributed. To evaluate this proposition we sort the u_{it} in increasing order u_k, with $k = 1, \ldots, N$ where N is the total number of observations across all dates and assets. We then graph the empirical distribution k/N against u_k to determine whether this graph is close to the identity function. One may use as a measure of performance the distance statistic

$$\kappa = \max_k \left(\left|u_k - \frac{k}{N}\right|\right).$$

4.2 Estimation Results

Tests for the physicality of the implied physical distribution extracted from option prices post depreferencing were reported in Madan et al. (2018) for the monthly maturity. The data covered 30 underlying assets over 133 nonoverlapping months across the time period January 2007 to December 2017. It was observed that the BG model came closest to being an adequate implied physical distribution. Since August 2014 there are many underlying assets for which numerous strikes trade at frequencies of 3 to 10 days. Here we report on

Table 4.1 *Preference parameters for bilateral gamma physical candidate. For the long and short side subscripted by L, S the table presents quantized means and variances for distributions of risk premia along with estimates for the proportion short. Also presented are the proportions represented by each preference set.*

μ_L	σ_L	μ_S	σ_S	a	prop.
0.9971	1.0029	0.9922	0.9956	0.05005	33.40
0.3885	1.1081	0.5684	0.6281	0.5591	3.30
2.2596	0.7707	0.1552	0.2461	0.9862	7.66
1.7538	1.0612	1.1935	1.1995	0.3776	4.51
2.6425	0.7669	1.0523	1.4760	1.1365	1.17
1.3821	0.9332	0.6216	0.6929	0.2104	21.27
0.8453	1.0311	1.0473	1.0176	0.8291	5.11
1.6332	1.0128	3.2595	1.8547	0.7612	1.79
1.8252	0.8740	0.3467	0.4620	0.1466	19.62
0.8099	0.3899	0.6233	0.2434	9.2487	2.19

48 such underlying assets for around 170 nonoverlapping 5-day intervals between August 2014 and December 2017. The total number of estimations is around 8, 000. The calibration procedure follows the methodology laid out in Madan et al. (2018) of first employing GPR (Rasmussen and Williams, 2006) to build volatility surface summaries as functions of moneyness and maturity, with the former defined by the log price relative of strike to the forward. The GPR functional summaries are used to extract option prices at exact 5-day maturities to avoid maturity variations, and prespecified levels of moneyness between negative to positive 20% in steps of 2% that yield 81 options prices maturing in exactly 5 days. The depreferenced physical model is calibrated to these option prices. A summary of the preference parameters is presented in §4.2.1 while tests of physicality are reported on in §4.2.2.

4.2.1 Summary of Preference Parameter Estimates

The two candidate models employed for potential five-day physical distributions were the BG and the BGG with four and six parameters respectively. In both cases the depreferencing model has five parameters. Across the 48 underliers and 170 nonoverlapping five-day intervals between August 2014 and December 2017 for the approximately 8, 000 estimations of both models we report here on the values for the preference parameters. The preference parameters are $c_L, \gamma_L, c_S, \gamma_S$, and a, but with a view toward reporting the mean and volatility of risk aversion on each side we report the mean γ/c and the volatility $\sqrt{\gamma}/c$ on the two sides and a the proportion short. The 8, 000 or so points were quantized into 10 sample reference points presented in Tables 4.1 and 4.2 for the BG and BGǴ respectively. Generally for the representative cases the premia on the long side are larger and more volatile, reflecting both a greater diversity and fear level on this side of the market. The other side being generally more professional has lower and less volatile risk premia.

Table 4.2 *Preference parameters for bilateral double gamma physical candidates. For the long and short side subscripted by L, S the table presents the quantized means and variances for risk premia along with the proportions short. Also presented are the proportion of points represented by each preference set.*

μ_L	σ_L	μ_S	σ_S	a	prop.
1.2127	1.0117	1.0429	1.0390	0.5642	17.47
0.6078	1.5113	0.4660	0.5445	1.8918	4.15
2.0235	0.8012	0.1333	0.2048	0.7837	12.25
1.6896	0.8955	1.5638	1.9991	0.3317	6.46
2.3281	0.7617	3.5737	1.4454	2.1136	0.90
0.6767	0.7429	0.3682	0.2284	2.3226	3.21
5.5704	0.5987	1.0223	1.0712	1.1900	12.51
1.3066	0.6306	0.3299	0.3142	6.7972	4.98
1.3717	0.8648	1.3554	1.2385	0.6862	13.96
1.7936	0.8974	0.3111	0.4199	0.1857	24.11

Table 4.3 *Representative option-implied bilateral gamma parameter sets quantized from estimates for 48 stocks and five-day maturities. Also shown are the proportion of points represented by each set.*

b_p	c_p	b_n	c_n	prop.
0.0137	5.3015	0.0221	4.3398	6.94
0.0212	1.5946	0.0426	0.8468	30.04
0.0212	1.2578	0.0357	0.8266	7.70
0.0693	0.3924	0.0677	1.0083	3.01
0.0232	0.5789	0.0462	0.3278	7.90
0.0511	0.7735	0.0693	1.0730	2.64
0.0196	0.9905	0.0213	0.9551	19.82
0.0248	0.8558	0.0596	0.4167	3.78
0.0238	1.0289	0.0291	1.0021	13.43
0.0422	0.3474	0.0689	0.2075	4.73

4.2.2 Summary of Physical Parameter Estimates

The quantized sets of option implied physical parameter estimates for both the BG and BGG physical candidates are presented in Tables 4.3 and 4.4.

4.2.3 Tests of Physicality from Option Data

With regard to assessing the quality of physicality in the implied candidate physical distribution we present for both models the empirical cumulative distribution function for the candidate distribution functions evaluated on the immediate forward five-day return. The distribution functions are across underliers and five-day intervals for all approximately

Table 4.4 *Representative option implied bilateral double gamma parameter sets quantized from estimates for 48 stocks and five-day maturities. Also shown are the proportion of points represented by each set.*

b_p	β_p	η_p	b_n	β_n	η_n	prop.
0.0210	1.3769	1.4077	0.0720	0.8902	1.1929	5.95
0.5673	0.7966	0.4790	0.2918	0.8076	0.7853	8.80
0.0170	0.7059	1.4315	0.0279	1.3303	0.5711	11.74
0.0039	2.1491	7.4249	0.0045	4.3395	4.0649	3.26
0.0301	1.9317	3.1933	0.0576	1.6661	2.7374	6.40
0.0138	0.7458	2.5331	0.0554	0.8120	1.0686	13.42
0.0248	0.6352	0.9550	0.1092	0.2907	0.9486	13.95
0.0294	0.9904	1.0819	0.1019	0.8233	0.9985	15.25
0.0214	0.5424	4.2439	0.0428	0.8573	1.1603	5.37
0.0173	0.7537	1.7457	0.0780	0.6481	1.0119	15.85

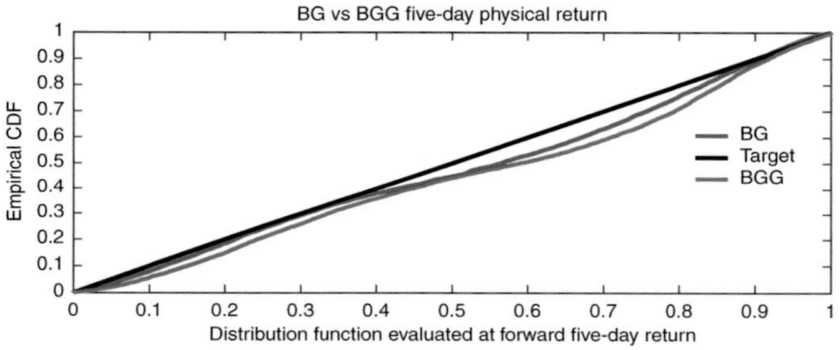

Figure 4.1 Empirical distribution function for candidate physical distributions evaluated at immediately following forward five-day returns. Shown in blue is the result for the bilateral gamma model and shown in red is the result for the bilateral double gamma model.

8,000 estimations for both models. Figure 4.1 presents these curves along with the identity that is the target.

4.2.4 Tests of Physicality from Time Series Data

For the 207 equity underliers reported on in §3.3.1 over the period 2007–2017 the BG parameters were estimated on 402 nonoverlapping five-day returns for a total of $83,214 = 207 \times 402$ estimations. Each of these estimations used a set of 150 immediately prior five-day returns on which digital moment estimations with Anderson–Darling weights were implemented. For each estimate on each underlier on each particular estimation day the estimated distribution function was evaluated on the immediately following five-day return. The empirical distribution function for the candidate uniformly distributed variates

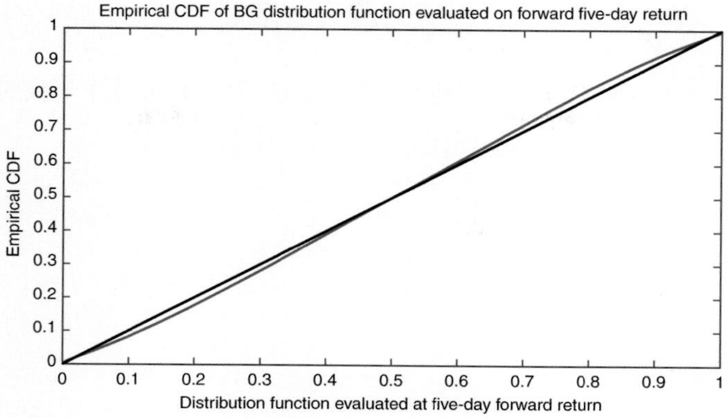

Figure 4.2 Full sample empirical candidate distributed variates for 207 underliers on approximately 400 estimation days

is presented in Figure 4.2 for the full sample of 82,795 sequence of candidate uniform variates. Formally the uniformity rejected at the critical value for a 5% test is 47 basis points and the ks statistic is 0.0260.

5

Multivariate Arrival Rates Associated with Prespecified Univariate Arrival Rates

One of the attractive features of the multivariate normal model is that it provides us with a joint distribution consistent with prespecified marginal distributions that are all in the class of normally distributed random variables. The multivariate model thereby maintains the structure of marginal univariate distributions that may have already been estimated. For the larger dimensions one may invoke factor structures to maintain invertibility of estimated covariance matrices. Furthermore, the analysis of multidimensional issues requires attention to the modeling of dependence between the various univariate returns with distributions in particular parametric classes. Examples include the multivariate VG of Madan and Seneta (1990) or the multivariate T-distribution as described, for example, by Kotz and Nadarajah (2004). Both of these constructions constrain the higher moments of marginal distributions.

There is an extensive literature on dependence modeling consistent with putting together arbitrary prespecified marginal distributions via the use of copulas. Classic references on this approach include Joe (1997) and McNeil et al. (2005). The interest here, however, is not in probability models at some specific horizon but in the modeling of multivariate arrival rates that are consistent with pre-estimated marginal arrival rates. The approach of copulas has been extended to arrival rates by the method of Lévy copulas developed in Kallsen and Tankov (2006). This method is not well suited to dealing with higher dimensions and does not provide analytical access to the embedded multivariate arrival rate function. Specific constructions in high dimensions are possible using pairwise combinations of what are called vine copulas (Grothe and Nicklas, 2013). They deliver access to high-dimensional tail measures. The arrival rate function itself will then involve a multidimensional differentiation.

When univariate risk is represented by exposure to instantaneous changes in value, as opposed to the aggregate effect of shocks on final positions at a horizon, the natural multivariate extension is a multivariate arrival rate function $k(x)$ announcing the arrival rates of simultaneous moves $x = (x_1, x_2, \ldots, x_n)$ for n underlying returns. Some remarks are in order with respect to such an agenda.

The first refers to the recent contribution by Madan and Schoutens (2018) where it is argued that in continuous time one should consider the possibility of zero covariations or no simultaneous moves by any two asset returns. Covariations over longer horizons may arise via the stochasticity of zero covariation arrival rates, and Madan and Schoutens model this using the dependence of univariate arrival rate parameters on the space variable or prices directly. However, if the instants of continuous time are not taken with such great precision, then for small intervals of time there are numerous observations of simultaneous movements. Here we seek multivariate arrival rate functions that do permit simultaneous moves.

More recently, Buchmann et al. (2019) introduced a multivariate model consistent with arbitrary VG marginals yielding a multivariate arrival rate function with full support in n-dimensional space. The model was used in a portfolio theory context by Madan (2018a). This approach is here extended to arbitrary BG marginals.

5.1 Multivariate Model for Bilateral Gamma Marginals

The construction of multivariate arrival rates consistent with prespecified VG marginals introduced in Buchmann et al. (2019) and implemented in Madan (2018a) was based on the αVG process for an n-dimensional vector α and the multivariate VG model (mvg) introduced in Madan and Seneta (1990). That model was defined by taking multidimensional Brownian motion with drift $\mu = (\mu_1, \mu_2, \ldots, \mu_n)$ and covariance matrix Σ evaluated at a random time given by a gamma process with unit mean rate and variance rate $\nu > 0$. The resulting n-dimensional process $X(t)$ has unit time characteristic function given by

$$\Phi_{\text{mvg}}(u) = \left(\frac{1}{1 - iu'\mu\nu + \frac{\nu}{2}u'\Sigma u}\right)^{\frac{1}{\nu}}, \qquad u = (u_1, \ldots, u_n). \tag{5.1}$$

When the underlying n-dimensional Brownian motion process is evaluated on each of its components at a separate set of random times, the resulting conditional characteristic function, conditional on the random times, loses the structure of linearity in the time variates, and as a consequence the composite process is no longer a Lévy process. Weak subordination was introduced in Buchmann et al. (2019) with a view to preserving a Lévy process on employing independent random time subordinators for each component. Buchmann et al. showed that one may add separate and independent VG processes for each component to obtain a multivariate process consistent with prespecified VG marginals. This result is here extended to being consistent with prespecified BG marginals. However, we work here with just the unit time random variables and not the processes.

Suppose the prespecified BG marginal on the jth component has the characteristic function

$$\phi_j(u_j) = \left(\frac{1}{1 - iu_j b_{\text{p}j}}\right)^{\widetilde{c}_{\text{p}j}} \left(\frac{1}{1 + iu_j b_{\text{n}j}}\right)^{\widetilde{c}_{\text{n}j}}. \tag{5.2}$$

Proposition 5.1 *Given a choice of a correlation matrix C and*

$$\nu > \frac{1}{\min_j(\min(\widetilde{c}_{\text{p}j}, \widetilde{c}_{\text{n}j}))},$$

there exists an mvg process and independent bilateral gamma processes for each component with the sum being consistent with the prespecified bilateral gamma marginals with characteristic functions (5.2) *and the given correlation matrix C. The mvg process is defined by*

$$\sigma_j^2 = \frac{2b_{\mathrm{p}j}b_{\mathrm{n}j}}{\nu}, \qquad \Delta(\sigma) = \mathrm{diag}(\sigma_j),$$

$$\Sigma = \Delta(\sigma)\, C\Delta(\sigma), \qquad \mu_j = \frac{b_{\mathrm{p}j} - b_{\mathrm{n}j}}{\nu}.$$

The independent BG marginal scale parameters the same as those of the prespecified marginals with speeds given by

$$\widetilde{c}_{\mathrm{p}j} - \frac{1}{\nu}, \widetilde{c}_{\mathrm{n}j} - \frac{1}{\nu}.$$

Proof The mvg marginal characteristic function in the jth component may be written as

$$\left(\frac{1}{1 - iu_j\nu\mu_j + \frac{\nu}{2}\sigma_j^2 u_j^2}\right)^{\frac{1}{\nu}}.$$

Now consider the difference of two gamma processes with scale coefficients

$$\sigma_j\eta_j\sqrt{\nu/2}, \qquad \sigma_j\theta_j\sqrt{\nu/2},$$

speed $1/\nu$, and characteristic function

$$\left(\frac{1}{1 - iu_j\sigma_j\eta_j\sqrt{\nu/2}}\right)^{\frac{1}{\nu}}\left(\frac{1}{1 + iu_j\sigma_j\theta_j\sqrt{\nu/2}}\right)^{\frac{1}{\nu}}.$$

The characteristic function may also be written as

$$\left(\frac{1}{1 - iu_j\sigma_j\sqrt{\nu/2}(\eta_j - \theta_j) + \frac{u_j^2\sigma_j^2\nu}{2}\eta_j\theta_j}\right)^{\frac{1}{\nu}}.$$

To identify with the mvg marginals make the identifications

$$\mu_j = \frac{\sigma_j(\eta_j - \theta_j)}{\sqrt{2\nu}},$$
$$1 = \eta_j\theta_j.$$

Consider further the addition of an independent BG process with scales $b_{\mathrm{p}j}, b_{\mathrm{n}j}$ and speeds $c_{\mathrm{p}j}, c_{\mathrm{n}j}$ and resulting characteristic function

$$\left(\frac{1}{1 - iu_j\sigma_j\eta_j\sqrt{\nu/2}}\right)^{\frac{1}{\nu}}\left(\frac{1}{1 + iu_j\sigma_j\theta_j\sqrt{\nu/2}}\right)^{\frac{1}{\nu}}\left(\frac{1}{1 - iu_jb_{\mathrm{p}j}}\right)^{c_{\mathrm{p}j}}\left(\frac{1}{1 + iu_jb_{\mathrm{n}j}}\right)^{c_{\mathrm{n}j}}.$$

Set

$$\eta_j = \sqrt{\frac{2}{\nu}}\frac{b_{\mathrm{p}j}}{\sigma_j}, \qquad \theta_j = \sqrt{\frac{2}{\nu}}\frac{b_{\mathrm{n}j}}{\sigma_j}.$$

The condition $\eta_j \theta_j = 1$ requires that

$$\sigma_j^2 = \frac{2 b_{\mathrm{p}j} b_{\mathrm{n}j}}{\nu}.$$

The marginals are now

$$\left(\frac{1}{1 - iu_j b_{\mathrm{p}j}}\right)^{c_{\mathrm{p}j}+1/\nu} \left(\frac{1}{1 + iu_j b_{\mathrm{n}j}}\right)^{c_{\mathrm{n}j}+1/\nu} = \left(\frac{1}{1 - iu_j b_{\mathrm{p}j}}\right)^{\widetilde{c}_{\mathrm{p}j}} \left(\frac{1}{1 + iu_j b_{\mathrm{n}j}}\right)^{\widetilde{c}_{\mathrm{n}j}}.$$

These are BG marginals. Therefore we may form a joint law that is mvg plus independent BG processes given a correlation matrix C and ν with

$$c_{\mathrm{p}j} = \widetilde{c}_{\mathrm{p}j} - \frac{1}{\nu}, \qquad c_{\mathrm{n}j} = \widetilde{c}_{\mathrm{n}j} - \frac{1}{\nu}, \qquad \mu_j = \frac{b_{\mathrm{p}j} - b_{\mathrm{n}j}}{\nu},$$

$$\sigma_j^2 = \frac{2 b_{\mathrm{p}j} b_{\mathrm{n}j}}{\nu}, \qquad \Sigma = \Delta(\sigma) C \Delta(\sigma).$$

The multivariate characteristic function is

$$\left(\frac{1}{1 - iu'\mu\nu + \frac{\nu}{2}u'\Sigma u}\right)^{\frac{1}{\nu}} \prod_{j=1}^{M} \left(\frac{1}{1 - iu_j b_{\mathrm{p}j}}\right)^{c_{\mathrm{p}j}} \left(\frac{1}{1 + iu_j b_{\mathrm{n}j}}\right)^{c_{\mathrm{n}j}}. \qquad \square$$

The multidimensional arrival rate function is

$$k(z) = \widetilde{m}(z) + \sum_{j=1}^{M} k_j(z_j) \prod_{i=1, i\neq j}^{M} \mathbf{1}_{z_i = 0}$$

where

$$\widetilde{m}(z) = \frac{2\exp\left(\mu^T \Sigma^{-1} z\right)}{\nu (2\pi)^{n/2} \sqrt{|\Sigma|}} \left(\sqrt{\mu^T \Sigma^{-1} \mu + \frac{2}{\nu}}\right)^{\frac{n}{2}} \left(\sqrt{z^T \Sigma^{-1} z}\right)^{-n/2}$$
$$\times K_{n/2}\left(\sqrt{\left(\mu^T \Sigma^{-1} \mu + \frac{2}{\nu}\right) z^T \Sigma^{-1} z}\right),$$

$$k_j(z_j) = \frac{c_{nj}}{|z_j|} \exp\left(-|z_j|/b_{nj}\right) \mathbf{1}_{z_j < 0} + \frac{c_{pj}}{z_j} \exp\left(-|z_j|/b_{pj}\right) \mathbf{1}_{z_j > 0}.$$

The arrival rate function may be written as

$$k(z) = \widetilde{m}(z) + \sum_{j=1}^{M} \left(\frac{\widetilde{c}_{nj} - \frac{1}{\nu}}{|z_j|} \exp\left(-|z_j|/b_{nj}\right) \mathbf{1}_{z_j<0} + \frac{\widetilde{c}_{pj} - \frac{1}{\nu}}{z_j} \exp\left(-|z_j|/b_{pj}\right) \mathbf{1}_{z_j>0}\right) \prod_{\substack{i \neq j \\ i=1}}^{M} \mathbf{1}_{z_i=0}.$$

Apart from the marginal arrival rates the multivariate arrival rate may be further simplified by isolating its dependence on ν. Define $\eta = b_p - b_n$ and note that $\mu = \eta/\nu$. Further note that

$$Q^{-1} = \Delta(\sqrt{\nu}\mathbf{1}_n)\Sigma\Delta(\sqrt{\nu}\mathbf{1}_n)$$

is independent of ν and

$$\Sigma = \Delta\left(\frac{1}{\sqrt{\nu}}\mathbf{1}_n\right)Q^{-1}\Delta\left(\frac{1}{\sqrt{\nu}}\mathbf{1}_n\right).$$

Next, observe that

$$\mu^T\Sigma^{-1} = \eta^T Q = \theta^T,$$
$$\sqrt{|\Sigma|} = \frac{1}{\nu^{n/2}\sqrt{|Q|}},$$
$$\sqrt{z^T\Sigma^{-1}z} = \sqrt{\nu}\sqrt{z^T Qz},$$
$$\left(\mu^T\Sigma^{-1}\mu + \frac{2}{\nu}\right)\left(z^T\Sigma^{-1}z\right) = (\theta^T Q\theta + 2)\left(z^T Qz\right).$$

Define

$$\delta = (\theta^T Q\theta + 2)$$

and write

$$m(z) = \frac{2\sqrt{|Q|}}{\nu(2\pi)^{n/2}}\left(\frac{\theta^T Q\theta + 2}{z^T Qz}\right)^{\frac{n}{4}} \exp\left(\theta^T x\right) K_{n/2}\left(\sqrt{\delta z' Qz}\right). \tag{5.3}$$

5.2 The Role of Dependency Parameters in the Multivariate Bilateral Gamma Model

The multivariate BG model has two dependency parameters: ν and the correlation matrix C. These parameters control different aspects of dependency in returns. Here we show by illustration in a simple case that C delivers return correlations while the parameter ν allows further control of correlations in squared returns.

Consider for this purpose the simple case of two multivariate VG variables (x_1, x_2) with

$$x_1 = \sqrt{g}z_1, \qquad x_2 = \sqrt{g}z_2,$$

where z_1, z_2 are multivariate normal and g is gamma distributed with unit mean and variance ν.

It is clear that returns are uncorrelated if the multivariate normal correlation is zero. So return correlations come from the correlation matrix C. Now suppose C is diagonal. Consider squared correlation with

$$x_1^2 = gz_1^2, \qquad x_2^2 = gz_2^2.$$

The variance of x_1 equals that of x_2 equals unity:

$$\mathrm{Cov}(x_1^2, x_2^2) = \mathrm{E}[x_1^2x_2^2] - \mathrm{E}[x_1^2]\mathrm{E}[x_2^2] = \mathrm{E}[g^2] - 1 = \nu.$$

However the variance of x_1^2 is

$$\mathrm{Var}(x_1^2) = \mathrm{E}[g^2z_1^4] - 1 = 2 + 3\nu.$$

So

$$\mathrm{Corr}(x_1^2, x_2^2) = \frac{\nu}{2 + 3\nu}.$$

When ν goes to infinity squared return correlation goes to unity.

Hence the parameter ν calibrates dependence in squared returns and differentiates it from dependence in returns. The return correlation coefficient ρ controls return correlation and ν controls squared correlation. More generally,

$$\mathrm{Corr}(x_1, x_2) = \mathrm{E}[x_1 x_2] = \mathrm{E}[g z_1 z_2] = \rho.$$

But also

$$\begin{aligned}\mathrm{Cov}(x_1^2, x_2^2) = \mathrm{E}[g^2 z_1^2 z_2^2] - 1 &= (1+\nu)\left(1 + 2\rho^2\right) - 1 \\ &= 2\rho^2 + \nu(1 + 2\rho^2).\end{aligned}$$

Hence

$$\mathrm{Corr}(x_1^2, x_2^2) = \frac{2\rho^2 + \nu(1 + 2\rho^2)}{2 + 3\nu}.$$

If ρ goes to ± 1, squared correlation goes to unity. For $\rho = 0$ the maximal squared correlation is $1/3$. For $\nu = 0$ squared correlation is ρ^2. Squared correlation lies in the interval

$$\rho^2 \le \mathrm{Corr}(x_1^2, x_2^2) \le \frac{1}{3} + \frac{2}{3}\rho^2.$$

Or alternatively

$$1 \le \frac{\mathrm{Corr}(x_1^2, x_2^2)}{\rho^2} \le \frac{2}{3} + \frac{1}{3\rho^2}.$$

A further analysis of squared correlations may be found in Madan (2020) and Madan and Wang (2020b).

5.3 Multivariate Bilateral Gamma Lévy Copulas

Every joint probability distribution function $f(x)$ for $x \in \mathbb{R}^n$ with marginal distributions $F_i(x_i)$ delivers a copula $C_f(u_1, \ldots, u_n)$ via the construction

$$\begin{aligned}C_f(u_1, \ldots, u_n) &= F\left(F_1^{-1}(u_1), \ldots, F_n^{-1}(u_n)\right), \\ F(x) &= \int_{-\infty}^{x_1} \cdots \int_{-\infty}^{x_n} f(y_1, \ldots, y_n) dy_1 \cdots dy_n.\end{aligned}$$

This copula may then be used to build dependence models for arbitrary marginal distribution functions $G_i(x_i)$, with $x_i \in \mathbb{R}$, by

$$H(x_1, \ldots, x_n) = C_f(G_1(x_1), \ldots, G_n(x_n)).$$

Kallsen and Tankov (2006) define Lévy copulas that construct a multivariate Lévy measure that can be used to build dependence between marginal Lévy measures. Multivariate signed tail measures were defined in equation (2.16) as

$$K(x) = \prod_i \operatorname{sign}(x_i) \int_{A(x)} k(y)dy,$$
$$A(x) = \left\{ y \in \mathbb{R}^d | \operatorname{sign}(x_i)y_i \geq \operatorname{sign}(x_i)x_i \right\}.$$

We observed in §2.3.2 that any multidimensional Lévy measure uniquely defines the complex exponential variation that is uniquely related to its multivariate signed tail measure. Furthermore the marginal measures are given by $k_i(y_i) = k(e_i y_i)$, where e_i is the ith column of the $n \times n$ identity matrix and

$$K_i(x_i) = -\mathbf{1}_{x_i<0} \int_{-\infty}^{x_i} k_i(y)dy + \mathbf{1}_{x_i>0} \int_{x_i}^{\infty} k_i(y)dy.$$

Consider now standard BG marginal tails $B_1(x), \ldots, B_n(x)$ given by

$$B_i(x) = -\mathbf{1}_{x<0} \int_{-\infty}^{x} \frac{e^{-|y|}}{|y|} dy + \mathbf{1}_{x>0} \int_{x}^{\infty} \frac{e^{-y}}{y} dy, \; i = 1, \ldots, n.$$

Further, for correlation matrix C and parameter $\nu > 0$ let $m_{\mathrm{sbg}}(x)$ be the multivariate VG Lévy measure with standard BG marginals, correlation matrix C, and gamma time change variance ν. Let $B(x)$ be the associated multivariate signed tail measure

$$B_{C,\nu}(x) = \prod_i \operatorname{sign}(x_i) \int_{A(x)} m_{\mathrm{sbg}}(y)dy,$$
$$A(x) = \left\{ y \in \mathbb{R}^d | \operatorname{sign}(x_i)y_i \geq \operatorname{sign}(x_i)x_i \right\}.$$

Now let $G_1(x), \ldots, G_n(x)$ for $x \in \mathbb{R}\backslash\{0\}$ be arbitrary marginal Lévy tail measures with $G_i(x) < 0$ for $x < 0$ and $G_i(x) > 0$ for $x > 0$. Define

$$L_i(x) = B_i^{-1}(G_i(x)), \text{ for } x \neq 0.$$

If we set $L_i(x) = y$, then

$$x = G_i^{-1}(B_i(y)),$$

and the original variates are nonlinear transforms of standard BG marginals. For correlation matrix C and gamma time change parameter ν the multivariate signed tail measure is given by

$$H(x) = B_{C,\nu}(L_1(x_1), \ldots, L_n(x_n)),$$

where the L_i are as defined above and B is the multivariate Lévy tail of the multivariate standard BG with parameters ν and C.

The multivariate standard BG Lévy measure is

$$m_{\text{sbg}}(x) = \frac{(\nu/2)^{n/2-1/2} \exp\left(-\sqrt{(x^{\mathrm{T}} C^{-1} x)}\right)}{\nu\,(2\pi)^{n/2-1}\sqrt{|C|}\sqrt{x^{\mathrm{T}} C^{-1} x}}.$$

It is constant on the ellipse $x^{\mathrm{T}} C^{-1} z = y$ at the level

$$\frac{(\nu/2)^{n/2-1/2}}{\nu\,(2\pi)^{n/2-1}\sqrt{|C|}} \frac{e^{-\sqrt{y}}}{\sqrt{y}}.$$

5.4 Multivariate Model for Bilateral Double Gamma Marginals

In the case of BGG marginals the log characteristic function of the prespecified marginals takes the form

$$\ln\left(\phi_j(u)\right) = -\widetilde{\eta}_p^j \ln\left(1 + \widetilde{\beta}_p^j \ln(1 - iu\widetilde{b}_p^j)\right) - \widetilde{\eta}_n^j \ln\left(1 + \widetilde{\beta}_n^j \ln(1 + iu\widetilde{b}_n^j)\right).$$

On splitting the VG into the difference of two gamma processes with the same speed we get the marginal characteristic function

$$-\frac{1}{\nu}\ln\left(1 - iuw_j^{(p)}\right) - \frac{1}{\nu}\ln\left(1 + iuw_j^{(n)}\right),$$

where

$$w_j^{(p)} - w_j^{(n)} = \mu_j \nu, \qquad w_j^{(p)} w_j^{(n)} = \frac{\sigma_j^2 \nu}{2}.$$

One may solve for the coefficients μ_j and σ_j^2 by

$$\sigma_j^2 = \frac{2 w_j^{(p)} w_j^{(n)}}{\nu}, \tag{5.4}$$

$$\mu_j = \frac{w_j^{(p)} - w_j^{(n)}}{\nu}. \tag{5.5}$$

Consider each side separately and note that on the positive side we have the target

$$\left(\frac{1}{1 + \widetilde{\beta}_p^{(i)} \ln(1 - iu\widetilde{b}_p^{(i)})}\right)^{\widetilde{\eta}_p^{(i)}}.$$

However, on adding independent BGG variates we attain the characteristic function

$$\left(\frac{1}{1 - iuw_j^{(p)}}\right)^{\frac{1}{\nu}} \left(\frac{1}{1 + \beta_p \ln(1 - iub_p)}\right)^{\eta_p}.$$

If we identify the two scale coefficients β_p, b_p with the corresponding prespecified ones we get the condition

$$\left(\frac{1}{1-iuw_j^{(p)}}\right)^{\frac{1}{\nu}} = \left(\frac{1}{1+\widetilde{\beta}_p^j \ln(1-iu\widetilde{b}_p^j)}\right)^{\widetilde{\eta}_p^j-\eta_p}.$$

Choosing

$$\eta_p = \widetilde{\eta}_p^j - \frac{1}{\nu},$$

we are left with

$$1-iuw_j^{(p)} = 1+\widetilde{\beta}_p^j \ln(1-iu\widetilde{b}_p^j).$$

Expanding in u on the right we have

$$1-iu\widetilde{\beta}_p^{(i)}\widetilde{b}_p^{(i)} + \frac{1}{2}\widetilde{\beta}_p^{(i)}\widetilde{b}_p^{(i)2}u^2.$$

Equating first-order terms we take

$$w_j^{(p)} = \widetilde{\beta}_p^j\widetilde{b}_p^j,$$

and ignore the second-order term as an approximate result.

The proposed solution is to define the scale coefficients of the independent BG as identical with the prespecified targets. Define

$$w_j^{(p)} = \widetilde{\beta}_p^j\widetilde{b}_p^j, \qquad w_j^{(n)} = \widetilde{\beta}_n^j\widetilde{b}_n^j,$$

and set the mvg parameters in accordance with (5.4) and (5.5) and the speed parameters as the prespecified ones less $1/\nu$. In this case mvg plus independent BG components is an approximation to the prespecified marginals.

5.5 Simulated Count of Multivariate Event Arrival Rates

For multivariate arrival rate functions (5.3) the evaluation of arrival rates events defined by subset A of $\mathbb{R}^n\backslash\{0\}$ requires the performance of multidimensional integrals of the form

$$\int_A m(z)dz.$$

Already in the one-dimensional case we have considered simulation for such integrals especially when the set A may be complicated. In the multidimensional case for the higher dimensions this will in fact be the only tractable path. This requires the introduction of a multidimensional density $g(z)$ that may be simulated from, with

$$\int_A m(z)dz = E^g\left[1_A\frac{m(z)}{g(z)}\right].$$

Abstracting from the exponential tilt $\theta' z$ the arrival rate function 5.3 is constant on the ellipse with radius

$$r = \sqrt{z'Qz},$$

with the structure of a gamma arrival rate for the radius. In fact, for $\theta = 0$ we may write

$$m(z) = \frac{\nu^{n/2-3/2}}{(2\pi)^{n/2-1} r\sqrt{c}} \exp(-r\sqrt{\delta}).$$

Such a structure suggests the use for $g(z)$ of a gamma-distributed elliptical radius.

Define

$$g(z) = \frac{\delta^{\gamma/2}\sqrt{z'Qz}^{\gamma-1}}{\Gamma(\gamma)} \exp(-\sqrt{\delta}\sqrt{z'Qz}).$$

One may write

$$z = ru,$$

where u is uniform on the ellipse $z'Qz = 1$ and r, u has the joint density

$$h(r,u) = \frac{\Gamma(n/2)\delta^{\gamma/2} r^{\gamma-1}}{(2\pi)^{n/2}\sqrt{|Q|}\Gamma(\gamma)} \exp(-\sqrt{\delta} r).$$

The normalizing constant for the surface area of the ellipse $z'Qz = 1$ is

$$\frac{2\pi^{n/2}\sqrt{|Q|}}{\Gamma(n/2)}.$$

Let $w = \sqrt{\delta} r$ have a standard gamma density $l(w)$ with shape parameter γ,

$$l(w) = \frac{w^{\gamma-1}\exp(-w)}{\Gamma(\gamma)}.$$

We then have

$$z = \frac{1}{\sqrt{\delta}} wu.$$

The measure change is explicitly given by

$$\rho(z) = \frac{m(z)}{g(z)} = \frac{2\pi\sqrt{\delta}\nu^{n/2-3/2}\exp(\theta' z)\Gamma(\gamma)}{\Gamma(n/2)\delta^{\gamma/2}(z'z)^{\gamma/2}}.$$

Notice that the small constant $c = |Q|$ is part of the surface area correction for the ellipse and is no longer present in the measure change. The use of precisely this ellipse in the definition of g helps in developing a relatively bounded measure change function.

Arbitrary multidimensional variations $c(z)$ may then be evaluated by

$$\int_{-\infty}^{\infty} c(z)m(z)dz = E^g\left[c(z)\rho(z)\right],$$

where E^g refers to expectations under the distribution g for a gamma-distributed elliptical radius. Simulation of a gamma-distributed elliptical radius is easily accomplished in relatively high dimensions. With respect to the independent bilateral components one may follow the procedures for univariate counts already presented. The selection of γ should be a small positive value as already noted in the univariate case.

6

The Measure-Distorted Valuation As a Financial Objective

The maximization of wealth has long been recognized and advocated as a sound objective for the decisions facing economic enterprises. A classic piece on its advocacy is Milton Friedman's article in the *New York Times*, dated September 13, 1970, and entitled, "The Social Responsibility of Business Is to Increase Its Profits." A more detailed advocacy is also presented in Jensen (2002). However, the law of one price coupled with market efficiency or the absence of arbitrage creates problems for such an objective and its advocacy.

It is demonstrated, for example in Ross (1978), that the valuation of risky streams in the absence of arbitrage must be done by a linear function of the streams. As a consequence of linearity, directions of increase in value maintain this status no matter how far one goes in this direction. Basically linear functions do not have interior extreme values and one must seek guidance from constraints in defining optimal solutions. For many problems, like the design of hedging strategies or the selection of value-maximizing portfolios, one often seeks optimal outcomes in the interior of a continuum of possibilities. The linearity of valuation dismisses all such solutions from the class of value-maximizing outcomes.

Yet another problem from a value-maximizing perspective is that the linear objective may be flat in numerous directions. For example, with zero cost hedges the value of the hedged position is invariant to the hedge and is the value of the unhedged position no matter what the hedge. Similarly the value of a portfolio is the sum of component values no matter the allocation between the components.

These value-maximizing failures due to the linearity of value have led the literature to pursue other objective functions for solving these problems. Primary among these has been the use of expected utility maximization with its long history in economics and financial economics. Many financial applications follow Merton (1969, 1971). With respect to hedging, we cite, for example, Duffie et al. (1997). A more recent summary of research along these lines is presented in Skiadas (2008). Nonetheless, we note that expected utility is an evaluation of personal welfare as opposed to economic or social welfare, and the advocacy of the latter was based on these wider concerns of corporate risk management in contrast to the narrower personal considerations embedded in utility theory. In this regard it may also be noted that the attraction of market value maximization lay to a considerable extent in the relative objectivity of market opinions over that of individual persons. By modifying value to be nonlinear, the aim is to restore the objectivity of markets in directing economic and financial decision-making. The affairs of corporations and nations are not to be directed by the interests of the few but the many, and this is why we need a foundation in a theory of nonlinear market valuation.

The theoretical support for value maximization also lies in the welfare theorems of general equilibrium theory as expounded by Arrow and Hahn (1971). In full generality recognizing a variety of externalities the case for a constitutional objective function for firms and enterprises falls apart as argued in Milne (1981). The considerations raised in that paper point toward the view that welfare theorems are, in any case, an extreme position of questionable practical relevance. They are not and cannot be the main reason supporting value maximization. Probably a better argument is the democratization of market values away from too personalized an objective that takes account of just a single possibly misdirected evaluation of a mix of beliefs and preferences inherent in good decision-making. Conservative market valuation aims at such a democratization.

The theory of competitive markets with its presumed underlying law of one price is after all an abstraction with an associated extreme view of trades that will be acceptable to market participants. The extreme view is represented by the assumption of a large supply or demand available at prices slightly above or below the equilibrium price. The offer to sell at a slightly lower price is presumed to be easily taken up by market participants, as is the offer to buy at a slightly higher price. The zero-cost risky streams acceptable to the market are thereby viewed as those with a positive market value and in the financial literature are termed positive alpha trades. With the focus on the return to an invested dollar the price is unity, and any offer below or above unity generates an excess return or alpha for the buyer or seller accordingly. At the equilibrium price (Sharpe, 1964; Lintner, 1965; Mossin, 1966; Ross, 1976) of unity for the invested dollar expected returns equal risk-adjusted required returns with deviations delivering alpha on one side or the other.

With a linear valuation operator this set of market acceptable trades is very large. In fact it is the largest convex set possible containing the nonnegative cash flows. It is what is technically referred to as a half-space, and no bigger set is convex. An alternative abstraction for the set of market-acceptable trades models the zero-cost tradeable claims as a convex set that, however, is not closed under negation, as the terms of trade are allowed to change with the direction of trade. Such a model is considered in Madan (2015a). Market efficiency and the absence of arbitrage result in a two-price economy with the lower price reflecting the terms at which market participants may sell risky streams and the upper price is one at which they may buy such streams. The two prices are, however, not the same and the law of one price fails. Fortuitous trades may occur at terms between the upper and lower prices, but all the economy may solve for or determine are the two separate prices. The two pricing operators are, however, nonlinear, and concave for the lower price and convex for the upper one. The former represents a conservative valuation of assets and may be maximized. The latter may be employed to value liabilities conservatively and may be minimized. They may be employed in solving for optimal hedges, portfolio allocations, and derivative design, among other risk management problems. As they turn out to be infima and suprema of classical linear valuation operators, they continue to represent the democratization of market valuations mentioned earlier in support of value maximization.

The resulting nonlinear conservative valuation principles yield easily implementable objective functions for financial decision-making when they are further related to the maximization of distorted expectations as opposed to expectations. The distortions incorporate our natural uncertainty with the probabilities of tail events in particular. The lack of experience with tail events is reflected in distortions by raising the probabilities attached to losses while simultaneously lowering gain probabilities to attain value conservatism. Much of this material was covered in Madan and Schoutens (2016). Here the approaches are extended to distorting directly unnormalized arrival rates of events given that we work with variation measures on variation spaces and there is no underlying probability or expectation to be evaluated. The result is a measure-distorted valuation operator as a financial decision-making objective function modeling conservative valuation in the context of variation spaces.

6.1 Linear Valuation Issues

A two-date one-period context is sufficient to highlight the issues. For the purposes of a general discussion one may also restrict attention to a finite probability space denoted by $(\Omega, \mathcal{F}, P)$ for a base physical probability measure P. Let random outcomes $X : \Omega \to \mathbb{R}$ have a value denoted by $V(X)$.

In the absence of arbitrage and under the law of one price it must be the case that

$$V(\alpha X + \beta Y) = \alpha V(X) + \beta V(Y),$$

and hence V is a linear function whereby there exists $m = m(\omega)$ such that

$$V(X) = \int m(\omega) X(\omega) P(d\omega).$$

Consider in particular a positive outcome $X_0(\omega) > 0$ with value $V_0 > 0$. We then have

$$V_0 = \int m(\omega) X_0(\omega) P(d\omega).$$

Furthermore, as positive outcomes have positive values $m(\omega) \geq 0$.

One may then write

$$\frac{V(X)}{V_0} = \frac{\int m(\omega) X_0(\omega) \frac{X(\omega)}{X_0(\omega)} P(d\omega)}{\int m(\omega) X_0(\omega) P(d\omega)} = \int q(\omega) \frac{X(\omega)}{X_0(\omega)} P(d\omega) = E^Q \left[\frac{X(\omega)}{X_0(\omega)} \right],$$

$$q(\omega) = \frac{m(\omega) X_0(\omega)}{\int m(\omega) X_0(\omega) P(d\omega)}.$$

Here $q(\omega)$ is referred to as a change of probability, X_0 is called a numeraire or store of value asset, and outcomes relative to the store of value have an expectation under a change of probability equal to that of the value of the outcome expressed in units of the store of value asset.

It is customary to use dollars as the numeraire and let $X_0(\omega) = 1$. In this case $V_0 = e^{-rt}$ for an interest rate r relevant to the time period t. One then writes

$$V(X) = e^{-rt}E^Q[X],$$

or one equates market value to the discounted expected value under the pricing probability Q.

The problems with value maximization as a criterion for financial decision-making begin at this point. Clearly if one is constrained to choose between two outcomes X_1, X_2, the criterion offers the guidance to make the choice X_1 if $V(X_1) > V(X_2)$ and the other way, otherwise. If, however, we have a liability Y that may be hedged by positions α in some hedging instruments H that are financed or alternatively have a zero cost, then the hedged position is $\alpha' H - Y$ and its value is $-V(Y)$ independent of the position α. Similarly if α is the proportion invested in X_1 and $(1 - \alpha)$ is invested in X_2, the portfolio return is

$$\alpha \frac{X_1}{V(X_1)} + (1 - \alpha)\frac{X_2}{V(X_2)},$$

with a portfolio value of unity independent of α. Value maximization fails to deliver, as an objective function, solutions to important problems in financial decision-making.

One may resort to other considerations for such problems, like expected utility maximization or mean variance considerations. The 200 years of work in economics and finance cited, for example, by Jensen (2002) as supporting value maximization leave us short on answering important financial questions. A position often taken in this situation is that financial questions have no real economic implications as demonstrated by the invariance of value to alternative financial plans. The famous Modigliani and Miller (1958) theorem is a case in point. However, one must recognize that theorems have no scientific content and only reflect how the conclusions have been assumed. Essentially, if all financial decisions may be reversed and unwound at zero cost, then this explains why they do not matter. The problem, however, lies not in the conclusion but in the assumption of financial irrelevance.

Madan (2015b) shows that in the absence of arbitrage when the space of zero-cost tradeable claims is not closed under negation, the law of one price fails and all a free market can deliver are two prices or valuations. The lower price is the terms at which one may deliver risks to the market. The higher one is the terms at which one may acquire risks from the market. Related to the structure of zero-cost tradeable claims is a set of risks acceptable to the market.

Under the law of one price for any $a > 0$ the market will accept $V(X) + a - X$ or $X - (V(X) - a)$ as positive-value trades. This is a large, convex set of trades with no larger set being convex. Figure 6.1 presents an example of such a set in two dimensions. The set of positive-value trades is to the right of the blue line.

Acceptable risks are not explicitly modeled under the law of one price, but are just a consequence of such an assumption. One may always buy for more or sell for less than the market price. However, there is another view of markets that is not so generous, and for a range of prices the market is not willing to either buy or sell the associated risk. One

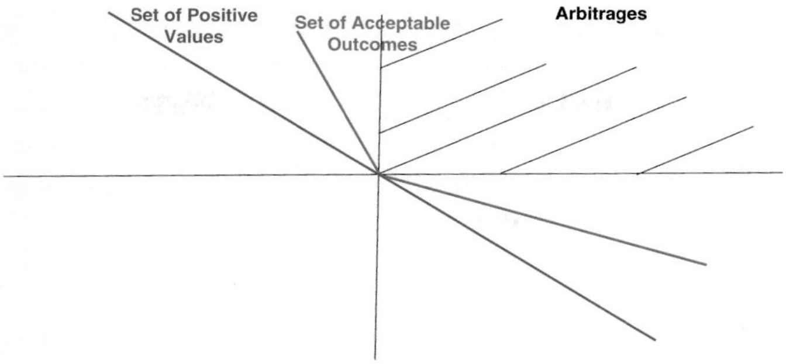

Figure 6.1 Arbitrages, positive values, and acceptable risks

can buy, but at a sufficiently high price, and one may sell at a substantial price reduction. Such two-price markets are an alternative abstraction to the law of one price in modeling markets.

We develop value maximization as a financial decision-making criterion for such two-price markets. The desire to go to markets for value is the same as in classical theory. We wish to disengage economic decisions from personal preference considerations that can be polluted by a variety of biases and can even border on corruption when viewed from a market perspective. The failure of value maximization in delivering guidance on important financial decisions was just an anomalous consequence of the abstraction associated with the law of one price that may not be relevant in more realistic models of market activity.

6.2 Modeling Risk Acceptability

The law of one price was observed to implicitly model the set of acceptable risks as in Figure 6.1 as among the largest convex sets possible. Actual markets may fail to be this generous. However, nonnegative outcomes offered at zero cost, being arbitrages, are always acceptable. A natural model for acceptable risks then leads one to consider convex sets containing the nonnegative outcomes. The convexity follows from recognizing that one may combine acceptable trades like packaging a quarter of one and three quarters of the other.

A simpler model of acceptable risks arises on permitting scaling of trades. The market being large relative to most participants, one anticipates no real difficulty in raising the size of a trade by 5 or 10%. Arbitrary scaling is a problem as it can test the sanity of both parties when seriously considered. Theoretically it is simpler to allow for arbitrary scaling, recognizing that in practice only small scales will be involved. When scaling is permitted at all levels, the set of acceptable risks simplifies to a convex cone containing the nonnegative outcomes. This approach to risk acceptability was adopted by Artzner et al. (1999) in their paper on coherent measures of risk. Figure 6.1 presents an example of such a cone as the area to the right and above the red lines. When each of the two red lines are extended to form lines, then they each define half-spaces of acceptable risks in the classical sense, but now we have to intersect the two half-spaces to form the acceptable outcomes.

Artzner et al. showed that all such convex cones of acceptable risks are defined as the intersection of numerous half-spaces. More formally, for any convex cone of acceptable risks or outcomes denoted by $X \in \mathcal{A}$ there exists a set of supporting probability measures $Q \in \mathcal{M}$ equivalent to the base probability P such that

$$X \in \mathcal{A} \text{ if and only if } E^Q[X] \geq 0 \text{ for all } Q \in \mathcal{M}.$$

In classical equilibrium theory two cones of acceptable risks are involved. At the aggregate level we ask for positive excess supply in all states, and this is the smallest cone of just the nonnegative outcomes. Market participants accept any trade with a positive market value, and this is the largest convex cone of the half-space described and discussed in Figure 6.1. At the aggregate level the set of acceptable risks being considered is not generous at all, requiring excess supplies to belong to a very small cone. At the level of market participants it is very generous. Demanding nonnegative excess supplies in all states may be too restrictive, leading to economies that are too small and politically unacceptable. One may therefore choose to work with a larger candidate for the small cone. But likewise we consider a smaller candidate for the larger cone. Some of these matters are addressed in an equilibrium context in Madan (2018b). Leaving aside equilibrium and social welfare considerations and the interactions between the two cones, a larger small cone and a smaller large cone, we go on to modeling market participation defined in terms of a proper subcone that is not a half-space. From this perspective the set of supporting probabilities $\mathcal{M}$ cannot just consist of a single probability Q. Risk acceptability in markets then requires a nonnegative expectation under a number of supporting probabilities. These probabilities are a mix of beliefs and preference considerations and work toward the democratization of value in markets.

It is interesting to inquire into the status of the law of one price once the cone of market-acceptable risks is no longer a half-space. For an outcome X to be associated with one price, both X and $-X$ must be acceptable. But then we must have that

$$E^Q[X] = 0, \text{ all } Q \in \mathcal{M}.$$

One may denote this by

$$\mathcal{N} = \left\{X \mid E^Q[X] = 0, \text{ all } Q \in \mathcal{M}\right\}.$$

The set $\mathcal{N}$ is the null space of outcomes with zero valuation by all supporting probabilities. It is an intersection of subspaces that is a linear subspace consisting of the set of claims for which the law of one price equal to zero holds. If $\mathcal{N}$ is just $\{0\}$, then the law of one price fails for all nonzero outcomes. If $\mathcal{N}$ is not empty, one may define an equivalence relation of outcomes by $X \sim X'$ if and only if $X - X' \in \mathcal{N}$. In the space of equivalence classes or the outcomes modulo $\mathcal{N}$, the law of one price fails universally. For a rich enough collection of supporting probabilities one anticipates $\mathcal{N} = \{0\}$. In this context, marking to market fails as markets do not and cannot assign a value to any outcome. What markets can do is to assign conservative valuations that may be used for marking purposes, assets at the lower valuation and liabilities at the upper valuation. The structure of these valuations is considered next.

Before taking up valuation in such conservative contexts it is important to note that in many applications one will have to monitor dynamic movements in the set $\mathcal{M}$ of supporting probabilities. Periods like the financial crisis of 2008 clearly witnessed a sizable expansion in $\mathcal{M}$ while the preceding boom may well have seen substantial decreases in this set. The size of the real economy is also significantly impacted across time by variations in this set. All risky production and transaction plans have to pass risk acceptability tests to come to fruition and the size of the set $\mathcal{M}$ controls what may or may not be approved. There are also a variety of interactions taking place in the definition of $\mathcal{M}$ with regulators proposing equity standards while equity markets ultimately define what the levels are. These issues are taken up later in the book.

6.3 Nonlinear Conservative Valuation

Consider an outcome X offered to the market for a candidate price $V(X)$. The market has to evaluate the acceptability of its position that is

$$X - V(X).$$

From the perspective of an existing cone of risk acceptability $\mathcal{A}$, this requires that $X - V(X) \in \mathcal{A}$ or equivalently that

$$V(X) \leq \inf_{Q \in \mathcal{M}} E^Q[X].$$

The best or highest lower valuation is then given by

$$L(X) = \inf_{Q \in \mathcal{M}} E^Q[X]. \tag{6.1}$$

Similarly if one offers a candidate upper price $V(X)$ for the acquisition of X, then acceptability of $V(X) - X$ leads to the best price as

$$U(X) = \sup_{Q \in \mathcal{M}} E^Q[X]. \tag{6.2}$$

The upper and lower valuation operators (6.2) and (6.1) are respectively convex and concave and suitable for minimization and maximization.

In particular we have in general that

$$L(X+Y) \geq L(X) + L(Y), \qquad U(X+Y) \leq U(X) + U(Y),$$

with sharp inequalities reflecting the benefits of risk diversification. Unlike the classical case, value-maximizing hedge and portfolio considerations may now be developed. Value optimization is not restricted to the boundaries of constraint sets. As the valuations scale with the size of positions, value optimization provides guidance on the direction of trades or positions and not the absolute scale. Issues of scale are to be settled by other considerations.

6.4 Risk Reward Decompositions of Value

The valuation operator $L(X)$ may quite generally be given a risk reward representation as follows. Suppose the collection $\mathcal{M}$ of supporting probabilities is viewed as perturbations of a base probability $P \in \mathcal{M}$. The reward may be taken as usual to be the expectation under the base, reference, or true probability, $E^P[X]$. Removing the reward, we are left with pure risk, given by

$$X^c = X - E^P[X].$$

To eliminate this pure risk component the long position holding the risk would have to access from the market its negative at a cost of the upper valuation

$$U(-X^c) = U\left(E^P[X] - X\right).$$

The reward less the cost of eliminating the pure risk is

$$E^P[X] - U\left(E^P[X] - X\right) = -U(-X) = L(X),$$

where the last equality is a consequence of the relation between suprema and infima. Hence the conservative value for long position in an outcome is always representable as reward less risk where the risk is the upper valuation for the negated pure risk component.

With respect to $U(X)$, this upper valuation operator is applied to liabilities and the representation is in the form of expectation or reward plus a risk add-on. The pure risk component is again X^c, a liability that can be eliminated at the cost of the upper valuation of X^c. The expectation plus risk add-on is then

$$E^P[X] + U(X^c) = U(X).$$

Alternatively one may define risk charges for the long position as

$$\text{RC}_\text{L} = E^P[X] - L(X) = U\left(E^P[X] - X\right),$$

and hence the risk charge is the upper valuation for the centered and negated outcome. Furthermore, we may write

$$L(X) = E^P[X] - \text{RC}_\text{L}.$$

For the short position the risk charge or add-on is

$$\text{RC}_\text{S} = U(X - E^P[X]),$$

and

$$U(X) = E^P[X] + \text{RC}_\text{S}.$$

6.5 Remarks on Modigliani–Miller Considerations

By the concavity of the lower valuational operator the value of a package can be greater than the sum of the value of the components. A consequence of this property is a possible violation of the Modigliani–Miller result with the lower valuation of an unlevered firm exceeding the value of debt plus equity for an equivalently levered firm. A simple example illustrates.

Let S be the value of the stock of an unlevered firm at one year for an initial value of 100. Let $\mathcal{M}$ consist of just two probabilities that are both geometric Brownian motion with volatilities $\sigma_1 = 0.2$ and $\sigma_2 = 0.4$. With a continuously compounded interest rate of 2%, the value of unlevered equity is 100 under both probabilities and the lower valuation is also 100. With leverage and debt at 75, the lower value of levered equity is 26.9436, attained at the lower volatility of σ_1 as a call option struck at 75. The lower value of debt is 69.2141 attained at the higher volatility of σ_2. This debt value is the present value of the bond face 75 less the upper value of a put struck at 75.

The lower value of the levered firm taken as the sum of the lower value of the component securities is then 96.1577, and it is below that of the unlevered firm. In this simple two-price economy the Modigiliani–Miller result fails for the lower valuation operator. We shall later exhibit a set of sufficient conditions restoring this result for the lower valuation operator.

Given that the Modigiliani–Miller result fails, there is a cost to issuing debt that may be offset by the tax benefits accruing to debt issuance. For a tax rate of 15% we present in Figure 6.2 a graph of the value of the levered firm as a function of the debt issued. The optimal debt in this case is around \$40.

6.6 Probability Distortions

A relatively large class of lower valuation operators for outcomes on a base probability space with probability P has been introduced, for example, in Kusuoka (2001) based on distortions of probability. The construction of each of these valuations requires a prior selection of a probability distortion function. This distortion function is a concave distribution $\Psi(u)$, with $0 \leq u \leq 1$, defined on the unit interval. For any potential outcome X the lower valuation is defined in terms of its distribution function $F_X(x) = P(X \leq x)$ and density $f_X(x)$. Specifically

$$L(X) = \int_{-\infty}^{\infty} x d\Psi(F_X(x))dx \tag{6.3}$$

$$= \int_{-\infty}^{\infty} x\Psi'(F_X(x)) f_X(x)dx. \tag{6.4}$$

In the absence of a distortion or when $\Psi(u) = u$ the valuation $L(X)$ is just the expectation of X. In the presence of a proper concave distortion equation (6.4) shows that one is evaluating a reweighted expectation. The lower quantiles have their weights lifted by the concavity of Ψ' while the higher quantiles have their weights reduced. As a result the lower valuation $L(X)$ is below the expectation of X. The greater the concavity of Ψ, the greater the distortion

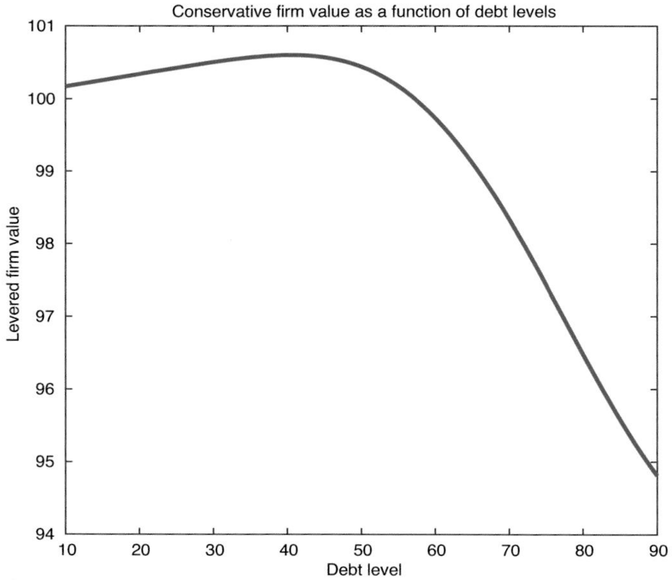

Figure 6.2 Optimal debt when levered firms are valued using a conservative lower valuation operator. The level of optimal debt is around 40. The tax rate used was 15%.

and the more extensive the reweighting. The reduction in the lower valuation thus depends on the concavity of the distortion. Conservatism in valuations is induced by the concavity of the distortion.

An equivalent formulation of equation (6.3) can be obtained by integrating by parts to obtain the Choquet (1953) representation for $L(X)$ as

$$L(X) = -\int_{-\infty}^{0} \Psi(F_X(x))dx + \int_{0}^{\infty} (1 - \Psi(F_X(x)))\,dx. \tag{6.5}$$

One may introduce the complementary convex distortion

$$\widehat{\Psi}(u) = 1 - \Psi(1 - u)$$

that lies below the identity just as Ψ lies above it to write with $\widehat{F_X}(x) = 1 - F_X(x)$ that

$$L(X) = -\int_{-\infty}^{0} \Psi(F_X(x))dx + \int_{0}^{\infty} \widehat{\Psi}\left(\widehat{F_X}(x)\right) dx. \tag{6.6}$$

It is clear from the representation of equation (6.6) that the left tail is lifted upward by Ψ and the right tail is shifted downward by $\widehat{\Psi}$. The greater the concavity of Ψ the further are both shifts and the more conservative the lower valuation.

One may further replace distribution functions and write directly in terms of the base probability P that

$$L(X) = -\int_0^\infty \Psi(P(X^- > x))\,dx + \int_0^\infty \widehat{\Psi}\left(P\left(X^+ > x\right)\right)dx, \tag{6.7}$$

where one is directly evaluating the integral of distorted tail probabilities, lifting the negative contributions and deflating the positive ones.

From expression (6.7) it is clear that

$$\begin{aligned} L(-X) &= -\int_0^\infty \Psi(P((-X)^- > x))\,dx + \int_0^\infty \widehat{\Psi}\left(P\left((-X)^+ > x\right)\right)dx \\ &= -\int_0^\infty \Psi\left(P(X^+ > x\right)dx + \int_0^\infty \widehat{\Psi}(P(X^- > x))\,dx. \end{aligned}$$

Hence $U(X) = -L(-X)$ is

$$U(X) = -\int_0^\infty \widehat{\Psi}(P(X^- > x))\,dx + \int_0^\infty \Psi\left(P(X^+ > x\right)dx, \tag{6.8}$$

where the roles of Ψ and $\widehat{\Psi}$ have been reversed in the evaluation of the distorted tail probability integrals.

The set of probabilities supporting distorted expectations as a lower valuation operator is identified in the following proposition.

Proposition 6.1 *The set $\mathcal{M}$ of probabilities Q for which the distorted expectation of equation (6.7) satisfies*

$$L(X) = \inf_{Q \in \mathcal{M}} E^Q[X],$$

is defined by

$$\mathcal{M} = \{Q | Q(A) \leq \Psi(P(A)), \textit{ for all } A\}.$$

Proof Suppose first that $Q(A) \leq \Psi(P(A))$ and note that

$$E^Q[X] = -\int_0^\infty Q(X^- > x)\,dx + \int_0^\infty Q\left(X^+ > x\right)dx.$$

Since

$$Q(X^- > x) \leq \Psi(P(X^- > x)),$$

we get that

$$-\int_0^\infty \Psi(P(X^- > x))dx \leq -\int_0^\infty Q(X^- > x)\,dx.$$

Furthermore, since

$$Q\left(X^+ \leq x\right) \leq \Psi\left(P\left(X^+ \leq x\right)\right) = 1 - \widehat{\Psi}\left(1 - P\left(X^+ \leq x\right)\right),$$

we obtain

$$\int_0^\infty \widehat{\Psi}\left(P\left(X^+ > x\right)\right) dx \le \int_0^\infty Q\left(X^+ > x\right) dx.$$

It follows that

$$L(X) \le E^Q[X]$$

for all such probabilities Q and hence their infimum over Q.

We also see that

$$L(X) = \int_{-\infty}^\infty x\Psi'\left(F_X(x)\right) f_X(x)dx = \int P(d\omega)X(\omega)\Psi'\big(P\left(X \le X(\omega)\right)\big).$$

Also

$$\int P(d\omega)\Psi'\big(P\left(X \le X(\omega)\right)\big) = \int P(d\omega)\Psi'(u(\omega)) = \int \Psi'(u)du = 1.$$

Hence define $\overline{Q}$ by

$$\frac{d\overline{Q}}{dP} = \Psi'\left(P\left(X \le X(\omega)\right)\right) = \Psi'(u(\omega)).$$

We wish to show that $L(X)$ is $E^Q[X]$ for $Q \in \mathcal{M}$. As a distorted expectation

$$L(X) = \int_{-\infty}^\infty x\Psi'\left(F_X(x)\right) f_X(x)dx = \int P(d\omega)X(\omega)\Psi'\big(P\left(X \le X(\omega)\right)\big).$$

Define

$$u(\omega) = P\left(X \le X(\omega)\right).$$

Now for an arbitrary set A we have

$$\overline{Q}\left(A\right) = \int_A P(d\omega)\Psi'(u(\omega)).$$

On the other hand

$$\Psi\left(P(A)\right) = \Psi\left(\int_A P(d\omega)\right).$$

One may write

$$A = \bigcup_u A(u),$$
$$A(u) = \left\{\omega \in A | u(\omega) = u\right\}.$$

We have that

$$P(A) = \int_0^1 P(A(u))du.$$

It is also the case that

$$\overline{Q}\left(A(u)\right) = \Psi'(u)P(A(u)),$$

and that

$$\overline{Q}(A) = \int_0^1 \overline{Q}(A(u))\,du,$$

as well as

$$\Psi(P(A)) = \int_0^1 \Psi(P(A(u)))du.$$

The result follows on observing

$$\Psi'(u)P(A(u)) \leq \Psi\left(P(A(u))\right).$$

Now it is the case that

$$\Psi\left(P(A(u)) = \int_0^{P(A(u))} \Psi'(w)dw \geq \Psi'(P(A(u))P(A(u)).\right.$$

Observe that

$$P(\{\omega \mid u(\omega) = u\}) \geq P\left(\{\omega \in A | u(\omega) = u\}\right) = P(A(u)).$$

So we see that

$$P(A(u)) \leq P(\Omega(u)) = P\left(u(\omega) \leq u\right) = u.$$

Hence

$$\Psi'(P(A(u)) \geq \Psi'(u),$$

and so

$$\Psi\left(P(A(u)) \geq \Psi'(u)P(A(u)).\right. \qquad \square$$

The lower valuation operators related to distorted expectation have two defining properties. The first is that they are determined purely as functions of the risk distribution function and ignore how outcomes may depend on underlying events. The second is that they are comonotone additive. Two outcomes X, Y are said to be comonotone if they have no negative comovements or if $(X(\omega_1) - X(\omega_2))\,(Y(\omega_1) - Y(\omega_2)) \geq 0$. In general $L(X + Y) \geq L(X) + L(Y)$ and the inequality is strict if there are diversification benefits. Comonotone additivity requires $L(X+Y) = L(X)+L(Y)$ when X, Y are comonotone. Under comonotone additivity and with distorted expectations in particular the lower value of the unlevered firm equals the sum of the lower values of debt plus equity for levered firms, as debt and equity are comonotone securities.

A variety of distortions have appeared in the literature (Cherny and Madan, 2009) and include minvar, maxvar, minmaxvar, and maxminvar, along with the Wang distortion (see Wang, 2000). A two-parameter distortion is also provided by minmaxvar2. These are now described along with their properties.

The distortion minvar is defined as

$$\Psi^{(\gamma)}_{\text{minvar}}(u) = 1 - (1 - u)^{1+\gamma}.$$

Its derivative tends to zero at unity but is finite at $1 + \gamma$ when u is equal to zero. The absence of gain enticement is strong, but large, losses are only reweighted upward by a limited amount. The distorted may be easily described as evaluating the expectation of the minimum of γ independent draws from the distribution when γ is an integer.

The distortion maxvar is defined as

$$\Psi^{(\gamma)}_{\text{maxvar}}(u) = u^{\frac{1}{1+\gamma}}.$$

Its derivative tends to infinity at zero but is finite at $(1 + \gamma)^{-1}$ when u is equal to unity. The absence of gain enticement is limited, but large losses are reweighted upward to infinity. The distorted may be easily described as evaluating the expectation from a distribution with the property that the maximum of γ independent draws from this distribution matched the original distribution, when γ is an integer.

The distortion minmaxvar is defined as

$$\Psi^{(\gamma)}_{\text{minmaxvar}}(u) = 1 - (1 - u^{\frac{1}{1+\gamma}})^{1+\gamma}.$$

Its derivative tends to infinity at zero and to zero at unity. The absence of gain enticement is strong as is the reweighting of large losses.

The distortion maxminvar is defined as

$$\Psi^{(\gamma)}_{\text{maxminvar}}(u) = \left(1 - (1 - u)^{1+\gamma}\right)^{\frac{1}{1+\gamma}}.$$

Its derivative tends to infinity at zero and to zero at unity. The absence of gain enticement is strong as is the reweighting of large losses.

The Wang distortion is defined as

$$\Psi^{(\gamma)}_{\text{Wang}}(u) = N\left(N^{-1}(u) + \gamma\right),$$

with $N(x)$ the standard normal distribution. The derivative tends to infinity and zero at zero and unity and both gain enticement and loss aversion are strong.

The distortion minmaxvar2 is defined by

$$\Psi^{(\gamma_1,\gamma_2)}_{\text{minmaxvar2}}(u) = 1 - (1 - u^{\frac{1}{1+\gamma_1}})^{1+\gamma_2}.$$

The parameters γ_1, γ_2 control the rates or speed at which the derivatives approach infinity and zero at zero and unity respectively.

Figure 6.3 presents a graph of the minmaxvar distortion and its complementary distortion $\widehat{\Psi}$.

For an example of a levered and unlevered firm in §6.5 one may take the stock price as a geometric Brownian motion with a 25% volatility, an initial stock price at 100, and a one-year maturity. With leverage and debt at 75 one may evaluate the lower valuation for the unlevered firm using *minmaxvar* with $\gamma = 0.25$. The lower value of the unlevered firm is 90.1237. The lower value of equity for the levered firm is 19.0578 and the lower value of debt is 71.0659. The total lower value of the levered firm is unchanged at 90.1237. Lower valuations remain consistent with the Modigliani–Miller result for distorted expectation as a consequence of comonotone additivity.

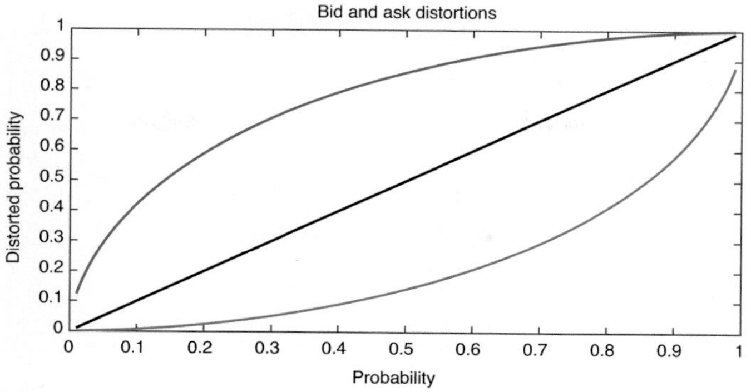

Figure 6.3 Graph of the minmaxvar distortion for distortion parameter γ. The distortion Ψ lies above the identity while the complementary distortion $\widehat{\Psi}$ lies below the identity.

6.7 Measure Distortions Proper

The motivations behind the distortion of probabilities reflect the view that as one does not have experience with rare tail events, their probabilities should not have the same status as those for which the experience is substantial. As a consequence for conservatism, tail probabilities associated with losses should be inflated while those associated with gains should be deflated. However, experience is related to one's familiarity with events or how often they happen. Probability, on the other hand, is a relative frequency obtained on normalizing event frequency by the frequency or all other possible events. The collection of all other possible events may not even be enumerable or describable let alone have knowledge of their aggregate frequencies. If we agree that we have familiarity with the frequency of events and little knowledge of the set of all possible events, then perhaps one should consider directly distorting event counts and not focus on probabilities. For short, we know what we know, the counts on events that have been observed, and we also know that we don't know what we do not know, which is the inability to describe the entire collection of what has not been experienced but could be.

Measure distortions technically arose as limits of probability distortions in discrete time models as we let the size of the time step go to zero to approach continuous time. This work was presented in Madan et al. (2017a). Here we focus directly on measure distortions as distortions of variation measures that count event frequencies and take values in the positive half-line, as opposed to probabilities that take their values in the unit interval. Two measure distortions mirroring the work of Ψ and $\widehat{\Psi}$ for probability distortions are introduced that we denote by G^+, G^-. These are positive increasing functions on $\mathbb{R}^+$, respectively concave/convex and above/below the identity. Figure 6.4 presents a typical example.

Lower and upper measure-distorted valuations $\mathcal{L}(X)$, and $\mathcal{U}(X)$ for variation outcomes on a variation space with base variation measure ν in analogy with the distorted tail integrals (6.7) and (6.8) are defined as follows.

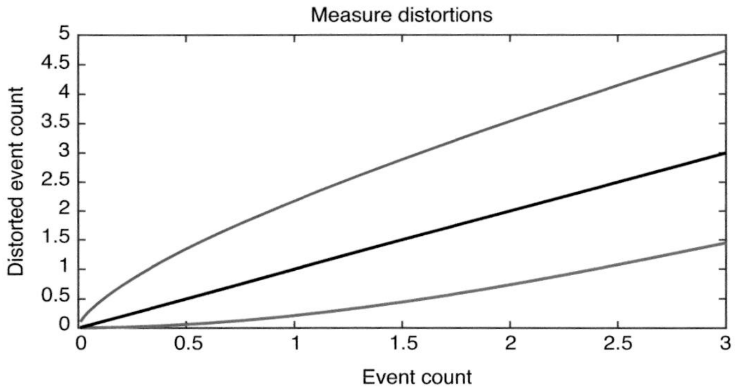

Figure 6.4 The measure distortion $G^{+}(x)$ is concave and above the identity. The measure distortion $G^{-}(x)$ is convex and below the identity.

$$\mathcal{L}(X) = -\int_0^{\infty} G^{+}\left(\nu\left(X^{-} > x\right)\right) dx + \int_0^{\infty} G^{-}\left(\nu(X^{+} > x)\right) dx, \tag{6.9}$$

$$\mathcal{U}(X) = -\int_0^{\infty} G^{-}\left(\nu\left(X^{-} > x\right)\right) dx + \int_0^{\infty} G^{+}\left(\nu(X^{+} > x)\right) dx. \tag{6.10}$$

In the absence of the distortions one is merely evaluating the variation. The loss tails are lifted by G^{+} and by the concavity of G^{+} the lift is higher for lower frequencies. The gain tail is dropped by G^{-} and by the convexity this is higher for the smaller frequencies. The measure distorted variation may be seen as a reweighted variation with weights proportional to the derivatives of the distortions, and they are higher at lower frequencies for G^{+} by concavity and they are lower for G^{-} by its convexity.

If the variation is seen as a reward, then the measure-distorted variation may be seen as reward less risk on adding and subtracting the variation and writing

$$\mathcal{L}(X) = V(X) - \int_0^{\infty} \Gamma^{+}\left(\nu\left(\{\omega | X(\omega) < -a\}\right)\right) da - \int_0^{\infty} \Gamma^{-}\left(\nu\left\{\omega | X(\omega) > a\right\}\right) da, \tag{6.11}$$

$$\Gamma^{+}(z) = G^{+}(z) - z, \tag{6.12}$$

$$\Gamma^{-}(z) = z - G^{-}(z). \tag{6.13}$$

The two integrals with respect to Γ^{+} and Γ^{-} may be seen as concave risk charges for tail risk that are positive penalties subtracted from the variation to get the lower conservative measure-distorted variation. The proposed objective function for risk management in variation space is the maximization of a measure-distorted variation.

Proposition 6.2 *There are a set of supporting variation measures $\widetilde{\nu} \in \widetilde{\mathcal{M}}$ such that*

$$\mathcal{L}(X) = \inf_{\widetilde{\nu} \in \widetilde{\mathcal{M}}} \int X(\omega) \widetilde{\nu}(d\omega),$$

$$\mathcal{U}(X) = \sup_{\widetilde{\nu} \in \widetilde{\mathcal{M}}} \int X(\omega) \widetilde{\nu}(d\omega)$$

with $\widetilde{\nu} \in \widetilde{\mathcal{M}}$ only if, for all A, $\nu(A) < \infty$, it is the case that

$$G^{-}(\nu(A)) \leq \widetilde{\nu}(A) \leq G^{+}(\nu(A)).$$

Proof Let

$$K(y) = \{\omega \mid X(\omega)\text{sign}(y) > \text{sign}(y)y\}.$$

We may then write

$$\mathcal{L}(X) = -\int_0^\infty G^{+}(\nu(K(-y)))dy + \int_0^\infty G^{-}(\nu(K(y)))dy.$$

Consider now $\widetilde{\nu}$ absolutely continuous with respect to ν such that for all A with $\nu(A) < \infty$ we have

$$G^{-}(\nu(A)) \leq \widetilde{\nu}(A) \leq G^{+}(\nu(A)).$$

It follows that

$$-G^{+}(\nu(K(-y))) \leq -\widetilde{\nu}(K(-y)),$$
$$G^{-}(\nu(K(y))) \leq \widetilde{\nu}(K(y)),$$

and hence we have that

$$\mathcal{L}(X) \leq \int_\Omega X(\omega)\widetilde{\nu}(d\omega).$$

Therefore it is the case that

$$\mathcal{L}(X) \leq \inf_{\widetilde{\nu} \in \mathcal{M}} \int_\Omega x(\omega)\widetilde{\nu}(d\omega),$$

where

$$\widetilde{\mathcal{M}} = \left\{\widetilde{\nu} \mid \widetilde{\nu} \ll \nu \quad \text{and} \quad G^{-}(\nu(A)) \leq \widetilde{\nu}(A) \leq G^{+}(\nu(A)) \text{ for all } A,\ \nu(A) < \infty\right\}.$$

For the other direction note that

$$\mathcal{L}(X) = -\int_0^\infty G^{+}(\nu(\{\omega \mid X(\omega) > -y\}))\,dy + \int_0^\infty G^{-}(\nu(\{\omega \mid X(\omega) > y\}))\,dy.$$

Equivalently we may write

$$\mathcal{L}(X) = -\int_{-\infty}^0 yG^{+\prime}(\nu(\{\omega \mid X(\omega) < y\}))\,d\nu(\{\omega \mid X(\omega) < y\})$$
$$-\int_0^\infty yG^{-\prime}(\nu(\{\omega \mid X(\omega) > y\}))\,d\nu(\{\omega \mid X(\omega) > y\}),$$

or that

$$\mathcal{L}(X) = \int_{\Omega \cap (X(\omega) \leq 0)} X(\omega)G^{+\prime}(\nu(\{\omega' \mid X(\omega') < X(\omega)\}))\,d\nu(\{\omega' | X(\omega') < X(\omega)\}$$
$$-\int_{\Omega \cap (X(\omega) > 0)} X(\omega)\,G^{-\prime}(\nu(\{\omega' \mid X(\omega') > X(\omega)\}))\,d\nu(\{\omega' \mid X(\omega') > X(\omega)\}).$$

So

$$\mathcal{L}(X) = \int_\Omega X(\omega)\widetilde{\nu}(d\omega),$$

where

$$\begin{aligned}\widetilde{\nu}(d\omega) &= G^{+\prime}\left(\nu\left(\{\omega' \mid X(\omega') < X(\omega)\}\right)\right) d\nu(\{\omega' \mid X(\omega') < X(\omega)\}\,\mathbf{1}_{x(\omega)\le 0}\\ &\quad -G^{-\prime}\left(\nu\left(\{\omega' \mid X(\omega') > X(\omega)\}\right)\right) d\nu\left(\{\omega' \mid X(\omega') > X(\omega)\}\right)\mathbf{1}_{x(\omega)>0}.\end{aligned}$$

We now show that for all A, $\nu(A) < \infty$ that

$$G^{-}(\nu(A)) \le \widetilde{\nu}(A) \le G^{+}(\nu(A)).$$

It then follows that

$$\mathcal{L}(X) \ge \inf_{\widetilde{\nu}\in\mathcal{M}} \int_\Omega X(\omega)\widetilde{\nu}(d\omega),$$

and hence that

$$\mathcal{L}(X) = \inf_{\widetilde{\nu}\in\mathcal{M}} \int_\Omega X(\omega)\widetilde{\nu}(d\omega).$$

First observe that

$$\begin{aligned}\widetilde{\nu}(A) &= \int_{A\cap(X(\omega)\le 0)} G^{+\prime}\left(\nu\left(\{\omega' \mid X(\omega') < X(\omega)\}\right)\right) d\nu(\{\omega'|X(\omega') < X(\omega)\}\\ &\quad + \int_{A\cap(X(\omega)>0)} G^{-\prime}\left(\nu\left(\{\omega' \mid X(\omega') > X(\omega)\}\right)\right)\left(-d\nu\left(\{\omega' \mid X(\omega') > X(\omega)\}\right)\right)\\ &\le \int_{A\cap(X(\omega)\le 0)} G^{+\prime}\left(\nu\left(\{\omega' \mid X(\omega') < X(\omega)\}\right)\right) d\nu(\{\omega' \mid X(\omega') < X(\omega)\}\\ &\quad + \nu(A\cap X(\omega) > 0)\\ &= \nu(A) + \int_{A\cap(x(\omega)\le 0)} \Gamma^{+\prime}\left(\nu\left(\{\omega' \mid X(\omega') < X(\omega)\}\right)\right) d\nu(\{\omega' \mid X(\omega') < X(\omega)\}.\end{aligned}$$

On the other hand we have that

$$\begin{aligned}G^{+}(\nu(A)) &= \nu(A) + \Gamma^{+}(\nu(A)) \qquad (6.14)\\ &\ge \nu(A) + \Gamma^{+}(\nu(A\cap(x(\omega) < 0))\\ &= \nu(A) + \int_{A\cap(X(\omega)\le 0)} \Gamma^{+\prime}\left(\nu\left(\{\omega'|X(\omega') < X(\omega)\}\right)\right) d\nu(\{\omega' \mid X(\omega') < X(\omega)\}\\ &\ge \widetilde{\nu}(A).\end{aligned}$$

For the last equality we write

$$g(y) = \Gamma^{+}(\nu(A\cap(X(\omega) < y)),$$

and note that, as

$$g(0) - g(-\infty) = \int_{-\infty}^{0} g'(y)dy,$$

we obtain

$$\Gamma^{+}(\nu(A \cap (X(\omega) < 0)) = \int_{-\infty}^{0} g'(y)dy$$
$$= \int_{A\cap(X(\omega)\leq 0)} \Gamma^{+\prime}(\nu(\{\omega' \mid X(\omega') < X(\omega)\}))\, d\nu(\{\omega' \mid X(\omega') < X(\omega)\}\,.$$

Similarly we observe that

$$\begin{aligned}
\widetilde{\nu}(A) &= \int_{A\cap(X(\omega)\leq 0)} G^{+\prime}(\nu(\{\omega' \mid X(\omega') < X(\omega)\}))\, d\nu(\{\omega' \mid X(\omega') < X(\omega)\} \\
&\quad + \int_{A\cap(X(\omega)>0)} G^{-\prime}(\nu(\{\omega' \mid X(\omega') > X(\omega)\}))(-d\nu(\{\omega' \mid X(\omega') > X(\omega)\})) \\
&\geq \nu(A \cap X(\omega) \leq 0) \\
&\quad + \int_{A\cap(X(\omega)>0)} G^{-\prime}(\nu(\{\omega' \mid X(\omega') > X(\omega)\}))(-d\nu(\{\omega' \mid X(\omega') > X(\omega)\})) \\
&= \nu(A) - \int_{A\cap(X(\omega)>0)} \Gamma^{-\prime}(\nu(\{\omega' \mid X(\omega') > X(\omega)\}))(-d\nu(\{\omega' \mid X(\omega') > X(\omega)\}))\,.
\end{aligned}$$

On the other hand

$$\begin{aligned}
G^{-}(\nu(A)) &= \nu(A) - \Gamma^{-}(\nu(A)) \qquad (6.15)\\
&\leq \nu(A) - \Gamma^{-}(\nu(A \cap (X(\omega) > 0)) \\
&= \nu(A) - \int_{A\cap(X(\omega)>0)} \Gamma^{-\prime}(\nu(\{\omega' \mid X(\omega') > X(\omega)\}))(-d\nu(\{\omega' \mid X(\omega') > X(\omega)\})) \\
&\leq \widetilde{\nu}(A).
\end{aligned}$$

One may also note that on reversing the roles of G^{+}, G^{-} and defining

$$\mathcal{U}(X) = -\int_{0}^{\infty} G^{-}(\nu(K(-y)))dy + \int_{0}^{\infty} G^{+}(\nu(K(y)))dy,$$

then equivalently one has that

$$\mathcal{U}(X) = \sup_{\widetilde{\nu}\in\mathcal{M}} \int_{\Omega} X(\omega)\widetilde{\nu}(d\omega).$$ □

When the variation involved is an asset held, one maximizes $\mathcal{L}$, but for liabilities to be paid out as opposed to being received, the objective switches to minimizing $\mathcal{U}$. The former is a concave functional while the latter is a convex functional.

The risk charge is by construction

$$\mathcal{R}_{\mathrm{L}}(X) = \int_{\Omega} X(\omega)\nu(d\omega) - \inf_{\widetilde{\nu}\in\mathcal{M}} \int_{\Omega} X(\omega)\widetilde{\nu}(d\omega) = \sup_{\widetilde{\nu}\in\mathcal{M}} \int_{\Omega} X(\omega)(\nu(d\omega) - \widetilde{\nu}(d\omega))\,.$$

It is computed as a measure-distorted variation from

$$\mathcal{R}_{\mathrm{L}}(X) = \int_{0}^{\infty} \Gamma^{+}(\nu(\{\omega \mid X(\omega) < -a\}))\, da + \int_{0}^{\infty} \Gamma^{-}(\nu\,\{\omega \mid X(\omega) > a\}),$$

using the concave distortions Γ^+, Γ^- introduced in equations (6.12) and (6.13). The measure-distorted variations are

$$\mathcal{L}(X) = V(X) - \mathcal{R}_L(X), \tag{6.16}$$

while

$$\mathcal{U}(X) = V(X) + \mathcal{R}_U(X), \tag{6.17}$$

where

$$\mathcal{R}_U(X) = \int_0^\infty \Gamma^-\left(\nu\left(\{\omega \mid X(\omega) < -a\}\right)\right) da + \int_0^\infty \Gamma^+\left(\nu\left\{\omega \mid X(\omega) > a\right\}\right).$$

In both constructions the risk charge trades with the variation on a one-to-one basis.

An important difference in managing variation risk from that of position risk is the focus of attention. In the latter the focus is on the position as an outcome to be valued, taking account of its probabilities. With variation risk the focus is on the motion or change in value. The risk charges are precisely charges for such exposures incorporated into the conservative valuation of variation exposures.

Examples of measure distortions employed in the literature model the risk charges Γ^+, Γ^- by

$$\Gamma^+(x) = a\left(1 - e^{-cx}\right)^{\frac{1}{1+\gamma}}, \tag{6.18}$$

$$\Gamma^-(x) = \frac{b}{c}\left(1 - e^{-cx}\right). \tag{6.19}$$

For symmetric risk charges at infinite counts we may take $a = b/c$. The derivative of $\Gamma^{-\prime}(0) = b \leq 1$ and a maximal choice is $b = 1$. The reweighting has a component of the form e^{-cx} and this is effectively zero, say, for $cx = 10$. Hence c may be set at 10 divided by the count at which reweighting stops. The parameter γ controls the rate at which $\Gamma^{+\prime}(x)$ tends to infinity near zero. The behavior of $\Gamma^{+\prime}$ near zero is that of $x^{-\frac{\gamma}{1+\gamma}}$ and $\gamma = 1$ is associated with behavior like the reciprocal of the square root.

Measure distortions were applied to derive bid and ask prices for the stock and to conservatively value insurance losses in Eberlein et al. (2014a). Applications to the valuation of portfolios of derivatives may be found in Eberlein et al. (2014b). Portfolio theory applications were studied in Madan (2018a). Risk charges based on measure distortions are estimated in Madan (2017b). Applications to regulatory capital requirements are in Madan (2017a). The estimation of measure distortion parameters from market data on bid and ask prices is undertaken in Elliot et al. (2020).

6.8 Dual Formulation of Measure Distortions

For probability distortions a dual formulation was presented in Cherny and Madan (2009) and employed in Madan and Wang (2016a). The primal optimization problem may not be a disciplined convex program, but this may be true for the dual formulation. Madan and

Wang exploited this property for forming bounds on forward-starting options. For a variety of problems it will turn out to be useful to work with such a dual formulation. First define the conjugate duals of the distortions by

$$\Phi(a) = \sup_{x\geq 0} G^{+}(x) - ax, \qquad \widetilde{\Phi}(a) = \inf_{x\geq 0} G^{-}(x) - ax.$$

Let an alternative variation measure $\widetilde{\nu}$ have density with respect to ν given by

$$\frac{d\widetilde{\nu}}{d\nu} = Z(\omega).$$

Proposition 6.3 *The set supporting variation measures $\widetilde{\nu} \in \widetilde{\mathcal{M}}$ have densities Z satisfying*

$$\int (Z(\omega) - a)^{+}\, \nu(d\omega) \leq \Phi(a),\ a > 1,$$
$$\int (a - Z(\omega))^{+}\, \nu(d\omega) \leq -\widetilde{\Phi}(a),\ a < 1.$$

Furthermore, in the region $Z(\omega) \neq 1$ we also have that

$$G^{-\prime}(0) < Z(\omega) < G^{+\prime}(0).$$

Proof Consider an alternate variation measure $\widetilde{\nu}$ with a density $Z(\omega)$ with respect to ν and write the constraint on Z as

$$G^{-}(\nu(A)) \leq \int_{A} Z(\omega)\nu(d\omega) \leq G^{+}(\nu(A)).$$

Next write for $A = \{\omega \mid Z(\omega) > a\}$ that

$$Z(\omega)\mathbf{1}_A = (Z(\omega) - a)^{+} + a\mathbf{1}_A,$$

and observe that we have

$$G^{-}(\nu(A)) \leq \int (Z(\omega) - a)^{+}\nu(d\omega) + a\nu(A) \leq G^{+}(\nu(A)),$$

and equivalently that

$$G^{-}(\nu(A)) - a\nu(A) \leq \int (Z(\omega) - a)^{+}\nu(d\omega) \leq G^{+}(\nu(A)) - a\nu(A).$$

Now as

$$\Phi(a) = \sup_{x\geq 0} G^{+}(x) - ax, \qquad \widetilde{\Phi}(a) = \inf_{x\geq 0} G^{-}(x) - ax,$$

where $a > 1$ for G^{+} and $0 < a < 1$ for G^{-}, we obtain

$$\widetilde{\Phi}(a) \leq \int (Z(\omega) - a)^{+}\nu(d\omega) \leq \Phi(a).$$

Alternatively we note that $\Phi(a) = \infty$ for $a < 1$, and for $a > G^{+\prime}(0)$, we have $\Phi(a) = 0$. Hence $\Phi(a)$ places an upper bound for $1 < a < G^{+\prime}(0)$. Furthermore for $a \geq G^{+\prime}(0)$ we have

$$\int (Z(\omega) - a)^+ \nu(d\omega) = 0,$$

and so

$$Z(\omega) < G^{+\prime}(0).$$

For $a > 1$ we have $\widetilde{\Phi}(a) = -\infty$, and for $a > G^{-\prime}(0)$ we have $\widetilde{\Phi}(a) < 0$, while for $a < G^{-\prime}(0)$ it follows that $\widetilde{\Phi}(a) = 0$. We may consider the set $A = \{\omega \mid Z(\omega) < a\}$, for which we also have

$$Z(\omega)\mathbf{1}_A = -(Z(\omega) - a)^- + a\mathbf{1}_A,$$

and we observe that

$$G^-(\nu(A)) \leq -\int (Z(\omega) - a)^- \nu(d\omega) + a\nu(A) \leq G^+(\nu(A)),$$

and so

$$G^-(\nu(A)) - a\nu(A) \leq -\int (Z(\omega) - a)^- \nu(d\omega) \leq G^+(\nu(A)) - a\nu(A);$$

so the inequality on the right is satisfied, but we must now have

$$\widetilde{\Phi}(a) \leq -\int (Z(\omega) - a)^- \nu(d\omega),$$

or that

$$\int (a - Z(\omega))^+ \nu(d\omega) \leq -\widetilde{\Phi}(a).$$

We thus obtain upper bounds on puts and calls on $Z(\omega)$. For $a < 1$ we have

$$\int (a - Z(\omega))^+ \nu(d\omega) \leq -\widetilde{\Phi}(a).$$

Also for $a < G^{-\prime}(0)$, as $\widetilde{\Phi}(a) = 0$, it follows that

$$\int (a - Z(\omega))^+ \nu(d\omega) = 0,$$

and so

$$Z(\omega) > G^{-\prime}(0).$$

So we also have that

$$G^{-\prime}(0) < Z(\omega) < G^{+\prime}(0),$$

in the region $Z(\omega) \neq 1$. These bounds are vacuous when we take $G^{-\prime}(0) = 0$, $G^{+\prime}(0) = \infty$.

For the argument in the opposite direction, suppose we have for a density that

$$\int (Z(\omega) - a)^+ \, \nu(d\omega) \leq \Phi(a), \; a > 1,$$
$$\int (a - Z(\omega))^+ \, \nu(d\omega) \leq -\widetilde{\Phi}(a), \; a < 1.$$

We are interested in the object

$$\int_A Z(\omega)\nu(d\omega)$$

for a set A satisfying $\nu(A) < \infty$ where we wish to establish that

$$G^-(\nu(A)) \leq \int_A Z(\omega)\nu(d\omega) \leq G^+(\nu(A)).$$

On the set $Z(\omega) > a > 1$ we have

$$Z(\omega)\mathbf{1}_A = (Z(\omega) - a)^+ + a\mathbf{1}_A,$$

and hence

$$\int Z(\omega)\mathbf{1}_A\nu(d\omega) = \int (Z(\omega) - a)^+\nu(d\omega) + a\nu(A) \leq \Phi(a) + a\nu(A) \leq G^+(\nu(A)) + \varepsilon.$$

Therefore we have that

$$\int Z(\omega)\mathbf{1}_A\nu(d\omega) \leq G^+(\nu(A)).$$

Similarly for A given by $Z(y) < a < 1$ we have

$$G^-(\nu(A)) \leq \int_A Z(\omega)\nu(d\omega).$$

For an arbitrary A let $A^+ = \{\omega|\omega \in A, Z(\omega) > 1\}$ and let $A^- = \{\omega|\omega \in A, Z(\omega) < 1\}$ with $A_0 = \{\omega|\omega \in A, Z(\omega) = 1\}$. It follows that

$$\int_{A_0} Z(\omega)\nu(d\omega) = \nu(A_0),$$

and

$$G^-(\nu(A_0)) \leq \nu(A_0) \leq G^+(\nu(A_0)).$$

We also have

$$G^-(\nu(A^+)) \leq \nu(A^+) \leq \int_{A^+} Z(\omega)\nu(d\omega) \leq G^+(\nu(A^+)),$$
$$G^-(\nu(A^-)) \leq \int_{A^-} Z(\omega)\nu(d\omega) \leq \nu(A^-) \leq G^+(\nu(A^-)).$$

The inequalities hold for all three sets. Now furthermore

$$G^-(\nu(A)) = G^-\left(\nu(A^-) + \nu(A_0) + \nu(A^+)\right),$$
$$G^+(\nu(A)) = G^+\left(\nu(A^-) + \nu(A_0) + \nu(A^+)\right).$$

We have that

$$G^{+}(\nu(A^{+})) + \nu(A^{-}) + \nu(A_0) \leq G^{+}(\nu(A)).$$

This follows from

$$G^{+}(a+b) - G^{+}(a) = \int_a^{a+b} G^{+\prime}(u)du \geq b.$$

Hence

$$G^{+}(a+b) \geq G^{+}(a) + b.$$

But

$$\begin{aligned}\int_A Z(\omega)\nu(d\omega) &\leq \nu(A^{-}) + \nu(A_0) + \int_{A^{+}} Z(\omega)\nu(d\omega)\\ &\leq G^{+}(\nu(A^{+})) + \nu(A^{-}) + \nu(A_0)\\ &\leq G^{+}(\nu(A)).\end{aligned}$$

Similarly one observes that

$$G^{-}(\nu(A)) \leq \int_A Z(\omega)\nu(d\omega).$$

Hence the required inequalities hold for all A with $\nu(A)$ finite. □

6.9 Explicit Representation of Dual Distortions $\Phi, \widetilde{\Phi}$.

For the distortions introduced in §6.7, the form used for Γ^{-} has been

$$\Gamma^{-}(x) = \frac{b}{c}(1 - e^{-cx}),$$

where the condition of being dominated above by the identity requires that one take $b \leq 1$. The extreme choice would therefore be $b = 1$. If an arrival frequency of 20 times a day is viewed as sufficient to eliminate a risk charge on reweighting, then the choice of $c = 0.5$ yields a reweighting of exp(-10) or half a basis point. A symmetric maximal risk charge is accomplished with the choice of

$$\Gamma^{+}(x) = a\left(1 - e^{-cx}\right)^{\frac{1}{1+\gamma}},$$

and $a = b/c$. It remains to select γ, and low-loss frequencies being reweighted at the rate of the reciprocal of the square root of the frequency are attained by $\gamma = 1$. For this choice of distortions one may explicitly construct the required dual distortions.

For our examples

$$G^{+}(x) = x + a(1 - e^{-cx})^{\frac{1}{1+\gamma}},$$

and the first-order condition for $\Phi(\lambda)$ is

$$(1 - e^{-cx})^{-\frac{\gamma}{1+\gamma}} e^{-cx} = \frac{(\lambda - 1)(1+\gamma)}{ac}.$$

Let $u = e^{-cx}$ and write

$$\frac{u}{(1-u)^{\frac{\gamma}{1+\gamma}}} = \frac{(\lambda-1)(1+\gamma)}{ac}.$$

Let $g(u)$ be the function on the left. $g(0) = 0$, and $g(1) = \infty$. Further

$$g'(u) = \frac{1}{(1-u)^{\frac{\gamma}{1+\gamma}}} + \frac{\gamma}{1+\gamma}\frac{u}{(1-u)^{\frac{1+2\gamma}{1+\gamma}}} > 0.$$

Hence there is a unique value $u(\lambda)$ in the unit interval meeting the equality. Let

$$x(\lambda) = -\frac{\ln(u(\lambda))}{c},$$

and then set

$$\Phi(\lambda) = G^{+}(x(\lambda)) - \lambda x(\lambda).$$

This will give us

$$\Phi(\lambda) = -(1-\lambda)\frac{\ln(u(\lambda))}{c} + a\,(1-u(\lambda))^{\frac{1}{1+\gamma}}.$$

For $\widetilde{\Phi}$ the first-order condition is for

$$G^{-}(x) = x - \frac{b}{c}(1-e^{-cx}),$$

$$1 - be^{-cx} = \lambda,$$

or that

$$\frac{1-\lambda}{b} = u.$$

For $1-\lambda < b$ or $\lambda > 1-b$ we have

$$x = -\frac{\ln\left(\frac{1-\lambda}{b}\right)}{c},$$

and

$$\widetilde{\Phi}(\lambda) = -(1-\lambda)\frac{\ln\left(\frac{1-\lambda}{b}\right)}{c} - \frac{b}{c}\left(1 - \frac{1-\lambda}{b}\right).$$

For $b = 1$ and $c = 0.5$ we write

$$-\widetilde{\Phi}(\lambda) = 2\left[\lambda + (1-\lambda)\ln(1-\lambda)\right].$$

Figure 6.5 presents a graph of explicit bounds on the options written on the density $Z(y)$.

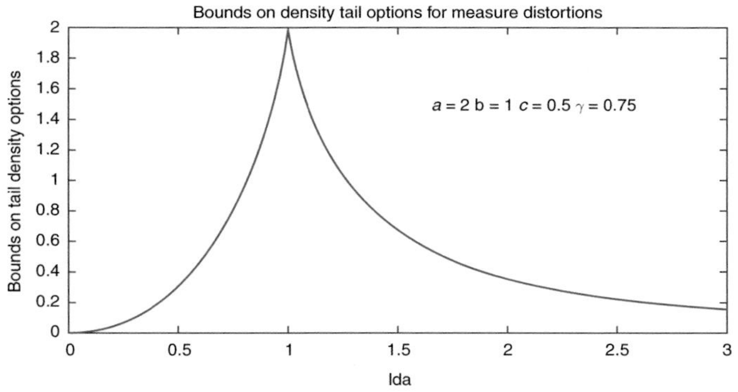

Figure 6.5 Bounds on density options

6.10 Generic Considerations in the Maximization of Market Valuations

At a general level when actions given by a vector a are associated with a market valuation denoted by $V(a)$, certain considerations are relevant for the objective of value maximization when the value function satisfies a scaling property. Similar considerations apply to minimization and the discussion is here restricted to the maximization problem. By the scaling property for every $\lambda > 0$, it is the case that $V(\lambda a) = \lambda V(a)$. Hence if for some admissible a it is the case that $V(a) > 0$, then clearly the maximization problem has no bounded solution. If $V(a)$ is an infimum of classical expectations over a set of measures $Q \in \mathcal{M}$, then for every admissible action a there must exist $Q \in \mathcal{M}$ such that under this measure the value of the action is negative. In this case $V(a) < 0$ for all admissible actions and the maximization may proceed. Raising stress levels enhances the set of measures over which an infimum is taken to define value, and maximization can then be feasible.

Alternatively one may seek a direction of action and not focus on the scale. There are of course a number of ways to constrain action directions. One may restrict them to satisfy conditions like $\mathbf{1}^{\mathrm{T}}a = b > 0$ coupled with a bound on how far negative the actions may get. One may instead regularize actions into the set $a^{\mathrm{T}}a \leq b$. The optimum action $a^{*}(b)$ will have a value $V^{*}(b)$. There will then be a frontier defined by the curve $V^{*}(b)$, and one is positioning on this curve by either fixing the level b or the gradient given by $\partial V^{*}(b)/\partial b$. Both types of solutions will be encountered in the applications being made.

7
Representing Market Realities

There is a tendency and a tradition to think of reality as represented by a true or real-world physical probability measure. This is often represented by a probability space $(\Omega, \mathcal{F}, P)$ where P is the true probability measure. It is then understood under the supposed law of one price that there must exist a pricing measure delivering the forward price of events that is often denoted by Q equivalent to P with density $Z = dQ/dP$. Market realities, abstracting from time value or numeraire issues, are then represented by the pair of probabilities P, Q. Much has been written on the interplay between P, Q and the associated studies of event risk premia describing when and how the P and Q probabilities of events differ or when the two expectations of random outcomes are different. For a general discussion on the pricing of risks at an abstract level we refer the reader to Atlan et al. (2007). For a more classical treatment see Campbell et al. (1997) and the references therein.

Quite generally for a claim delivering the random outcome $c = (c(\omega), \omega \in \Omega)$ the expected outcome is $E^P[c]$ and its price is $E^Q[c]$. The rate of return, assuming a positive price, is then

$$R_c = \frac{E^P[c]}{E^Q[c]} - 1.$$

When R_c is positive the claim c is viewed as earning a premium as compensation for the risk embedded in c. However, a negative R_c is a premium being paid for the insurance provided by c.

The description of market realities is significantly altered when dealing with conservative markets characterized by proper cones of risk acceptability. As already noted, for proper cones of risk acceptability with a sufficient and diverse set of supporting measures, the law of one price fails for all nontrivial claims and only holds for constants. With respect to variation measures the only relevant constant exposure is the identically zero exposure that then has a zero price. The law of one price then merely asserts that the price of the identically zero payoff is zero.

For all other claims, the conservative market is constitutionally incapable of delivering a price. There is then no such object to speak about, discuss, or develop. However, neither is the physical or true probability the reality to be described by markets. Physical probability as a reality is at best a statistical reality. It is not the object of attention of a market. For a market the economic reality is the valuation of risk exposures, given their statistical, probabilistic, or measure theoretic description.

The financial market reality we seek to deliver is that of a conservative two-sided valuation of risky exposures. In the context of two-price economies these are the lower and upper valuation operators. The market reality is then described on specifying the variation measure ν and the set $\mathcal{M}$ of measures supporting risk acceptability. In the special case when $\mathcal{M}$ is based on measure distortions, the task is accomplished on defining the functional form for the measure distortion and its associated parameter values.

In the univariate case with BG arrival rates and the measure distortion proposed in §6.7 the market reality is completed on specifying the physical BG return distribution parameters b_p, c_p, b_n, c_n, and the measure distortion parameters b, c, and γ. Seven parameters are required in this case to formulate the valuation operators of markets. These valuation operators then serve as the objectives to be employed for financial decision-making in a variety of risk management contexts. We therefore begin with strategies for describing these market realities.

7.1 Risk Charges and the Measure Distortion Parameters

The lower valuations may be seen as a risk charge subtracted from the variation, while the upper valuation adds a risk charge to the variation. In general the two risk charges that are subtracted from and added to the variation are different and are given in equations (6.16) and (6.17). The two risk charge distortions given by equations (6.18) and (6.19) in the symmetric case of $a = b/c$ are defined by three parameters: b, c, and γ.

The parameter b refers to the limiting percentage gain discount and must be less than unity. One may refer to the parameter c as controlling the count level at which reweighting stops. When the count x is such that $cx = 10$, then the reweighting factor e^{-cx} is half a basis point and reweighting has essentially stopped. For $c = 1$ this count is 10 and for $c = 0.5$ we have a count of 20. Reweighting stops at counts of 1,000 when $c = 0.01$. The parameter γ controls loss aversion or the rate at which the derivative of G^+ tends to infinity at zero. It may be related to acceptability indices that for hedge fund returns reach levels near 0.75, but for daily equity returns are generally below 0.25. For estimates of γ over time in a variety of economic sectors the reader is referred to Madan et al. (2017b).

With a view toward observing the level of risk charges as functions of these parameters as they are varied, consider the base case of BG parameters given by

$$b_p = 0.018172; \quad c_p = 1.0259; \quad b_n = 0.019889; \quad c_n = 0.9571.$$

For BG arrival rates for jump sizes x the return exposure to the invested dollar is $(e^x - 1)$. At these parameters the variation is 7.3545 basis points.

Consider a base measure distortion parameter setting of gain discount $b = 1.0$, with reweighting stopped at a count of 20 or $c = 0.5$, and loss aversion $\gamma = 0.75$. The risk charge embedded in the shave of lower valuation below the variation is 0.04332 while that embedded in the add-on for the upper valuation is 0.046294.

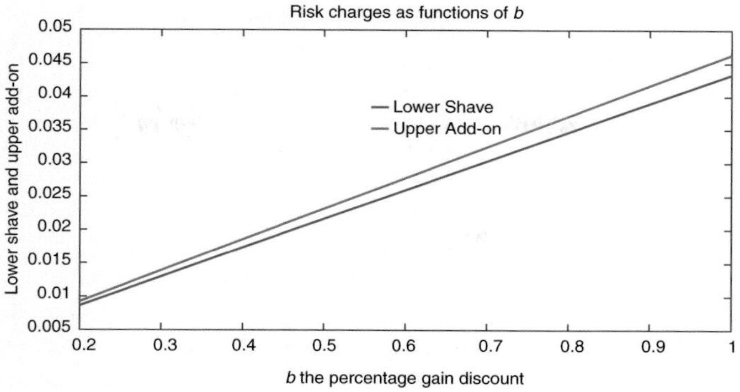

Figure 7.1 Risk charge as a function of b, the percentage gain discount

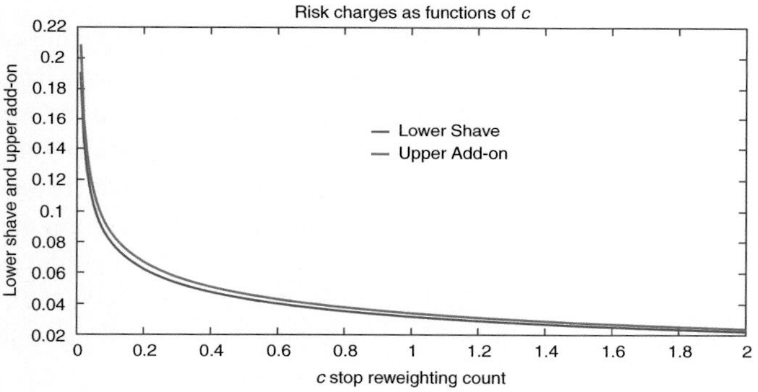

Figure 7.2 Risk charge as a function of c, the count level for reweighting to stop

Figures 7.1, 7.2, and 7.3 present the variations in the risk charge shave and add-on as functions of varying the parameters b, c, and γ from their respective base case levels of 1, 0.5, and 0.75.

We observe that risk charges are linear in the percentage gain discount b, sharply nonlinear as one goes to high count levels for reweighting stops, and nonlinear in the loss aversion.

7.2 Measure Distortions and Option Prices

For many applications of financial risk management the interest is in maximizing the market value of assets or minimizing the market value of liabilities. The former represents what we may safely get for the sale while the latter is an assessment of the potential cost of unwinding the liability. Classical valuation methodologies based on the absence of arbitrage and the law of one price have been observed to lead to linear valuation schemes that are not suitable for optimization. In two-price economies with no arbitrage the price or market value depends on the direction of trade and the lower value or price is a concave function of the risk while

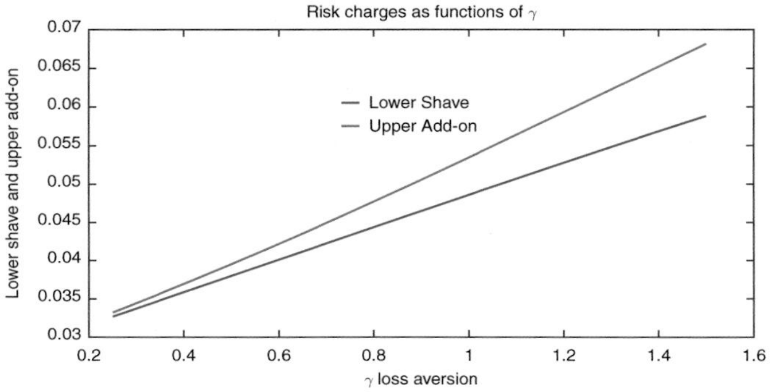

Figure 7.3 Risk charge as a function of γ, the loss aversion parameter

the upper price or value is convex. We therefore take as decision-making objectives such lower and upper price or valuation operators. When these operators are based on measure distortions, operational objectives require a determination of distortion parameters. The parameters may be likened to degrees of risk aversion prevailing in markets at the time of decision-making, or the current level of risk aversion to be found in markets. One would like to assess relevant values for these distortion parameters from current market data.

One possibility is to use option price data for the purpose. From option prices one may extract the market price of various tail events associated with stock prices at various future dates given by the option maturities. The risk-neutral Lévy measures embedded in the market prices of options then deliver the prices for instantaneous tail events. The principle behind the selection of distortion parameters is that of organizing risk-neutral prices or event valuations to be sandwiched between the distorted physical measures for the same events. More formally, for a physical measure $\nu(A)$ of event A we want the risk-neutral measure $\widetilde{\nu}(A)$ to satisfy

$$G^{-}(\nu(A)) \leq \widetilde{\nu}(A) \leq G^{+}(\nu(A)). \tag{7.1}$$

The distortions G^{-}, G^{+} have to be strong enough to drop physical measures distorted by G^{-} to be below the risk-neutral price while G^{+} lifts the physical measure above the risk-neutral price. For this purpose we also need to estimate the physical Lévy measures.

For the physical measure, time series data, overlapping or nonoverlapping for 5-, 10-, or 15-day returns may be used. From these data we extract information on currently relevant arrival rates of jumps $\nu(dx)$ of various sizes consistent with the various horizon-specific distributional return outcomes. Calibration of option prices taken at mid-quotes for these horizons delivers the comparable risk-neutral counterparts. For many underliers we now have a large number of such short maturity options being actively traded for a wide range of strikes. The distortion parameters may then be set at minimal levels, ensuring the validity of the inequalities (7.1). Here we illustrate the calibration of distortion parameters from estimates of physical and risk-neutral Lévy measures.

Consider as an example data on SPY as of February 8, 2018. For 150 random overlapping drawings of 5-, 10-, and 15-day returns taken from February 23, 2012, to February 8, 2018, the BG model was estimated by digital moment matching using Anderson–Darling weightings. The resulting physical arrival rate parameters are:

	b_p	c_p	b_n	c_n
5 day	0.0070	2.3501	0.0090	1.5667
10 day	0.0026	18.2449	0.0063	6.4195
15 day	0.0016	97.2516	0.0015	97.2515.

(7.2)

For any $x > 0$ one may evaluate three candidate tail probabilities using the parameters displayed in (7.2).

$$K_{\mathrm{BG},j}(x) = \int_{\mathrm{sign}(x)y \geq |x|} k_{\mathrm{BG},j}(y)dy \text{ for } j = 5, 10, \text{ or } 15 \text{ days}.$$

These are candidate values for $\nu(A(x))$ when $A(x) = \{y \mid \mathrm{sign}(x)y \geq \mathrm{sign}(x)x\}$.

One may also estimate the risk-neutral BG model on exact 5-, 10- and 15-day maturities using prices generated by implied volatility interpolation at these maturities from the raw data using a GPR summary of the option surface. The Gaussian process summary of option surfaces follows Spiegeleer et al. (2018). The resulting risk-neutral arrival rates are:

	b_p	c_p	b_n	c_n
5 day	0.0074	0.1230	0.0529	0.0175
10 day	0.0115	0.1779	0.0840	0.0246
15 day	0.0111	0.3179	0.0896	0.0397.

(7.3)

These parameter values may be used to evaluate $\widetilde{\nu}(A(x))$.

In general there are regions for x where $\widetilde{\nu}(A(x))$ exceeds $\nu(A(x))$ that may be used to assess the required lift in G^+. For regions where $\widetilde{\nu}(A(x))$ falls below $\nu(A(x))$ one may learn about the required drop in G^-. The analysis may be repeated for each maturity. Clearly if the reality were Lévy processes both physically and risk neutrally there would not be any horizon dependence. More generally one anticipates a term structure for the implied distortion parameters with the horizon chosen based on the specific application being conducted.

Consider first the five-day maturity for levels of x in absolute value between 5 and 10%. We present two graphs for $\widetilde{\nu}(A(x))$ against $\nu(A(x))$ for negative and positive x. Figure 7.4 presents the result for x on the negative side. We can see that for such tails there is a considerable lift upward to be accomplished by the distortion G^+.

Figure 7.5 presents the result for the comparable positive side. On this side the risk-neutral tail is in fact thin and we obtain information on how far G^- must fall.

Figure 7.6 presents the result for the absolute value of x below 5% on both sides. This situation comments on the drop of G^-.

Such analyses may be conducted for many underliers at a variety of maturities to build option-implied levels for measure distortions to be employed in the design of solutions to

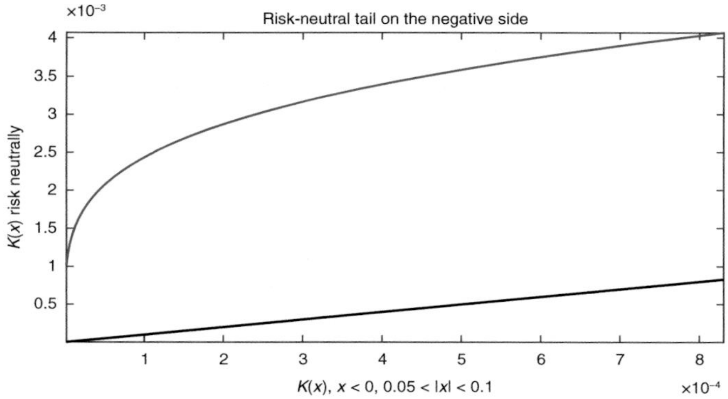

Figure 7.4 Risk-neutral tail versus the physical tail for $x < 0$ and in absolute value between 5 and 10%. Shown in black is the identity function.

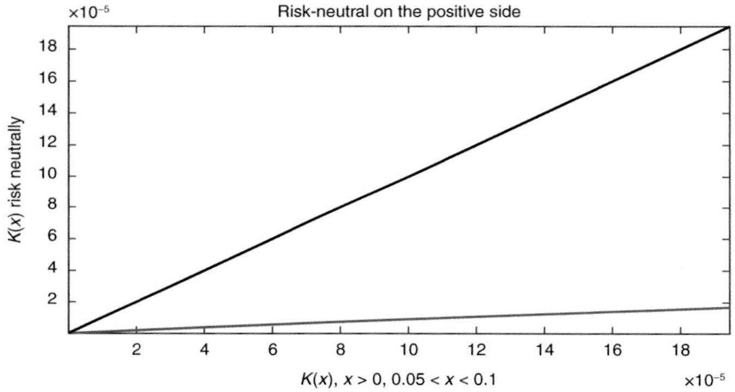

Figure 7.5 Risk-neutral tail versus the physical tail for $x > 0$ and in absolute value between 5 and 10%. Shown in black is the identity function.

risk management problems. It is also clear that the precise lifts and drops are sensitive to the range of jump sizes being monitored and one may employ information on trading liquidity on the two sides to calibrate such matters. For examples of measure-distortion calibrations along these lines we refer to Madan (2019) and Chapter 14.

7.3 Measure-Distorted Value-Maximizing Hedges for a Short Gamma Target

Consider the design of a hedge position in the stock for an exposure that is short gamma in the amount of a single dollar. We shall select the hedge to maximize a measure-distorted value. However, as measure-distorted values scale one may restrict attention to the short position of a single dollar gamma. For other dollar values one merely scales by the level of gamma. Furthermore, we assume a delta-neutral situation as the delta may be hedged out to begin with.

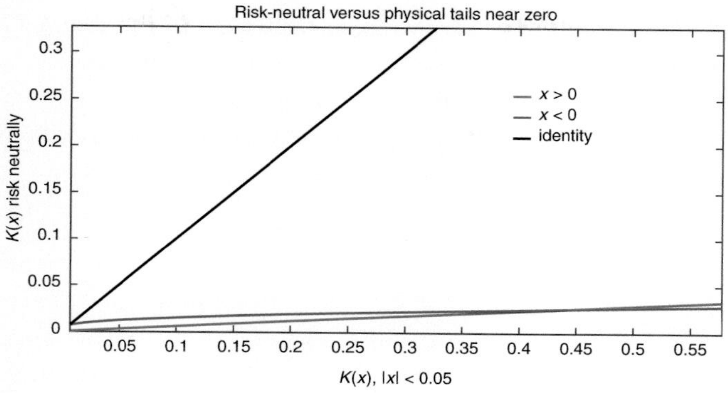

Figure 7.6 Risk-neutral tail versus the physical tail for $|x| < 0.05$ on both sides. Shown in black is the identity function.

For jumps x in the log price relative to the target exposure $c(x)$ of being short a single dollar, gamma is (the second derivative with respect to the stock price is unity and the position is $(S-1)^2/2$)

$$c(x) = -\frac{1}{2}\left(e^x - 1\right)^2. \tag{7.4}$$

Now consider investing a dollars in the stock to access the hedged cash flow of $c_a(x)$ given by

$$c_a(x) = -\frac{1}{2}\left(e^x - 1\right)^2 + a(e^x - 1). \tag{7.5}$$

The optimal hedge position may then be determined on finding the dollar investment that maximizes the lower valuation of the hedged exposure $c_a(x)$. For the evaluation of this lower measure-distorted value one has first to determine the tail counts on the base measure $\nu(dx)$ for the sets

$$A_a(w) = \{x| \ \mathrm{sign}(w)c_a(x) > \mathrm{sign}(w)w \}.$$

The lower valuation is then obtained as

$$\mathcal{L}(a) = -\int_0^\infty G^+\left(\nu\left(A_a(-w)\right)\right)dw + \int_0^\infty G^-\left(\nu\left(A_a(w)\right)\right)dw. \tag{7.6}$$

Let the base measure be given by BG arrival rates of jumps in the log price relative to x with BG tail measures

$$K(x) = \begin{cases} \int_x^\infty k_{\mathrm{BG}}(u; b_p, c_p)du & x > 0 \\ \int_{-\infty}^x k_{\mathrm{BG}}(u; b_n, c_n)du & x < 0 \end{cases}, \tag{7.7}$$

where the subscript BG on K and the dependence of K on the parameters has been suppressed for notational convenience. The integrations in equation (7.6) are conducted numerically. We use the tail measure (7.7) to determine the measure $\nu(A_a(w))$ of the sets $A_a(w)$ that have to be distorted in the expression (7.6) for integration.

The measures $\nu(A_a(w))$ are determined analytically as follows.

For this purpose let $y = (e^x - 1)$ and write the hedged cash flow (7.5) in terms of y as

$$c_a(y) = -\frac{y^2}{2} + ay,$$

with

$$x = \ln(1 + y).$$

Note that only values of y in excess of -1 are attainable.

7.3.1 Tail Measures for Hedged Outcomes

For $a > 0$. The maximum of the hedged exposure occurs at $y = a > 0$ and is $a^2/2$. The range for $w = V_a(y)$ on the positive side, $w > 0$, is $0 \le w \le a^2/2$. The general solution in y for $w > 0$ of $-\frac{y^2}{2} + ay = w$ is

$$y = a \pm \sqrt{a^2 - 2w}.$$

The larger positive solution is

$$y_2 = a + \sqrt{a^2 - 2w} > 0,$$

while the smaller positive solution is

$$y_1 = a - \sqrt{a^2 - 2w}.$$

The relevant upper tail in w is then

$$K(\ln(1 + y_1)) - K(\ln(1 + y_2)).$$

On the negative side or for $w < 0$ the general solution is

$$y = a \pm \sqrt{a^2 + 2|w|},$$

and one tail area is that of

$$K(\ln(1 + y_2)), \qquad y_2 = a + \sqrt{a^2 + 2|w|} > 0.$$

We add to this

$$K(\ln(1 + y_1)), \qquad y_1 = a - \sqrt{a^2 + 2|w|}, \text{ if } y_1 > -1.$$

For $a < 0$. In this case the maximum occurs at $y = a$ with value $a^2/2$. On the positive side $w > 0$ the tail area is zero unless $0 < w < a^2/2$. The solutions are

$$y_2 = a - \sqrt{a^2 - 2w} < y_1 = a + \sqrt{a^2 - 2w}.$$

If $y_2 > -1$ the tail measure is

$$K(\ln(1 + y_1)) - K(\ln(1 + y_2)).$$

If $y_2 < -1$ and $y_1 > -1$ the tail measure is

$$K(\ln(1 + y_1)).$$

The tail measure is zero if $y_1 < -1$.

For the negative side $w < 0$ the general solutions are

$$a \pm \sqrt{a^2 + 2|w|},$$

where

$$y_1 = a + \sqrt{a^2 + 2|w|} > 0,$$

with tail area

$$K\left(\ln(1 + y_1)\right).$$

The other point is

$$y_2 = a - \sqrt{a^2 + 2|w|} < 0,$$

and if $y_2 > -1$ we add to the tail area the value

$$K\left(\ln(1 + y_2)\right).$$

7.3.2 Implied Hedge Positions

Consider the BG five-day return parameter values reported on in §7.2 displayed in (7.2) with $b_p = 0.007$, $c_p = 2.35$, $b_n = 0.009$, and $c_n = 1.5667$. For the measure-distortion parameter values $b = 1$, $c = 0.5$, and $\gamma = 1.0$. The optimal hedge is to take a short position in stock of 17.77 basis points. With respect to the sensitivity of the optimal hedge position with respect to the distortion parameters we present Figures 7.7, 7.8, and 7.9 on the response in basis points for the variation in γ, b, and c respectively.

We observe that the sensitivity to c is small, while for both b and γ the greater short positions are taken at higher levels of b and γ or greater risk aversion on the two sides. However, for the BG parameters employed the hedge position is quite small, amounting to just an \$18 stock investment for a dollar gamma position of \$10,000. More substantial positive and negative positions may occasionally occur with zero delta being the usual situation. We present next some examples of more significant positive and negative hedge positions.

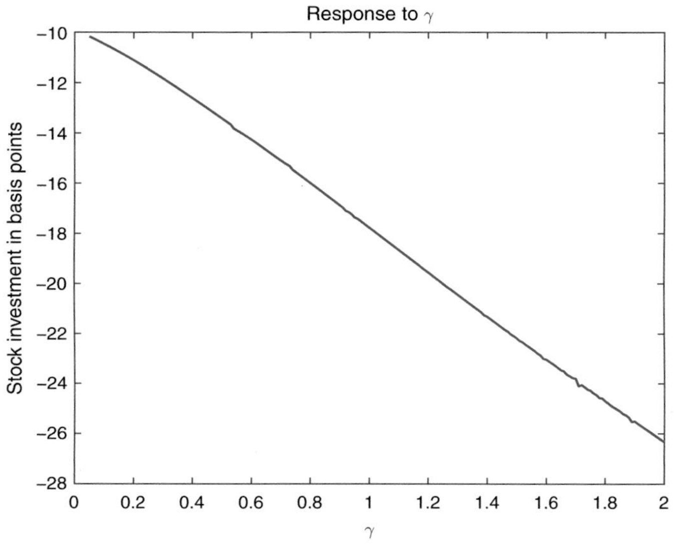

Figure 7.7 Sensitivity of hedge positions to variations in γ

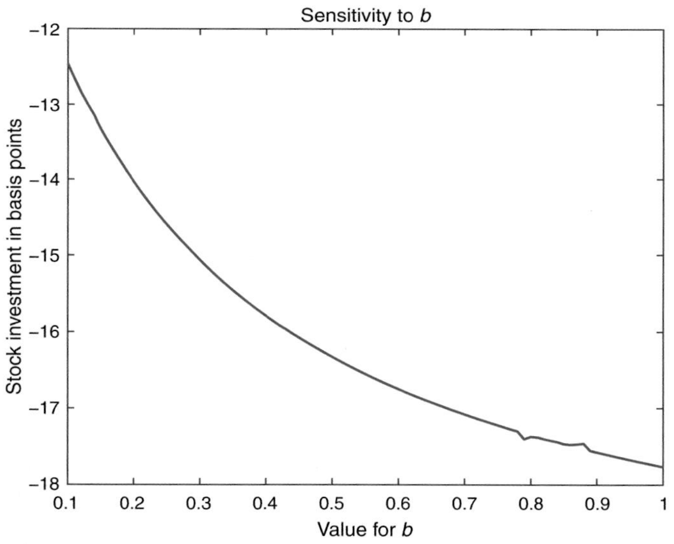

Figure 7.8 Sensitivity of stock investment to variations in b

For 48 equity underliers over 150 days BG parameters were extracted from data on five-day option prices for which optimal stock positions were constructed. From this set five parameter sets were randomly selected with stock positions in absolute value between 1 and

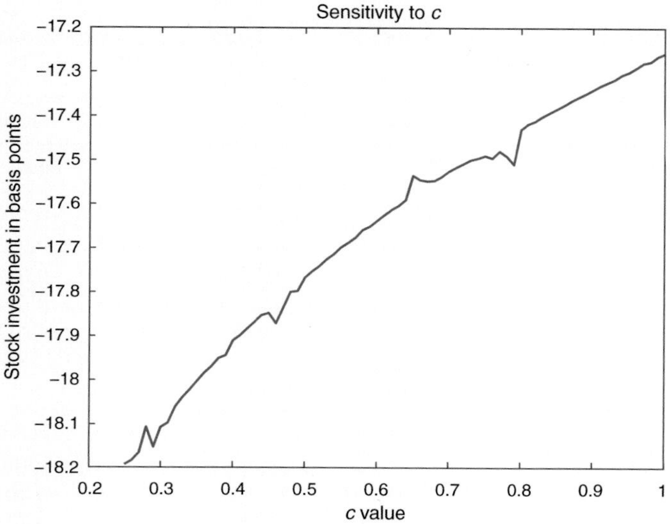

Figure 7.9 Sensitivity of stock investment to variations in c

5% and between 10 and 20% for both positive and negative positions. Table 7.1 presents the parameter values associated with such positions along with the position and the associated exponential variations.

For the larger positions the exponential variations are substantially negative and positive for positions that are respectively negative and positive. For the smaller positions the exponential variations may be small and of the opposite sign. Significant positions do arise depending on the monitoring of BG parameters.

7.4 Measure Distortions Implied by Hedges for a Long Gamma Target

Consider now a single-dollar long gamma target with the risk exposure $c(x)$ given by

$$c(x) = \frac{1}{2}\left(e^x - 1\right)^2.$$

We may write the outcome for the hedge position that shorts a dollars of stock as

$$c_a(x) = \frac{1}{2}\left(e^x - 1\right)^2 - a\left(e^x - 1\right). \tag{7.8}$$

The selection of the short position a that maximizes the lower measure-distorted valuation is equivalent to one that minimizes the negative of this valuation. But by the relationship between infima and suprema this is equivalent to minimizing the upper valuation for the negated outcome, which is identical to the outcome we have already considered for the short gamma target. Hence the undistorted measure evaluations are the same as those for the short dollar gamma target and we merely have to alter the distortions by applying G^-

Table 7.1 *Parameters associated with significant stock positions*

Small Negative

Sample	Bilateral Gamma Parameters				Exp. Var.	Stock Position
	b_p	c_p	b_n	c_n		
1	0.0658	0.9944	0.0973	0.9962	−245.4092	−0.0144
2	0.0795	0.4506	0.1039	1.0091	−605.2452	−0.0435
3	0.0128	1.5930	0.0298	0.6210	23.0835	−0.0126
4	0.0174	1.4697	0.0356	0.6802	19.8196	−0.0119
5	0.0130	2.0248	0.0340	0.6973	31.0437	−0.0155

Large Negative

Sample	Bilateral Gamma Parameters				Exp. Var.	Stock Position
	b_p	c_p	b_n	c_n		
1	0.0343	0.5111	0.1617	0.1613	−63.1984	−0.1012
2	0.0298	0.3470	0.1832	0.0889	−44.4294	−0.1313
3	0.0254	0.3955	0.1376	0.1172	−49.3686	−0.1064
4	0.0215	0.2242	0.1595	0.0461	−19.6030	−0.1348
5	0.0226	0.4252	0.1932	0.0921	−65.0855	−0.1657

Small Positive

Sample	Bilateral Gamma Parameters				Exp. Var.	Stock Position
	b_p	c_p	b_n	c_n		
1	0.0532	0.4681	0.0383	0.3229	135.5196	0.0449
2	0.0250	1.2876	0.0100	1.2172	206.9306	0.0350
3	0.0409	0.7481	0.0250	1.2968	−7.5198	0.0162
4	0.0963	0.5055	0.0698	1.0190	−174.1349	0.0185
5	0.0397	0.6246	0.0250	0.9531	18.2087	0.0178

Large Positive

Sample	Bilateral Gamma Parameters				Exp. Var.	Stock Position
	b_p	c_p	b_n	c_n		
1	0.1076	0.3623	0.0205	1.0514	201.4303	0.1798
2	0.0952	0.2698	0.0165	0.9523	114.3314	0.1562
3	0.1135	0.8019	0.0501	1.2410	365.9042	0.1149
4	0.0656	0.3305	0.0110	1.0319	112.1819	0.1152
5	0.0764	0.2052	0.0097	1.0406	62.6106	0.1372

to the loss outcomes and G^+ to the positive outcomes and then minimizing the associated upper value and negating the hedge position.

In general for positive gamma targets the hedge position is a small positive or negative one depending on the sign of the exponential variation. For the five samples with large negative and positive positions reported in §7.3.2 the corresponding long positions were as follows. For the large negative short gamma positions the long gamma positions were 1.42% for the first three and −1.42% for the remaining two. For the positive short gamma positions the corresponding long gamma positions were +1.83% for the first one and +1.42% for the other four. The measure-distortion parameters used were $b = 1$, $c = 0.5$, and $\gamma = 1.0$.

8

Measure-Distorted Value-Maximizing Hedges in Practice

The objective here is to report on the implementation of measure-distorted value-maximizing hedges for an enterprise over time. Actual enterprises are quite complex, involving many transactions undertaken for a variety of purposes with hedges put in place regularly at end of day for some enterprises or sporadically for others. Access to the variation exposure of an actual enterprise historically with a view to evaluating the impact of different hedging approaches is not really possible, though each proposed enterprise or hedging desk can backtest the implementation of such hedges and compare the results of different hedging schemes. To focus attention we consider well-defined enterprises that many banks construct as dynamic enterprise indexes. The enterprise is given a Bloomberg ticker and the enterprise value is reported to Bloomberg at market close each day. This is the market for structured products that stands at \$7 trillion, around 1% of the \$700 trillion derivatives market. It is, however, larger than the total exchange traded fund (ETF) market at \$5.3 trillion (Bloomberg Professional Services, October 16, 2019).

The structured products provide interested parties investment exposures to a variety of different asset classes with a view toward diversification. We report here on implementing different hedging strategies for such a structured product. Actual structured products reported in Bloomberg can be quite complex, taking positions in multiple markets across the globe. To simplify matters, for illustrative purposes, we restrict positions to the market for options on the S&P 500 index and the index itself. The structured product considered sells daily 20, 30, 40, and 50 delta strangles of an exact three-month maturity, marks to market all open positions, and implements daily a delta hedge for the variation exposure implied by the open positions. Three delta hedges are reported on: a standard Black–Merton–Scholes delta-neutral hedge and hedges that maximize the measure-distorted value of the post-hedge exposure. The latter requires an evaluation of the arrival rates of moves in the S&P 500 index, and this is accomplished by estimating on immediately past S&P 500 return data the BG model, or alternatively, the extraction of physical BG laws from option data on implementing the depreferencing technology described in Chapter 4. The actual exposure is approximated quadratically in constructing the measure-distorted value-maximizing hedge.

8.1 Hedging Overview

The classical hedging problem as presented, for example, in the context of hedging a call option in the context of the Black and Scholes (1973) and Merton (1973) models explains the procedure for replicating a call option by dynamically trading the stock continuously through time. Critical to the validity of this exercise is the martingale representation theorem associated with the filtration of a Brownian motion whereby all the martingales of this filtration may be represented as stochastic integrals with respect to the underlying Brownian motion (Elliot, 1982; Karatzas and Shreve, 1988; Protter, 2005). For many underlying process dynamics, including those of pure jump Lévy processes like the VG or the BG, such martingale representations are not valid and replication by dynamic trading in the stock is not possible. If the empirical observations supporting the local validity of such pure jump processes is entertained as the reality, then one has to ask how the hedge position is to be selected with a view to managing the quality of the residual or unhedged component that exists of necessity. As all uncertainty is resolved at maturity and any potential hedging strategy will be employed dynamically through time until maturity, one has to evaluate the terminal residual at maturity and assess its quality. The use of measure distortions in such a dynamic context with respect to the hedging of, say, a single call option could be considered using methods introduced later in this work. However, we recognize that in practice one does not hedge a single option to its maturity but all open positions at each hedging instance, taking account of the fact that the open positions are themselves dynamically evolving. The theoretical context of a single option hedged to its maturity is of limited interest. Hedge positions are constructed in practice for enterprises across all open positions in claims of multiple maturities. The enterprise must estimate the variation exposure of all open positions to an instantaneous move in the log price relative of the underlying in the amount x. Denote by $v(x)$ this variation exposure function. Dynamic valuation procedures may be employed to estimate this exposure and are considered later in this work. Here we take the function $v(x)$ as given. Typically a variety of models associated with a variety of products to be simultaneously hedged are used to estimate such exposure functions. For simple options one may employ valuations at implied volatilities that are either constant or move with the spot in some predetermined fashion to construct the exposure function.

With a dollars invested currently in the underlying asset one assesses the hedged position as leading to the post-hedge exposure of

$$h(x) = v(x) + a\left(e^x - 1\right),$$

and the hedging problem is that of choosing the optimal or desirable level for the dollar hedge position a given the instantaneous residual or post-hedge exposure h.

It is now clear and apparent that if a linear valuation operator is applied to this residual exposure, as is the case under applications of the law of one price, then the value of h is the value of v, as the value of the financed hedge position is zero. There is no sense of a value-maximizing optimal hedge position. For any valid optimality the post-hedge exposure must be valued by a nonlinear operator, say $\Lambda(h)$, that is capable of evaluating the interaction between the unhedged exposure $v(x)$ and the hedge exposure $a(e^x - 1)$. From the perspective of maximizing Λ it would be desirable that Λ be concave. In this case the set of h for which Λ is positive would be convex. If it is further supposed that the value of a positive h is positive, then this convex set contains the nonnegative outcomes. It is then a candidate for a set of acceptable outcomes (Artzner et al., 1999).

The conic perspective arises and the candidate set of acceptable outcomes is a convex cone, if it is additionally supposed that values scale with positions and the value of twice h is double the value of h. In support of scaling and a conic structure we observe that for small levels of scaling being contemplated one may view the market as a large enough counterparty to scale values with outcomes. Given a conic perspective the valuation must then be an infimum or supremum of classical valuations depending on whether we are dealing with an asset to be delivered to market or a liability to be accessed from the market. Chapter 6 then shows how the result is a distorted valuation, be it of probabilities or measures. Quite generally, then, the optimal hedging problem in the absence of replication is the optimization of a distorted valuation, and this is how we address the hedging issue at the enterprise level given an estimate of the variation exposure.

In the absence of scaling for valuations and the associated conic perspective the literature has considered personalized valuations for residual unhedgeable risk. These are based on expected utility maximization or equivalently the maximizations of certainty equivalents. The associated valuations are also referred to as indifference prices; see Henderson and Hobson (2009). The related set of acceptable risks is a convex set and not a cone with valuations that typically do not scale with outcomes.

Apart from value and utility maximization, somewhat more classically the literature has considered the objective of mean variance hedging both statically and dynamically. We cite in this regard Föllmer and Sondermann (1986), Duffie and Richardson (1991), Föllmer and Schweizer (1991), and Schweizer (1995, 2005). Yet other approaches include works on risk minimization (Davis and Lleo, 2015).

When working under candidate physical laws describing the motion of the asset, one has to be careful that the asset is not too attractive under this measure for the conic and expected utility objectives. For if this is the case an optimizer will switch from being a hedger to becoming a straight investor in the asset. One therefore has to ensure sufficiently high levels of distortions or risk aversions in these approaches. Of course these approaches do not arise for mean variance contexts as the variance dominates and enters the objective negatively. However, risk and reward are measured in different units in such approaches.

8.2 The Hedge-Implementing Enterprise

Hedges designed to attain a zero delta exposure are typically undertaken daily at market close by option desks. Sometimes such hedges may be undertaken at midday, but generally a position is taken in the underlying asset to zero out the delta currently implied by open positions on the underlier. The implied positions are determined by evaluating the derivative of the Black and Scholes (1973) and Merton (1973) option value with respect to the price of stock, where the derivative for each option in the book is computed at the implied volatility of the option. The derivatives are then added up across all open positions to arrive at the firm wide delta for the day. A position in the underlying stock to the extent of the negative of this delta is then taken in the stock. The result is a zero delta book at market close each day. We wish to compare this zero delta hedging strategy with alternatives that derive their positions by maximizing a conservative valuation of the post-hedge position. All hedging strategies will be executed daily from January 4, 2010, to December 21, 2018.

There is an extensive literature comparing a variety of strategies for hedging an option under various models and objectives, analytically when possible and numerically otherwise, but they amount to comments about performance in models and do not amount to outcomes that any operating enterprise would experience. With a view toward getting closer to what may be experienced in practice, we define a stylized enterprise that is a structured product. The specific structure product analyzed here sells options each day and then continuously marks open positions to market, and implements an aggregate hedge for all open positions. For simplicity we restrict attention to a single underlier and consider just the S&P 500 index as the underlying asset on which options are being traded.

The stylized enterprise index with a stylized ticker SPXSTR4 being reported on here sells four strangles every day on the S&P 500 index. The strikes on the strangles are chosen to be at delta levels of 20, 30, 40, and 50%. Both the call and put arms of each of the four strangles have these delta levels respectively. The options mature in exactly three months or 63 business days. At time of sale the index SPXSTR4 receives the market prices from the option sales. Additionally, SPXSTR4 pays out any cash flows due on maturity dates based on the level of the spot 63 days later. In addition all open positions are to be marked to market and supplemented by the daily cash flows earned from the implementation of a hedge using a variety of hedge designs, including the classical Black and Scholes (1973) and Merton (1973) delta hedges.

8.3 Summarizing Option Surfaces Using Gaussian Process Regression

Each day we have to find the 20, 30, 40, and 50 delta strikes for the day with a three-month maturity. Additionally, we have to mark all open positions to market. These activities are considerably enhanced if we have a decent functional summary of the traded option quotes for each day. The prices are best summarized by their Black and Scholes (1973) and Merton (1973) implied volatilities. A functional summary may be attempted using machine learning algorithms once it is clear such a function exists.

For this we turn to the consequences of the absence of static arbitrage. It has been shown in Carr and Madan (2005) or Davis and Hobson (2007) that in the absence of static arbitrage there exists a one-dimensional Markov martingale process for the discounted stock price such that the price of each option is the discounted expected payoff to the option with the expectation taken with respect to the law of this one-dimensional Markov martingale. Numerous parametric examples of such martingale models fitting option surfaces may be found in the literature. In fact there are examples of even additive processes or processes of independent increments that summarize the option surface. For examples one may refer to Carr et al. (2007). For any of these martingale models one may derive the option price and hence its implied volatility as just a function of the option maturity and the level of moneyness relative to the forward of the option strike. These considerations suggest that there exists a nonlinear function relating the option-implied volatility to its maturity and moneyness. Each day or at each moment, the martingale may change and the function may therefore move, but each day or at each moment there exists such a function. We therefore consider the application of machine learning methods to summarize for us this nonlinear function each day. In addition to the functional representation of implied volatilities we also construct functions for the forward price and the discount rate as just functions of the maturity.

The particular machine learning algorithm we employ is GPR that models the implied volatility as a Gaussian process over the moneyness maturity domains using an exponential kernel for the covariance functional and a zero mean function. For each underlier and each day we build three functional representations for the volatility surface, the forward curve, and the rate curve that are stored in separate files for each underlier and day. For purposes of marking and determining the 20, 30, 40, and 50 delta strikes we just call the appropriate function at the appropriate arguments to get the implied volatility and then the option price or delta as the case may be.

For an illustration of the results consider options on the S&P 500 index as of December 21, 2017. The options covered 12 maturities and there were 982 options in the training set. The maximum absolute error was 77 basis points between market and the GPR predicted implied volatilities. This is well within desired ranges. Figure 8.1 presents the market and GPR predicted implied volatilities at 6 maturities of 29, 85, 176, 365, 449, and 729 days.

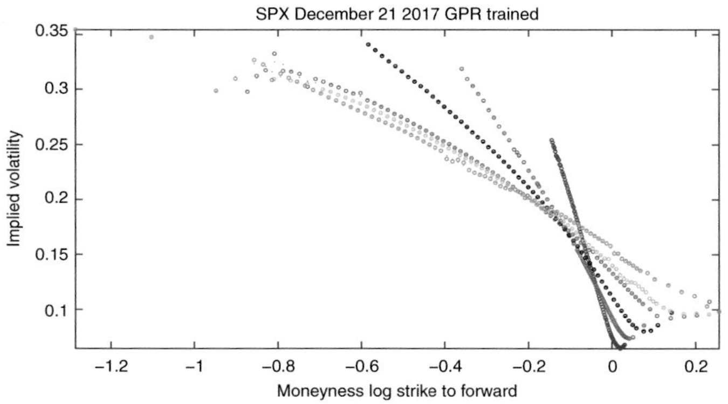

Figure 8.1 Market-implied volatilities are displayed by circles while the GPR predictions are displayed as dots. The maturities of 29, 85, 176, 365, 449, and 729 days are displayed in the colors blue, red, black, magenta, cyan, and green.

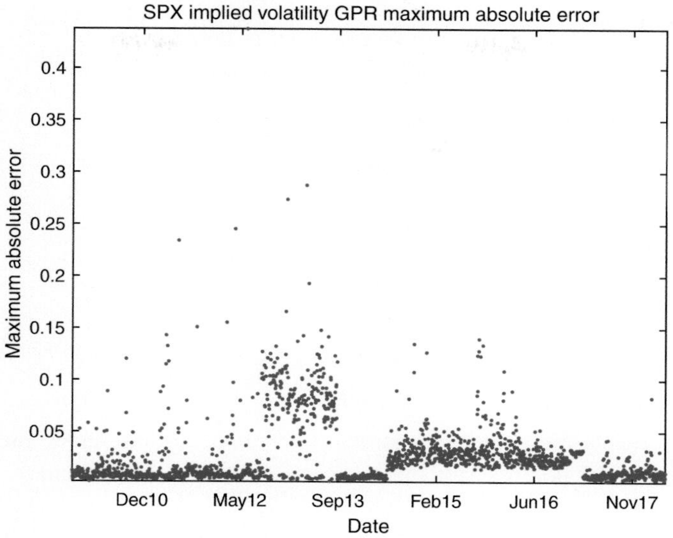

Figure 8.2 Maximum absolute error in GPR prediction of SPX implied volatilities across time from January 4, 2010, to April 17, 2018

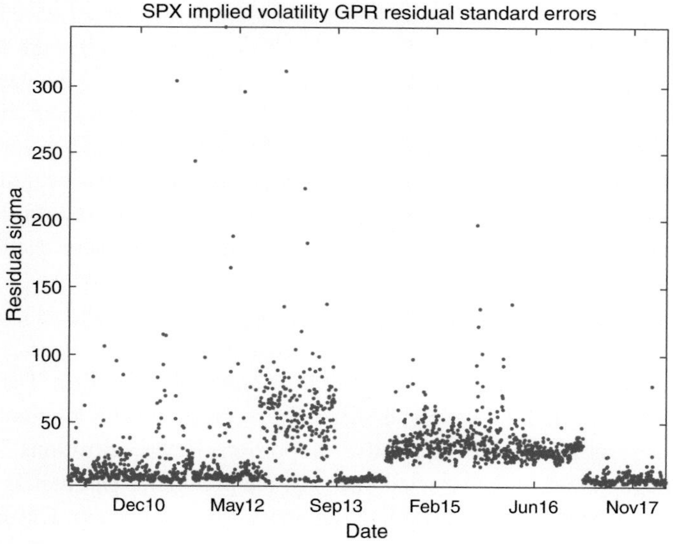

Figure 8.3 Residual standard error in basis points for GPR prediction of SPX implied volatilities across time from January 4, 2010, to April 17, 2018

Additionally two graphs of the maximum absolute error and the residual standard error across time from January 4, 2010, to April 17, 2018, are presented in Figures 8.2 and 8.3, with the latter in basis points.

Table 8.1 *Error percentiles in basis points*

Percentile	Max Absolute	Standard Deviation
1	34	5
5	43	6
10	50	7
25	69	9
50	155	18
75	304	34
90	673	52
95	908	68
99	1282	102

Table 8.1 presents the percentiles for the maximum absolute error and the residual standard error in basis points across this time period for the S&P 500 index.

8.4 Selecting the Hedging Arrival Rates

The hedge position is to be selected each day by maximizing the integrated measure-distorted tail arrival rates for gains and losses of different sizes experienced by the post-hedge exposure. This requires a determination of the tail arrival rates that are then distorted. One approach is to employ immediately past data on daily returns on the underlying asset to estimate the parameters of, say, a BG model for the return density and then obtain the arrival rates from the associated Lévy measure. This is a slow-moving approach that is also backward looking. A more forward-looking approach is to extract candidate physical densities from data on option prices via the depreferencing methods introduced in §4.1. There are, however, two possible different implementations of these methods, and we present and implement both methods.

The first of these is as presented in §4.1.1. Here one takes the candidate density $f(x)$ for a candidate traded maturity to be in the BG class with parameters b_p, c_p, b_n, and c_n that is impacted by the preference model with gamma-distributed risk aversions. The risk aversion distribution has parameters of scale and shape $c_{\mathrm{L}}, \gamma_{\mathrm{L}}, c_{\mathrm{S}}, \gamma_{\mathrm{S}}$ for the long and short sides. There is also a parameter a for the relative proportion of short positions. The resulting model is fit to option data at the selected maturity, the preference parameters are set aside, and the arrival rates are taken in line with the estimated BG parameters. The calibration has the task of determining all parameters to match the forward and option prices by choice of these nine parameters. The forward is not being penned exactly but approximately in such a calibration.

Since the forward is known up front and is critical to the calibration of option prices, an alternative is to apply the preference perturbation to a mean shifted physical candidate that pegs the forward using a standard convexity correction associated with the BG parameters. In this way the surface calibration works directly on the tail option prices with the forward

Table 8.2 *Quantiles for bilateral gamma with unpegged forward*

quantile	b_p	c_p	b_n	c_n
1	0.0021	0.0004	0.0018	0.0056
5	0.0084	0.0010	0.0105	0.0142
10	0.0157	0.0097	0.0488	0.0208
25	0.0569	0.0898	0.1212	0.0810
50	1.9019	0.7677	0.7014	0.6396
75	5.2237	1.7888	2.4622	1.2185
90	6.6128	8.6434	5.2269	4.5393
95	28.4066	33.0438	13.5863	22.9300
99	643.1100	78.6355	217.3280	91.7600

being exactly pegged. The estimated BG model will have its own estimated drift extracted from the option surface but the calibration has less work to do by virtue of pegging the forward.

We implement both these methods for access to arrival rates to be distorted in the hedge construction. The first we refer to as bguf for bilateral gamma with unpegged forward while the latter is referred to as bgpf for bilateral gamma with pegged forward. The data we are working with are from January 2010 onward, for which the first liquid maturity is nearly a month. The calibrations employed option prices for this maturity.

Option calibrations do suffer from randomness introduced by variations in the actual maturity to be employed in the raw data. An exact one-month maturity is not typically traded. There are also variations in the number of puts relative to calls employed in calibrations by movements in the spot price. These variations can cause variations in estimated parameters and thereby the implied arrival rates. With a view to minimizing these variations we use our GPR summarized surfaces to obtain option prices at an exact monthly maturity with strikes evenly distributed on both sides of the forward. Calibrations were conducted in two ways for each method bguf and bgpf for each day from January 4, 2010, to April 17, 2018, for a period of $2,082$ days.

Tables 8.2 and 8.3 present the quantiles for the four parameters under the two estimations. We observe that the estimates from the penned forward procedure are less variable and more stable.

8.5 Approximating Variation Exposures

Optimization routines were developed in Chapter 7 to determine dollar hedge positions in the stock to hedge a unit dollar short gamma position in the stock when the risk was BG and the objective was a measure-distorted valuation. The position may be scaled to arbitrary gamma levels as computed by Black and Scholes (1973) and Merton (1973) gamma computations. This optimization algorithm may be employed on approximating the actual variation exposure on account of open positions by a quadratic in the return. For a position in the stock price denoted by $c(S)$ one may employ the Black and Scholes

Table 8.3 *Quantiles for bilateral gamma with pegged forward*

quantile	b_p	c_p	b_n	c_n
1	0.0011	0.0042	0.0009	0.1980
5	0.0014	0.1298	0.0014	0.3666
10	0.0033	0.2485	0.0029	0.5383
25	0.0135	0.7437	0.0286	0.9547
50	0.0395	1.0912	0.0503	1.2530
75	0.9006	2.2409	0.2672	2.0113
90	1.8210	9.3059	0.7853	7.9303
95	2.2540	16.1874	1.1460	11.6597
99	3.8681	43.9790	2.1992	27.0496

(1973) and Merton (1973) delta Δ and Γ computations with a current stock price of S_0 to approximate the variation exposure by

$$c(S) - c(S_0) \approx \Delta(S - S_0) + \frac{1}{2}\Gamma (S - S_0)^2 .$$

With respect to the return

$$R = \frac{S - S_0}{S_0},$$

one may write

$$c(S) - c(S_0) = \Delta S_0 R + \frac{1}{2}\Gamma S_0^2 R^2.$$

For a traditional delta neutral hedge we short the return in the amount of ΔS_0 dollars to get to a zero delta exposure. For a short gamma position we scale the optimal short unit gamma hedge by ΓS_0^2.

Figure 8.4 shows the actual variation exposure and the quadratic hedge for the same for product SPXSTR4 on 4 days that are 100 days apart.

Figure 8.5 presents the intercept, delta, and gamma of a quadratic approximation to the variation exposure along with the quantization points for the same with 50 and 100 points.

8.6 Measure Distortion Parameters

The measure distortion we employ in the analysis of hedging was introduced in (6.18) and (6.19) with four parameters: a, b, c, and γ. For the asymptotic symmetry of risk charges as measures tend to infinity, we take $a = b/c$. The value of b was set at its largest level of unity. In keeping with the study of earlier examples we continue with $c = 0.5$. For the choice of γ we take a high value of 9 with a view to ensuring that the hedge asset is unlikely to be desirable on its own account and will be only used with a view to improving hedge quality. We verify that at this particular measure distortion the stock asset is not desirable for purchase or sale for both sets of BG parameters with the forward penned or unpenned.

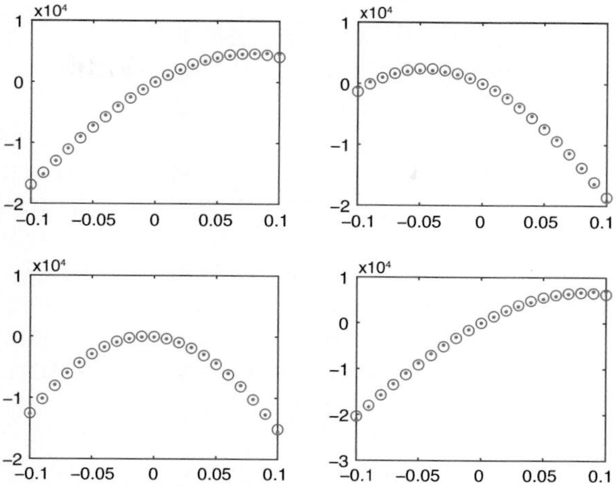

Figure 8.4 The actual variation exposure is presented by circles while the quadratic approximation is represented by dots. The exposures are as of May 26, 2010, October 19, 2010, March 14, 2011, and August 4, 2011.

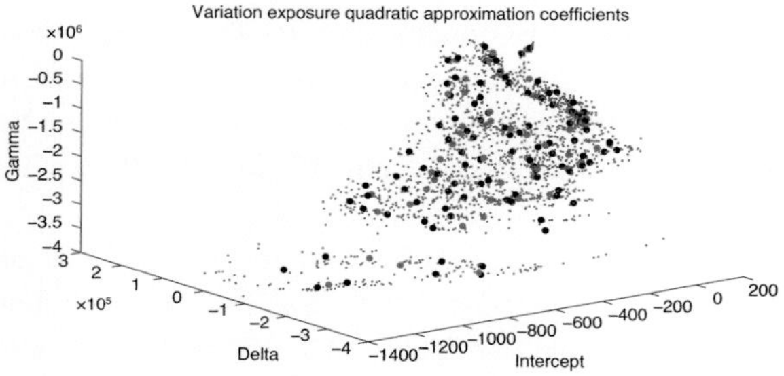

Figure 8.5 Quadratic approximation intercept, deltas, and gammas along with their quantization in red and black into 50 and 100 points

Table 8.4 presents the quantiles across time for the measure-distorted value of going long or short with the stock asset. The valuations are quite consistently negative, thereby ruling out any interest in the stock itself.

We note that the valuations for the case with penned forward are more stable than those for when the forward was unpenned. It would appear that the activity of penning the forward appears to assist the estimation by reducing the work being done by the calibration and focusing it on the structure of the tails as opposed to fixing the forward as well. There are also some cases for the unpenned estimations when the long position is attractive for the asset alone.

Other strategies for selecting measure distortion parameters in defining hedges can also be entertained, as opposed to using a constant level. One could, for example, first select

Table 8.4 *Quantiles for stock valuation*

	Long Position		Short Position	
quantile	bguf	bgpf	bguf	bgpf
1	−30.4112	−8.6354	−130.0162	−38.7782
5	−11.4154	−4.5572	−61.9074	−30.0821
10	−9.4500	−2.9779	−37.8309	−18.1038
25	−7.4473	−0.7394	−31.9885	−10.5802
50	−2.0944	−0.2834	−18.4232	−0.2302
75	−0.5686	−0.1771	−0.4397	−0.1085
90	−0.1213	−0.0093	−0.1272	−0.0218
95	−0.0341	−0.0074	−0.0752	−0.0211
99	0.8943	0.0000	−0.0186	−0.0148

active bid and ask prices on SPX options for the day. We could then calibrate distortion parameters to match these prices after extracting BG parameters from the option surface mid-quotes using depreferencing technology to identify the law to be distorted. These procedures may lead to better and more relevant hedges at an additional computational cost. The choice of constant distortion parameters is just an illustrative first step here.

Alternatively one may estimate BG parameters from time series data for the motion in major indices and then set distortion parameters at minimal levels consistent with the lower measure-distorted value of long short strategies on these assets being bounded above by a fixed negative reserve level. The reserve level could be selected to monitor the conservatism desired in valuation.

Such market-adapted time series of distortion parameters could be generated and saved to serve as inputs to the design of hedge strategies. The hedge strategies could then be backtested in the manner illustrated in Section 8.7. The actual hedge to be implemented could then be selected from the set of those currently delivering a good performance. A considerable set of research avenues is open in the formulation of competitive hedging systems.

8.7 Backtest Hedging Results for Multiple Strangles on SPX

The hedging strategies for the product SPXSTR4 were backtested on option and time series data on the S&P 500 index for the period January 4, 2010, to April 17, 2018. The last date on which the strangles were sold was January 11, 2018. The options expired in 63 days and cash flow consequences were associated with options sold on January 11, 2018, and maturing before April 17, 2018.

Cash flows were earned daily from the sale of four strangles and payouts were recorded at their maturities. In addition the GPR summaries of volatility surfaces were used to mark all open positions to market each day. For the hedge adopted on each day, BG parameters were extracted from prices of options maturing in a month at prespecified strikes using both unpenned and penned forward strategies on the candidate physical density employing

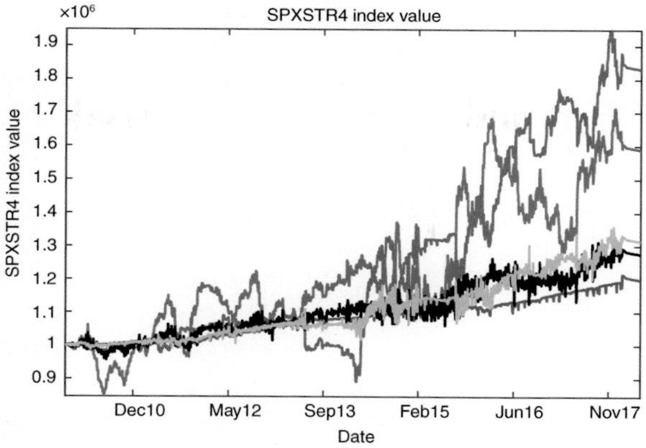

Figure 8.6 The SPXSTR4 index level through time using delta-neutral hedging and four hedges incorporating measure-distorted value (MDV) maximization. The delta-neutral hedge is presented in blue. The MDV hedge for unpenned forward estimation is presented in red while the penned forward estimation is in magenta. Black and cyan represent the unpenned and penned delta strategies with a 80% weight on the delta-neutral hedge.

the depreferencing technology for the parameter calibration. For the associated arrival rate function, hedges were selected that maximized the post-hedge exposure. The exposure was described using a quadratic in the log price relative derived from the Black and Scholes (1973) and Merton (1973) delta and gamma calculations for all open positions on the day. Also implemented was the Black and Scholes (1973) and Merton (1973) delta-neutral hedge strategy.

Figure 8.6 presents five mark-to-market index values for SPXSTR4 starting at $1 million. They are for the base delta-neutral hedge, the measure-distorted value-maximizing hedges for the arrival rates associated with the unpenned and penned forward BG estimations, and for hedging strategies that employ the base delta-neutral strategy with a weight of 80% and a weight of 20% on the measure-distorted value-maximizing strategy.

We observe that the penned hedge makes a substantial improvement over the delta-neutral hedge with some volatility and drawdowns. These may be somewhat controlled by weighting the delta-neutral hedge. In any case this is a first run on exploring measure-distorted value-maximizing hedges. Further research in these directions may make further improvements.

9
Conic Hedging Contributions and Comparisons

Dynamic hedging, and hedging more generally, took its beginnings (Black and Scholes, 1973; Merton, 1973) in the literature on replication where strategies are developed, under suitable assumptions, making it possible to create the hedged liability exactly. The final cash flow after implementing the hedge is exactly zero. Markets with this property are referred to as complete markets and numerous examples of such appear in the literature (see, for example, Cox et al., 1979, or Cvitanić and Zapatero, 2004). As a consequence any two hedges that accomplish replication are financially identical with no real basis to choose between them. Under the law of one price the cost of hedging under the two hedges must also be identical. Leaving aside such textbook examples of complete markets, one recognizes that perfect replication may not be possible in actual markets, and this places the task of hedging in the context of incomplete markets (Staum, 2008).

Most practical hedges have a post-hedge exposure, and the hedge design has to choose between these possibilities. A popular approach is to minimize this error or gap as far as possible in the spirit of attempting to replicate as best we can. A particular method is the implementation of a least-squares hedge that minimizes the sum of squared hedging errors. One may also consider minimizing the mean absolute deviation or exposure along with other loss functions applied to the exposures.

In this direction there are contributions on variance minimization, its time inconsistency, and related modifications to local variance minimization (Bertsimas et al., 2001; Heath et al., 2001; Basak and Chabakauri, 2012). But variance minimization, locally or globally, is burdened by its unrealistic symmetric treatment of gains and losses.

Alternatives associated with semi-variance or just the downside variance have been considered and developed (see, for example, Chen et al., 2001). These methods ignore an assessment of the risk beyond two moments and are suited, for example, to situations where it is relatively clear that the risk is close to a Gaussian one. Such hypotheses are questionable when we consider the hedged residuals of complex derivatives with a variety of embedded optionalities. We note in this regard that the risk-neutral distributions observed in markets are, however, highly non-Gaussian (Christoffersen et al., 2010) and reflect varying degrees of skewness and kurtosis for even just a position in the stock, let alone the addition of optionalities.

The residual post-hedge exposures, however, are not just hedging or prediction errors to be minimized, but actual cash outcomes that are either to be paid or received. As a result the positive cash flows are preferred while the negative ones are avoided. The valuation of exposures is sensitive to their direction or sign, and this fact is not reflected in the objective when they are treated as errors in usually symmetric loss functions. Conic hedging objectives assessing lower and upper market values for post-hedge claims are sensitive to the sign of the cash flow with distorted expectations reweighting upward losses that are to be avoided while simultaneously deflating gains in the calculation of the lower valuation that is then maximized.

So far we have focused attention in hedging on the comparison of conic revisions to traditional delta hedging where the hedge instrument is just the underlying stock with the hedge being implemented at daily close. A richer and more informative comparison may be conducted once the set of hedging assets is expanded to include options. However, this expansion has its own issues when working with a single underlying asset. Basically, the options-complete markets as all smooth, twice differentiable functions of the underlying stock price at the option maturity may be perfectly replicated by a portfolio of options in addition to the stock and bond payoffs (Carr and Madan, 1998). We are then back to zero post-hedge cash flows with nothing interesting left to discuss. In higher dimensions, for liabilities written as a function of multiple underlying asset prices at the option maturity, the incompleteness reappears as options fail to span functions of several variables. But before venturing into multidimensional issues taken up later, it is useful to explore and report on the one-dimensional possibilities. With a view toward limiting the span of the hedging assets we employ just two options of a six-month maturity. They are a call struck 5% up and a put struck 10% down, in addition to the stock itself. There are then just three hedging assets.

Unlike numerous hedging studies that hedge terminal cash flows given by some function of the final stock price at a future date, the target studied here is an exposure function modeling the difference between the value of claims subsequent to a jump in the stock price logarithm of x and the value prior to the jump. The exposure function is by design zero at zero. It is to be hedged in terms of the exposures delivered by positions in the stock and the hedging options. The latter requires a model for the valuation of options post the jump in the price of the stock. Though there are results (Carr and Madan, 1998) showing the completeness of options with respect to twice differentiable functions of the final stock price, such results do not extend directly to exposure functions and the result, if any, would depend on the valuation model being employed to determine the post-jump option prices and the associated hedging asset jump exposures.

With respect to hedging criteria, here we maximize the measure-distorted value of the post-hedge exposure. In addition, for comparative purposes, we consider the least-squares hedge that may be viewed as a mechanism for constructing the conditional expectation as projected onto the space of attainable spanning functions. This methodology will be generalized toward the construction of conditional distorted expectations that build in direction sensitivity into least-squares methodologies. These procedures will be later applied in the multivariate context.

9.1 Univariate Exposure Hedging Study

Say we have exposure $y(x)$ to a univariate jump in the log price of x. In addition there are assets with an exposure to a jump in the same price given by functions $g_i(x)$ for the ith hedging asset. By way of a hedge we wish to construct hedge positions a_i yielding the post-hedge exposure of

$$u(x) = y(x) - \sum_i a_i g_i(x).$$

One may implement objectives like least squares or mean absolute deviation or ε-insensitive optimization for the choice of positions. Such objectives are motivated by error reduction considerations that treat positive and negative deviations symmetrically. From an economic viewpoint the post-hedge exposure represents risk still left to be carried and one should be interested in maximizing its market value. From a measure distortion perspective the lower valuation for the position $a = (a_1, \ldots, a_N)$ is

$$L(a) = -\int_0^\infty G^+ \left(\nu\left(\{x \mid u^-(x) > w\}\right)\right) dw + \int_0^\infty G^- \left(\nu\left(\left\{x \mid u^+(x) > w\right\}\right)\right) dw.$$

The market value-maximizing hedge chooses a to maximize $L(a)$. The equivalent when dealing with probability distortions is

$$\widetilde{L}(a) = -\int_0^\infty \Psi\left(P\left(\{x \mid u^-(x) > w\}\right)\right) dw + \int_0^\infty \widehat{\Psi}\left(P\left(\left\{x \mid u^+(x) > w\right\}\right)\right) dw.$$

One may compare the results obtained from statistical and conic economic approaches toward hedge specification.

Consider hedging a shorter dated claim with longer dated out-of-the-money calls and puts. Option prices subsequent to a jump in the stock price may be taken from a calibrated risk-neutral model with given parameters. For example, one could employ the Sato process with BG density at unit time. This is the BGSSD model reported on in Madan and Schoutens (2019b). Alternatively valuations could be based on market-implied volatilities that are constant for each strike or defined by deterministic functions of the post-jump moneyness. Let $c_{\mathrm{T}}(S)$ be the value function of the target while $c_i(S)$, for $i = 1, \ldots, n$ are the values for the hedge instruments. The exposure function is

$$u(x) = -\left(c_{\mathrm{T}}\left(Se^x\right) - c_{\mathrm{T}}(S)\right) + \delta S(e^x - 1) + \sum_{i=1}^{n} a_i \left(c_i(Se^x) - c_i(S)\right).$$

One does not really want to minimize the residual in size via least squares, mean absolute deviation, or some other metric, but rather we would like to cover liabilities and be covered by assets so that in both cases we tend to have a positive cash flow if possible. These possibilities are investigated and the results reported.

We do, however, have the issue presented in §8.1 of ensuring that the hedge assets are not desirable on their own. In this case optimizers will forget hedging and become investors. Generally a high enough stress level can take care of this issue. As we are not hedging under a specific measure but allow for a rich collection of alternative valuation measures, the hedge assets should be negative in value under some alternative valuation measure. The problem is then well founded. These issues may be avoided by focusing attention on conditional expectations either distorted or not. For this purpose we introduce the concept of distorted least squares that makes least-squares procedures directionally sensitive as is appropriate when dealing in financial matters.

9.2 Distorted Least Squares

Consider a span of functions $f(x, \theta)$ with parameters θ that are attempting to reach to outcomes y_i observed at points x_i. An approach is to assert that the target y is the function output plus an error to define

$$y = f(x, \theta) + \varepsilon.$$

The nonlinear least-squares exercise selects θ with a view toward minimizing

$$\frac{1}{2} \sum_i (y_i - f(x_i, \theta))^2 .$$

Suppose the minimum occurs at θ_* defined by the first-order conditions

$$\sum_i (y_i - f(x_i, \theta)) \frac{\partial f(x_i, \theta)}{\partial \theta} = 0.$$

The minimum will occur somewhere as the objective is bounded below by zero. Say we have the global minimum at θ_*. We then have produced the function

$$f(x, \theta_*),$$

and we may inquire into its properties.

If the data were generated by calculating $f(x, \theta)$ and adding a Gaussian error ε with zero mean and some variance, then such an estimate is known to be a maximum likelihood estimate. But there is no basis for such a construction as the data in all probability are not generated in this manner.

An alternate argument supporting this procedure is that of supposing for each fixed x there were many values of y possible and our data were really

$$x, y(x)_1, y(x)_2, \ldots, y(x)_{N(x)},$$

representing $N(x)$ different outcomes for y occurring at the same x. So we postulate the situation whereby

$$y(x)_j = f(x, \theta) + \varepsilon(x)_j, \; j = 1, \ldots, N(x).$$

Now define

$$g(x) = \frac{1}{N(x)} \sum_{j=1}^{N} y(x)_j$$

as the conditional expectation of y given x.

One may then write more generally

$$\begin{aligned}
&\sum_{i,j \leq N(x_i)} \left(y(x_i)_j - f(x_i, \theta)\right)^2 \\
&= \sum_{i,j \leq N(x_i)} \left(y(x_i)_j - g(x_i) + g(x_i) - f(x_i, \theta)\right)^2 \\
&= \sum_{i,j \leq N(x_i)} \left(y(x_i)_j - g(x_i)\right)^2 + 2 \sum_{i,j \leq N(x_i)} \left(y(x_i)_j - g(x_i)\right)\left(g(x_i) - f(x_i, \theta)\right) \\
&\quad + \sum_i N(x_i)\left(g(x_i) - f(x_i, \theta)\right)^2 .
\end{aligned}$$

Recognizing that

$$N(x_i)g(x_i) = \sum_{j \leq N(x_i)} y(x_i)_j,$$

we have

$$\begin{aligned}
&\left[\sum_{i,j \leq N(x_i)} \left(y(x_i)_j - f(x_i, \theta)\right)^2\right] \\
&= \left[\sum_{i,j \leq N(x_i)} \left(y(x_i)_j - g(x_i)\right)^2\right] + \sum_i N(x_i)\left(g(x_i) - f(x_i, \theta)\right)^2,
\end{aligned}$$

and hence $f(x, \theta_*)$ is on the manifold defined by $f(x, \theta)$ that is closest in a weighted least-squares sense to the conditional expectation function $g(x)$.

Let us now generalize this construction toward the formation of distorted conditional expectations. In this regard we first construct

$$\begin{aligned}
\mathfrak{g}(x) &= \sum_{j=1}^{N(x)} y(x)_j w_j, \\
w_j &= \Psi\left(q_j\right) - \Psi\left(q_j - \frac{1}{N(x)}\right), \\
q_j &= \frac{1}{N(x)} \sum_{k=1}^{N(x)} \mathbf{1}_{y(x)_k \leq y(x)_j},
\end{aligned}$$

where $\mathfrak{g}(x)$ is the distorted expectation of y given x with distortions based on the quantiles of the residuals $y(x)_j - \mathfrak{g}(x)$ that are also the quantiles of $y(x)_j$. In this way we embed directional sensitivity with negative outcomes receiving a greater weight, thereby lowering the conservative conditional distorted expectation.

More generally one then writes using a conditional distorted expectation operator $\mathcal{E}$ as

$$\left[\sum_{i,j\leq N(x_i)} (y(x_i)_j - f(x_i,\theta))^2\right] = \left[\sum_{i,j\leq N(x_i)} (y(x_i)_j - \mathrm{g}(x_i))^2\right] + \sum_i N(x_i)(\mathrm{g}(x_i) - f(x_i,\theta))^2,$$

where we eliminate the cross-product term on noting that

$$N(x)\mathrm{g}(x) = \sum_{j\leq N(x)} y(x)_j w_j.$$

The distorted weights are now defined by

$$w_j = \Psi(q_j) - \Psi\left(q_j - \frac{1}{N}\right),$$
$$q_j = \frac{1}{N(x)}\sum_{k=1}^{N(x)} \mathbf{1}_{r(x)_k \leq r(x)_j},$$
$$r_j = y(x)_j - f(x_i,\theta).$$

The objective being minimized is

$$\sum_{i,j\leq N(x_i)} (y(x_i)_j - f(x_i,\theta))^2 w(x)_j.$$

This is the situation when the points x_i are viewed as occurring with equal probability of $1/M$ for the M points.

In the more general context, when x_i occurs with probability p_i, we define the objective as

$$\sum_i (y_i - f(x_i,\theta))^2 w_i,$$

with

$$w_i = \Psi(q_i) - \Psi(q_i - p_i),$$
$$q_i = \sum_{i=1}^{N} p_j \mathbf{1}_{r_j \leq r_i},$$
$$r_i = y_i - f(x_i,\theta).$$

For the upper distorted expectation one merely replaces the distortion Ψ by its complementary distortion

$$\widehat{\Psi}(u) = 1 - \Psi(1-u).$$

The strategy of minimizing squared errors weighted by functions of the residual quantiles leverages the fact that the minimum of squared errors weighted in this way occurs at the distorted expectation. To observe this, define

$$V(a) = \int (y(x) - a)^2 \, d\Psi(F(x)),$$

and note that $V'(a) = 0$ just if

$$a = \int y(x) d\Psi(F(x)).$$

The quantile weighting thus biases the choice of functions toward the distorted expectation as opposed to the expectation that is targeted by least squares. The use of $\widehat{\Psi}$ in place of Ψ lifts the outcome above the expectation while Ψ lowers it.

9.3 Example Illustrating Distorted Least-Squares Hedges

Consider, by way of an example, two hedging assets in addition to the stock, given by a call struck at 105 and a put struck at 90, both with a maturity of six months. For jump sizes between negative and positive 30% in steps of 10 basis points, we form the hedge matrix valuing these assets post jump less their value prior to a jump using BGSSD risk-neutral parameters as estimated for SPY as of December 17, 2018. These parameter values were

$$b_p = 0.0112, \qquad c_p = 39.1130, \qquad b_n = 0.3118,$$
$$c_n = 0.4086, \quad \gamma = 0.5726.$$

By way of an example consider a target that post a suitable hedge has a nonnegative exposure function. We design the target exposure function $T(x)$ as a nonnegative exposure function $c(x)$ plus a position holding two shares, one call struck at 105 and short 20 puts struck at 90. Both the call and put have a maturity of six months. Specifically we define

$$T(x) = 2\,(e^x - 1) + C(Se^x; 105, .5) - C(S; 105, .5)$$
$$- 20\,(P(Se^x; 90, .5) - P(S; 90, .5)) + c(x),$$

where $c(x)$ is zero at zero and otherwise positive on the sample space. Figure 9.1 presents a graph of the function $c(x)$. It was defined as

$$c(x) = 50000x^2 e^{-30|x|}.$$

Figure 9.2 presents the target cash flow as function of x the spot return.

The target cash flow is an asset for a party that can implement a least-squares hedge. In addition one may determine the hedge to maximize a distorted expectation with minmaxvar distortion at stress levels of 0.75 or 1.5. There are then three possible hedges. The hedge position attained on maximizing a distorted expectation may be modeled by a distorted conditional expectation with the latter attained on implementing a distorted least-squares prediction, and this is the result we present. Figure 9.3 presents the three post-hedge cash flows. We observe that the distorted conditional expectations raise the post-hedge outcomes, delivering outcomes closer to the targeted sample space arbitrage.

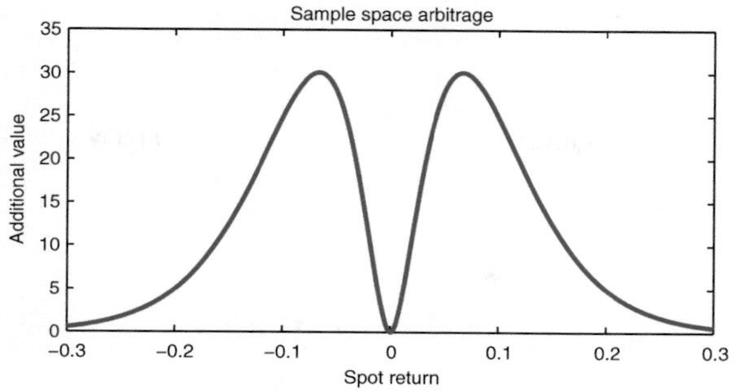

Figure 9.1 Sample space arbitrage cash flow possible post hedge with positions of two shares, one call at 105 and a short 10 puts at 90 both with maturity six month

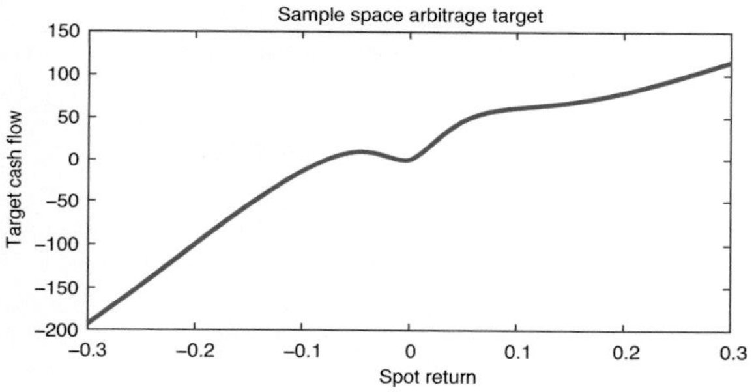

Figure 9.2 Target cash flow with access to a sample space arbitrage

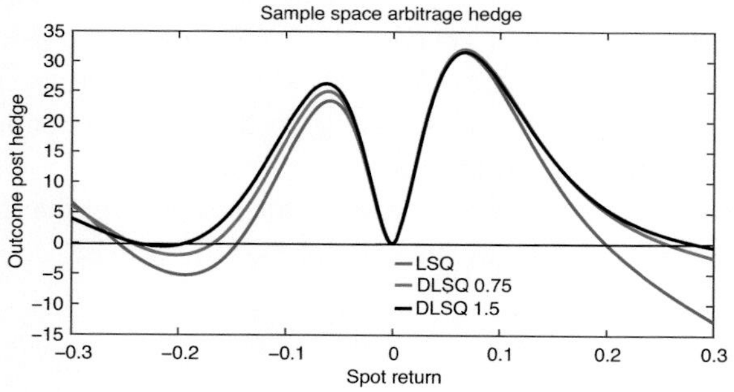

Figure 9.3 Post-hedge cash flow from a least-squares hedge in blue. The stress levels 0.75 and 1.5 are presented in red and black respectively.

Figure 9.4 addresses the hedge problem facing the counterparty that must deliver the target cash flow as its liability. The objective is then to over hedge and reduce the post-hedge cash flow to be delivered. The stress levels 0.75 and 1.5 deliver reductions relative to least squares.

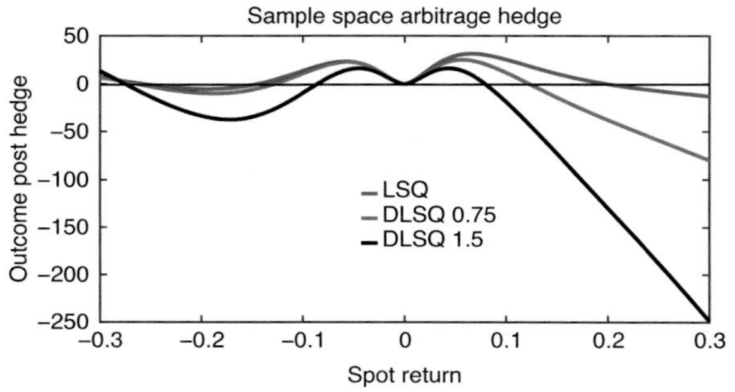

Figure 9.4 Post-hedge cash flows from the perspective of the counterparty that must deliver the target cash flow as its liability. Least squares is presented in blue. The stress levels of 0.75 and 1.5 are presented in red and black respectively.

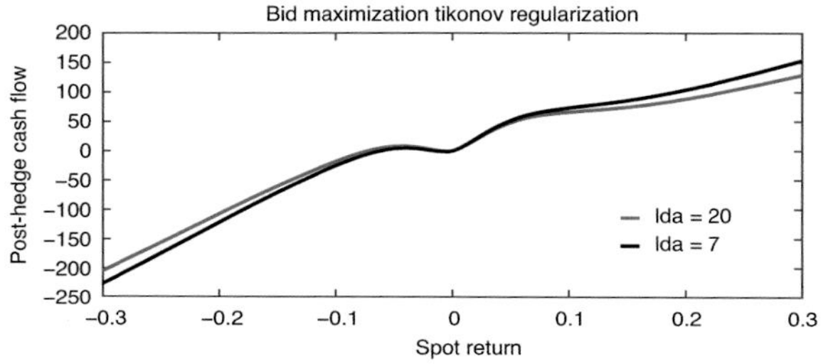

Figure 9.5 Tikonov regularization with penalties of 7 and 20

The directionally sensitive conic hedges move the hedge cash flows toward the delivery of post-hedge cash flows that are of greater value to the hedger, be they the recipients or payees of the target. The conic hedges vary with whether one is to receive or pay out the target. The least-squares hedge is invariant to the positioning direction in the target. From a value perspective a conic hedge is preferable.

One may also maximize directly the bid price for positions in the three hedge assets. Here we just maximize the post-hedge bid price or the bid price for

$$T(x) - \sum_i a_i g_i(x).$$

It turned out that the problem was unbounded as the hedge assets were attractive on the sample space given access to a sample space arbitrage. We then optimized using Tikonov regularization to get the result for two different penalty functions as shown in Figure 9.5.

9.4 Incorporating Weightings

Not all jump sizes are equally likely as represented on a grid. We employed the Lévy density to give the jump size x_i a weighting proportional to $k(x_i)$ using the BG parameters appropriate to the maturity of the hedging instruments. The weights were transformed to

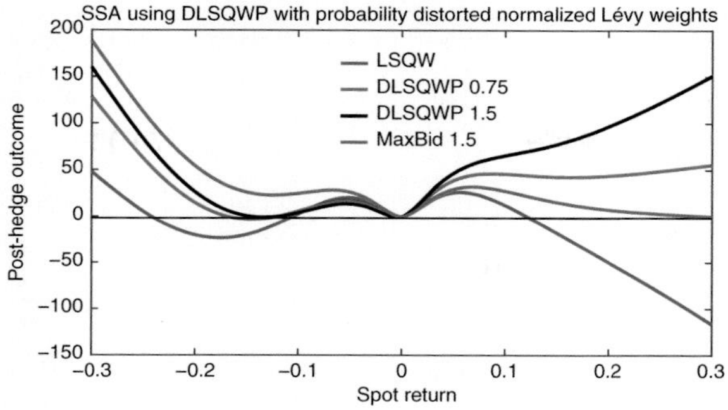

Figure 9.6 Post-hedge cash flows as a function of the jump size in returns. The least-squares solution is presented in blue. The two conditional distorted expectations for the stress levels of 0.75 and 1.5 are presented in red and black respectively. The maximal bid price for stress level 1.5 is presented in magenta.

probabilities on normalization. Suppose these probabilities are p_i for the point x_i. The program DLSQWP distorts the probabilities as follows. Define the quantile for the point x_i by

$$q_i = \sum_{u_j \le u_i} p_j .$$

The distorted least-squares objective is then

$$\sum_i u_i^2 \left(\Psi\left(q_i\right) - \Psi(q_i - p_i) \right).$$

Two stress levels of 0.75 and 1.5 are reported for the unweighted context presented earlier. The base position is taken to be long, and we employ the lower conditional distorted expectation for the hedge position.

Additionally we maximize the bid price obtained on distorting these probabilities. For the stress level of 0.75 this problem has an unbounded solution and is not reported. For the stress level of 1.5 the solution is interior. Figure 9.6 presents the least-squares solution, the two conditional distorted expectations, and the maximal bid price solution.

We observe that the conic hedges lift the cash flow outcomes uniformly, as expected by their preference for the positive direction embedded into the objective. The exercise reported supports the use of directionally sensitive hedging objectives. The use of conditional distorted expectations is particularly attractive as the problem is well founded with an expected interior solution. For maximal bid prices one has to ensure sufficient stress in the formulation of the hedge-designing objective function.

9.5 Measure-Distorted Value Maximization

For measure-distortion value-maximization parameters $b = 1$, $c = 0.5$, and $\gamma = 5$, the post-hedge cash flow is shown in Figure 9.7.

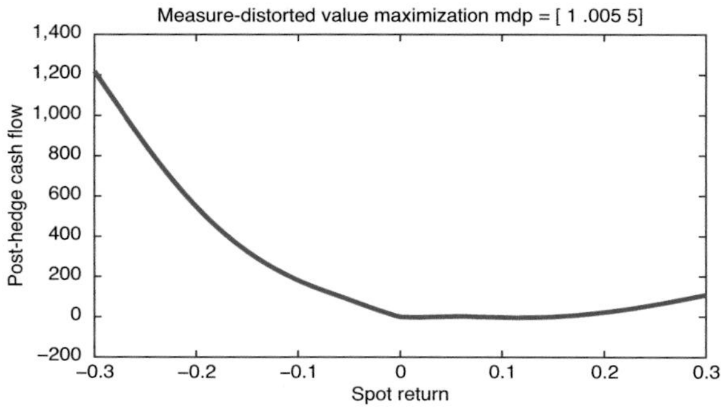

Figure 9.7 Post-hedge cash flow for measure-distorted value maximization for distortion parameters $b = 1$, $c = 0.005$, and $\gamma = 5$.

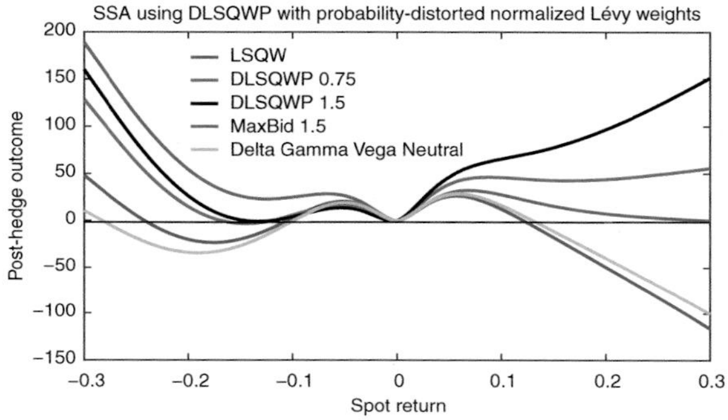

Figure 9.8 Additional greek-neutral hedge added

There are therefore a number of conic hedging approaches possible, all delivering directionally sensitive hedges that are different for long and short positions.

9.6 Greek Hedging

We may hedge the target exposure with a portfolio of options that have the same exposure. One may then evaluate the greeks of the hedging portfolio. The delta, gamma, and vega for the target were 3.5885, −0.02606, and −195.0105. The positions in the three assets attaining greek neutrality were 14.0705, −22.5218, and 36.3736. Figure 9.8 adds the greek-neutral hedge in the color cyan to the graph in Figure 9.6.

9.7 Theta Issues in Exposure Design

Suppose we have a set of hedging assets with exposure H in general to states that include a no-change state with zero exposure. Let y be any such target exposure. The least-squares approach seeks positions a such that we seek to minimize

$$(a'H - y')(H'a - y) = a'HH'a - 2y'H'a + y'y$$

with a solution defined by

$$a = (HH')^{-1}Hy.$$

However, in many cases the matrix HH' may be close to singular and not invertible. One may adopt a regularization scheme and impose a penalty whereby the objective becomes

$$a'HH'a - 2y'H'a + y'y + \lambda a'a.$$

The solution is now defined by

$$(HH' + \lambda I)a = Hy.$$

The smallest eigenvalue is now λ and the matrix is invertible with solution

$$a = (HH' + \lambda I)^{-1} Hy.$$

The resulting cash flow is

$$y'H'(HH' + \lambda I)^{-1}H.$$

We may now get directionally sensitive and construct positions with respect to minimizing

$$(a'H - y')W(H'a - y)$$

for a diagonal matrix W with diagonal w given by

$$\begin{aligned} w &= \Psi(q) - \Psi\left(q - \frac{1}{N}\right), \\ q &= \frac{1}{N}\sum_j \mathbf{1}_{r_j \le r_i}, \\ r &= H'a - y. \end{aligned}$$

We may adopt a regularization penalty and minimize

$$(a'H - y')W(H'a - y) + \lambda a'a.$$

The problem is, however, nonlinear and nonquadratic as w depends on r and a in a highly nonlinear fashion.

Consider now the target

$$y(x) = |x|$$

for jump sizes x in the logarithm in the spot price relative. Attaining such a nonnegative exposure at zero cost is possible, as is demonstrated in Figure 9.9. One is just accessing the curves shown in blue and red at zero cost. It is, however, not an arbitrage as such exposures are exposed to the cost of negative theta, as discussed later.

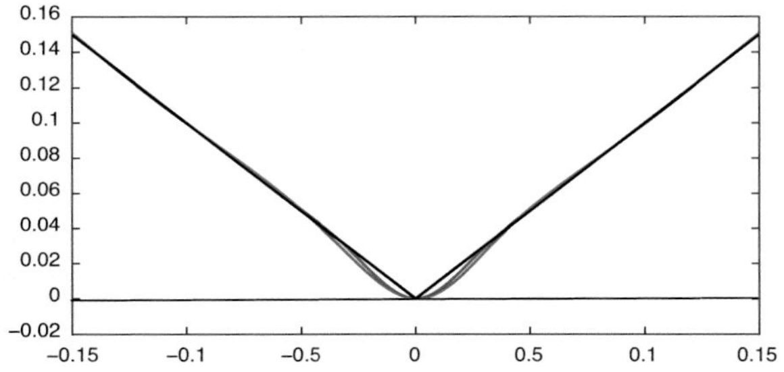

Figure 9.9 Access $|x|$ exposure using hedges based on option positions with valuations conducted using the BGSSD model in blue and using implied volatilities constant for each strike in red

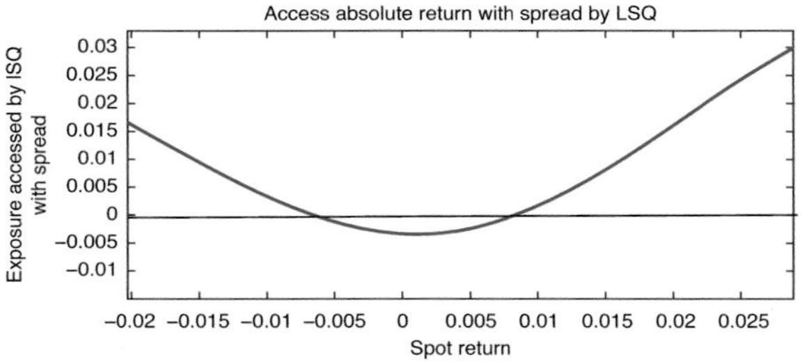

Figure 9.10 Least-squares hedge cash flow accessed by hedge with spreads for the target cash flow of the absolute value of x

9.8 Incorporating Spreads

We may recognize the presence of spreads in implied volatilities and incorporate the fact that when we go long, we buy at the ask or upper price and must unwind at the bid or the lower price. The role is switched when we go short, selling at bid and buying back at the ask. The hedge matrix may be expanded to HL and HS with nonnegative positions α_L, α_S and final exposures flows with a dollars in the stock of

$$aS_0(e^x - 1) + \alpha_L HL - \alpha_S HS.$$

We may still target the exposure $y(x) = |x|$ and try for a least-squares or distorted least-squares outcome. Figure 9.10 presents the result for least squares with spreads that represent the minimal distance of the target to the possibilities spanned by the hedge with spreads.

The hedge matrices were evaluated using bid and ask implied volatilities as appropriate. The least-squares hedge was obtained on minimizing the sum of squared residuals between the absolute value target and the exposure delivered by the hedge. We observe that outside a spot move of around 70 basis points the hedge delivers a positive outcome.

The gap between the upper distorted expectation or the upper least-squares solution and the least-squares solution is presented in Figure 9.11. We see that the upper hedge tends to

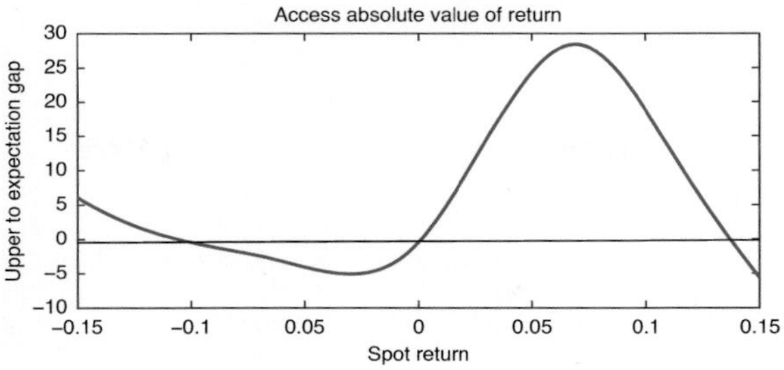

Figure 9.11 Gap between the upper least-squares solution and the least-squares solution

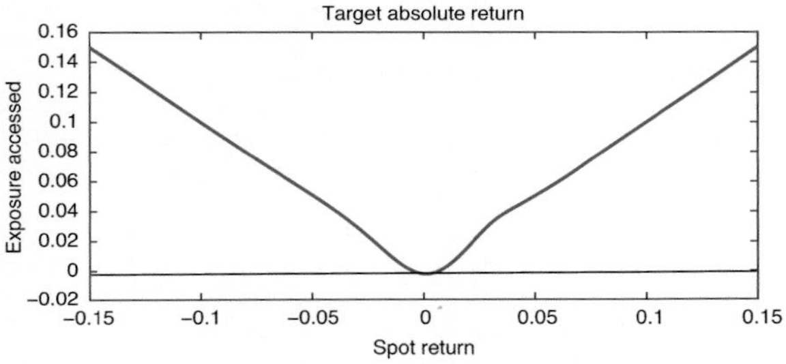

Figure 9.12 Least-squares exposure accessed by positions in the first two maturities with strike moneyness below 5%

deliver a more positive outcome, especially for an up move in the spot with a small negative cost associated with the down move.

The positions were near zero except for the first two maturities with moneyness below 5%. We removed all other options and reran the least-squares and upper least-squares programs. The result was as follows for SPY on January 2, 2013. Figure 9.12 presents the exposure accessed. The minimum value was negative 22 basis points at a spot return of zero. The exposure turns positive for spot returns above 50 basis points.

Figure 9.13 presents the gap in basis points between the upper least-squares exposure with stress level 1.5 for distortion minmaxvar and the least-squares exposure. The minimum upper exposure near at zero was 19 basis points. The exposure turns positive at moves of 30 basis points up or down. The spot was 145.7503.

The stock position was negative 165 basis points in both cases. The option positions were

Maturity	Strike	P/C	Position
0.0247	143	P	– 0.0092
0.0247	144	P	– 0.0026
0.0247	147	C	0.0599
0.0247	149	C	– 0.0467
0.0247	152	C	0.0165
0.0438	139	P	0.0032
0.0438	152	C	– 0.0070

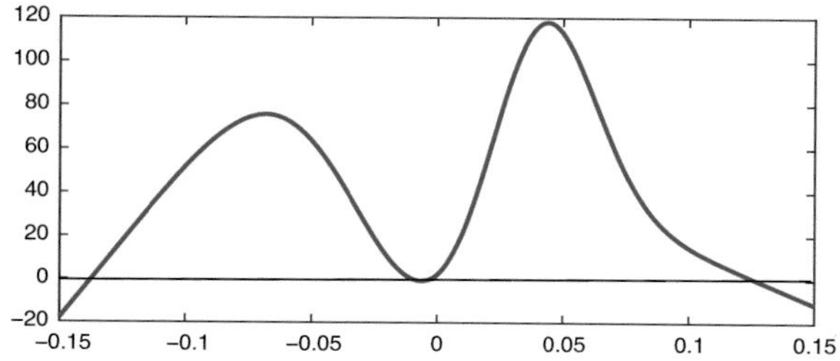

Figure 9.13 Gap between the upper least-squares exposure accessed and the least-squares exposure in basis points

The positions taken by the upper least-squares execution were

Maturity	Strike	P/C	Position
0.0247	140	P	– 0.0023
0.0247	143	P	– 0.0051
0.0247	144	P	– 0.0024
0.0247	147	C	0.0630
0.0247	149	C	– 0.0460
0.0438	153	C	0.0053

9.9 No Spread Access and Theta Considerations

We go back to pricing under the law of one price using the BGSSD model and implied volatility pricing with regularized hedges and $\lambda = 0.1$. There are options at two maturities. Figure 9.14 presents the results. The accessed cash flow was nonnegative for both hedges.

Under the law of one price with pure jump, risk-neutral arrival rates $k_Q(x)$ of finite variation, the martingale condition for values $V(S,t)$ implies that

$$V_t + \int_{-\infty}^{\infty} [V(Se^x, t) - V(S,t)]\, k_Q(x)dx = 0.$$

If the accessed exposure is that of $|x|$, then the theta of this portfolio should be

$$\theta_p = -\int |x| k_Q(x)dx.$$

There is then a loss in value over time if there is no movement in the stock price. When dealing with exposures the zero-cost condition translates to a zero-theta constraint.

Can we design better or optimal exposures? One may estimate a statistical arrival rate function $k_P(x)$ and seek to find good positions with respect to a risk-neutral arrival rate $k_Q(x)$. This problem is taken up in the next chapter as the problem of optimal exposure design. However, we impose the constraint on $y(x)$ of

$$\int_{-\infty}^{\infty} k_Q(x)y(x)dx = 0.$$

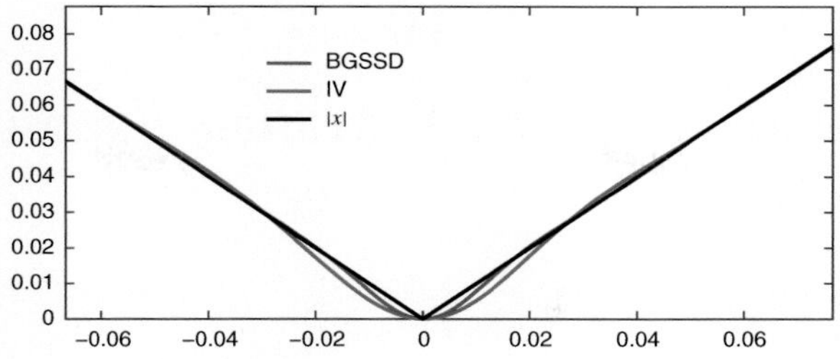

Figure 9.14 Accessing $|x|$ using options at the first two maturities

In this way we seek a zero-theta position. Such a constraint dismisses $|x|$ as a target as this cannot be a zero-theta exposure.

10

Designing Optimal Univariate Exposures

Suppose we have estimated both the physical and risk-neutral arrival rates $k_P(x)$ and $k_Q(x)$ respectively for jumps x in the log price relative. In such a context how should one design the desired exposure $y(x)$, of course satisfying $y(0) = 0$? At any time, every collection of positions in derivative securities on an underlying asset involving a variety of claims, including the stock, options of numerous strikes and maturities, and a variety of related structured products, will deliver an implied exposure function. This inherited exposure may be altered, subject to constraints, toward an optimal exposure, the determination of which is the question we address.

Much has been written on optimal positioning in derivatives of a specific maturity, and we cite as examples Leland (1980), Brennan and Solanki (1981) and Carr and Madan (2001). The question here, however, is divorced from maturities and considers instead the design of exposures that may amalgamate many maturities. Recognizing that many financial institutions are always holding some exposure often described by what are called spot slides, it is natural to inquire into the optimality of such.

It is useful to first summarize the result with respect to optimal claims as functions of the stock price at a single maturity. Suppose the risk-neutral density for this maturity is given by the probability density $q(S)$. Further suppose an individual has probability beliefs for the stock price at this maturity represented by the probability density $p(S)$. Let the cash flow accessed for this maturity be $c(S)$ with an expected utility for utility function $u(c)$, given by

$$\int_0^\infty u(c(S))p(S)dS. \tag{10.1}$$

The cost W of the attained cash flow $c(S)$ is in forward dollars

$$\int_0^\infty c(S)q(S)dS = W. \tag{10.2}$$

The first-order condition for the maximization of 10.1 subject to the constraint 10.2 with Lagrange multiplier λ for the constraint yields that

$$u'(c(S))p(S) = \lambda q(S). \tag{10.3}$$

The structure of the optimal cash flow accessed is then given by

$$c(S) = (u')^{-1}\left(\frac{\lambda q(S)}{p(S)}\right), \tag{10.4}$$

and the optimal cash flow is determined in terms of the ratio of the risk-neutral to physical densities. For the analysis of exposures, we anticipate that a critical role will be played by the two arrival rate functions k_Q, k_P.

It is reasonable to restrict attention to a manageable subclass of exposure functions $y(x)$. For example, discontinuities in exposure are not likely to be possible to create, so we certainly should impose continuity. However, even continuous functions can be highly oscillatory with infinite variation, as is the case with the paths of Brownian motion, for example. Such functions are also not likely to be accessible using traded securities. We restrict attention to exposures that are of finite variation and hence we write

$$y(x) = g(x) - h(x)$$

for two increasing functions g, h. With such a restriction it is natural to solve for the nonnegative derivatives of these functions that we write as

$$\begin{aligned} g(x) &= g_n(x)\mathbf{1}_{x<0} + g_p(x)\mathbf{1}_{x>0}, \\ h(x) &= h_n(x)\mathbf{1}_{x<0} + h_p(x)\mathbf{1}_{x>0}, \\ g_n', g_p', h_n', h_p' &\geq 0. \end{aligned}$$

As the functions are determined from their derivatives up to a constant, and given that $y(0) = 0$, we take $g(0) = h(0) = 0$ and define

$$\begin{aligned} g_n(x) &= -\int_x^0 g_n'(v)dv, & g_p(x) &= \int_0^x g_p'(v)dv, \\ h_n(x) &= -\int_x^0 h_n'(v)dv, & h_p(x) &= \int_0^x h_p'(v)dv. \end{aligned}$$

10.1 Exposure Design Objectives

The exposure design objective is a real-valued function $J(y)$ that one may seek to maximize or minimize. From a maximization viewpoint one would like J to be a concave functional. From a market value perspective one anticipates a scaling of value with $J(\lambda y) = \lambda J(y)$ for $\lambda > 0$. These considerations suggest the use of the lower and upper measure-distorted valuations with respect to the physical arrival rate function $k_P(x)$.

The dual characterization of measure-distorted valuations leads to the formulation

$$J(y) = \inf_{Z \in \mathcal{M}} \int_{-\infty}^{\infty} Z(x)k_P(x)(g(x) - h(x))dx.$$

The set $\mathcal{M}$ is defined by

$$\mathcal{M} = \left\{ \begin{array}{ll} Z \mid Z \geq 0, \int_{-\infty}^{\infty} (Z(x) - a)^+ \, k_P(x) dx \leq \Phi(a), & a > 1, \\ \int_{-\infty}^{\infty} (a - Z(x))^+ \, k_P(x) dx \leq -\widetilde{\Phi}(a), & a < 1. \end{array} \right.$$

The objective may be rewritten in terms of the derivative functions as

$$J(y) = \inf_{Z \in \mathcal{M}} \left[- \int_{-\infty}^{0} \int_{-\infty}^{x} Z(u) k_P(u) du \left(g_n'(x) - h_n'(x)\right) dx + \int_{0}^{\infty} \int_{x}^{\infty} Z(u) k_P(u) du \left(g_p'(x) - h_p'(x)\right) dx \right].$$

10.2 Exposure Design Constraints

We observed in §9.9 that one may access at zero cost a nonnegative exposure that must be paid for by a negative theta exposure. We eliminate these theta considerations by designing zero-theta exposures. This constraint then requires that

$$\int_{-\infty}^{\infty} k_Q(x)(g(x) - h(x)) dx = 0. \tag{10.5}$$

Equation (10.5) is the equivalent of a zero-cost constraint in the context of modeling exposures.

In terms of the derivatives we write

$$- \int_{-\infty}^{0} \int_{-\infty}^{x} k_Q(u) du \left(g_n'(x) - h_n'(x)\right) dx + \int_{0}^{\infty} \int_{x}^{\infty} k_Q(u) du \left(g_p'(x) - h_p'(x)\right) dx = 0.$$

10.3 Exposure Design Problem

The exposure design problem may then be formulated as maximizing b subject to

$$b \leq - \int_{-\infty}^{0} \int_{-\infty}^{x} Z(u) k_P(u) du \left(g_n'(x) - h_n'(x)\right) dx + \int_{0}^{\infty} \int_{x}^{\infty} Z(u) k_P(u) du \left(g_p'(x) - h_p'(x)\right) dx$$

$$\int_{-\infty}^{\infty} (Z(x) - a)^+ \, k_P(x) dx \leq \Phi(a), \qquad a > 1$$

$$\int_{-\infty}^{\infty} (a - Z(x))^+ \, k_P(x) dx \leq -\widetilde{\Phi}(a), \qquad a < 1$$

$$- \int_{-\infty}^{0} \int_{-\infty}^{x} k_Q(u) du \left(g_n'(x) - h_n'(x)\right) dx + \int_{0}^{\infty} \int_{x}^{\infty} k_Q(u) du \left(g_p'(x) - h_p'(x)\right) dx = 0$$

$$g_n', g_p', h_n', h_p', Z \geq 0.$$

10.4 Lagrangean Analysis of the Design Problem

The Lagrangean analysis of the exposure design problem may be written as follows:

$$\mathcal{L} = b - \lambda \begin{pmatrix} b + \int_{-\infty}^{0} \int_{-\infty}^{x} Z(u) k_P(u) du \left(g_n'(x) - h_n'(x)\right) dx \\ - \int_{0}^{\infty} \int_{x}^{\infty} Z(u) k_P(u) du \left(g_p'(x) - h_p'(x)\right) dx \end{pmatrix}$$
$$- \int_1^{\infty} \eta(a) \left(\int_{-\infty}^{\infty} (Z(x) - a)^+ k_P(x) dx - \Phi(a) \right) da$$
$$- \int_0^1 \eta(a) \left(\int_{-\infty}^{\infty} (a - Z(x))^+ k_P(x) dx + \widetilde{\Phi}(a) \right)$$
$$+ \zeta \begin{pmatrix} \int_{-\infty}^{0} \int_{-\infty}^{x} k_Q(u) du \left(g_n'(x) - h_n'(x)\right) dx \\ - \int_{0}^{\infty} \int_{x}^{\infty} k_Q(u) du \left(g_p'(x) - h_p'(x)\right) dx \end{pmatrix},$$

for dual variables $\lambda, \zeta, \eta(a)$.

The first-order conditions with respect to b yield $\lambda = 1$. As the objective is linear with respect to g_n', g_p', h_n', h_p' we have to implement a bang bang control with upper bounds for the derivatives g_n', g_p', h_n', h_p' and dual variables $\theta_n(x), \theta_p(x), \kappa_n(x), \kappa_p(x)$. We get respectively

$$- \int_{-\infty}^{x} Z(u) k_P(u) du + \zeta \int_{-\infty}^{x} k_Q(u) du - \theta_n(x) \leq 0$$
$$\text{with equality when } g_n' > 0,$$
$$\int_{x}^{\infty} Z(u) k_P(u) du - \zeta \int_{x}^{\infty} k_Q(u) du - \theta_p(x) \leq 0$$
$$\text{with equality when } g_p' > 0,$$
$$\int_{-\infty}^{x} Z(u) k_P(u) du - \zeta \int_{-\infty}^{x} k_Q(u) du - \kappa_n(x) \leq 0$$
$$\text{with equality when } h_n' > 0,$$
$$- \int_{x}^{\infty} Z(u) k_P(u) du + \zeta \int_{x}^{\infty} k_Q(u) du - \kappa_p(x) \leq 0$$
$$\text{with equality when } h_p' > 0.$$

With respect to $Z(x)$ for $x < 0$ we have

$$(g_n(x) - h_n(x)) - \int_1^{\infty} \eta(a) \mathbf{1}_{Z(x) > a} da + \int_0^1 \eta(a) \mathbf{1}_{Z(x) < a} da \leq 0$$
$$\text{with equality when } Z(x) > 0, \ x < 0.$$

Similarly with respect to $Z(x)$ for $x > 0$ we get that

$$(g_p(x) - h_p(x)) - \int_1^{\infty} \eta(a) \mathbf{1}_{Z(x) > a} da + \int_0^1 \eta(a) \mathbf{1}_{Z(x) < a} da \leq 0$$
$$\text{with equality when } Z(x) > 0, \ x > 0.$$

The problem is difficult to solve analytically, and we implement numerical solutions based on estimates of the two Lévy measures k_P, k_Q.

We observe, however, that positivity of derivatives is related to comparing Q tails with the corresponding tails of $Z(x)k_P(x)$, suggesting that Z values may be related to $k_Q(x)/k_P(x)$. This ratio is employed in truncating the infinite integral for calls on Z struck at a.

10.5 Discretization and Solution

We discretize the jump sizes from, say, −0.2 to 0.2 in steps of 10 basis points and exclude the zero outcome. The risk-neutral Lévy density is used to construct the tail integral at each point x denoted kkq. We also evaluate the physical Lévy densities at each x point to construct the vector kkp.

Next we define values for the vector $aaaa$ that will serve in truncating the calls on Z at $a > 1$. First define

$$aaaa = kkq./kkp.$$

The largest value for a considered was

$$mxa = \max(\min(\max(aaaa), 10), 1.5),$$

where we truncate the premium of k_Q over k_P at 10 and at least have it greater than 1.5. The minimal value for a was 0.01.

For calls on Z the strikes were 1.01 to mxa in steps of size 0.01. For puts on Z the strikes were from 0.01 to 0.99 in steps of 0.01. Measure-distortion parameters were used to evaluate the bounds $\Phi(a)$ and $-\widetilde{\Phi}(a)$ for the call and put strikes on Z at a. The sequence of $a's$ was truncated at $\Phi(a_c)$ and $-\widetilde{\Phi}(a_p)$ equal to 0.001. The constraints for call and puts on Z are on a grid of values below a_c and a_p.

Instead of solving for the derivatives g_n', g_p', h_n', h_p' and Z on the set of all grid values for jumps sizes that has a dimension of $401 * 5 = 2005$, we evaluate these derivatives at 21 points from −0.2 to 0.2 in steps of 0.02. The values at other points on the grid are obtained by linear interpolation at this coarser set of values. The optimization then has a dimension of $21 \times 5 = 105$.

There are no constant bounds on b. The lower bound on Z is zero and the upper bound was twice the maximum entry of $aaaa$. The upper bound on all the derivatives g_n', g_p', h_n', h_p' was unity. There was one equality constraint, and this was the zero-theta condition. The two nonlinear constraints on Z are the bounds on calls on Z struck at $a > 1$ by $\Phi(a)$ and the bound on puts on Z struck at $a < 1$ by $-\widetilde{\Phi}(a)$. In addition there is a nonlinear constraint given by the upper bound on b that is below the valuation of the exposure induced by $Z(x)k_P(x)$.

Implementation requires the choice of k_P, k_Q. We illustrate using BG physical Lévy measures implied from monthly option prices using the depreferencing technology introduced in Chapter 4. The risk-neutral Lévy measure is taken from fitting the BG model to the same option prices. The prices are taken at an exact one-month maturity at prespecified strikes inferred by GPR applied to market-implied volatility surfaces.

10.6 Details Related to Lévy Measure Singularities at Zero

We need to perform the following integrals in the neighborhood of the singularities in k_P, k_Q near zero,

$$\int_{-a}^{a} Z(x)k_P(x)(g(x) - h(x))dx; \qquad \int_{-a}^{a} k_Q(x)(g(x) - h(x))dx.$$

We partition at zero and work on the two segments.

Consider first

$$\int_{0}^{a} k_Q(x)(g(x) - h(x))dx.$$

We write this as

$$\begin{aligned}(g'(a) - h'(a)) \int_0^a xk_Q(x)dx &= (g'(a) - h'(a))c_p \int_0^a e^{-x/bp}dx \\ &= (g'(a) - h'(a))c_p b_p \left(1 - e^{-a/bp}\right).\end{aligned}$$

Similarly

$$\int_{-a}^{0} k_Q(x)(g(x) - h(x))dx = -(g'(-a) - h'(-a))c_n b_n \left(1 - e^{-a/b_n}\right).$$

Next we consider

$$\int_0^a Z(x)k_p(x)(g(x) - h(x))dx = Z(a)(g'(a) - h'(a))c_p b_p(1 - e^{-a/b_P})$$

using now the physical parameters. Similarly we take

$$\int_{-a}^{0} Z(x)k_P(x)(g(x) - h(x))dx = -Z(-a)(g'(-a) - h'(-a))c_n b_n \left(1 - e^{-a/bn}\right).$$

10.7 Sample Optimal Exposure Designs

A numerical constrained optimization problem was solved for the optimal exposure design for 29 underlying assets spread out over the years 2007 through 2017. Figure 10.1 presents nine representative exposure designs taken from the solutions for these names on a number of dates over the period. By design the exposure is zero at zero and tends to be generally positive for large moves, occasionally taking negative values in the negative tail and near zero to accomplish the zero-theta constraint. The desired exposure may be attained by taking positions in the shorter maturity options with a view toward exposure replication as opposed to replicating terminal cash flows. Of course the requisite valuations for post-jump derivative prices may be conducted using a model or implied volatilities, as reported on earlier.

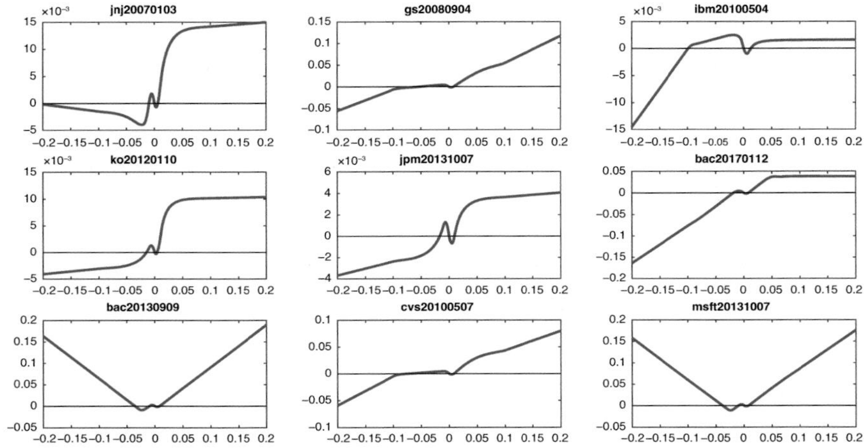

Figure 10.1 Sample of optimal exposure designs constructed for a variety of underlying assets on numerous days.

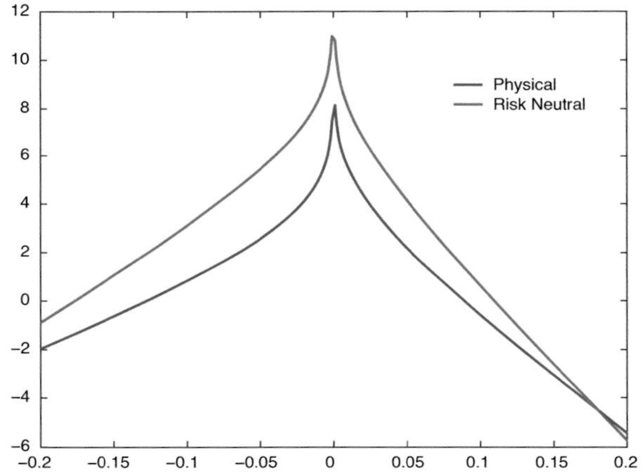

Figure 10.2 Logarithm of physical and risk-neutral arrival rates on MSFT on October 7, 2013. The densities were estimated from option prices with a maturity of one month after interpolation by a Gaussian process regression run on the raw data.

10.8 Further Details about Some Particular Cases

We present details here for MSFT as of October 7, 2013. The physical and risk-neutral BG parameters on this date were estimated as follows.

	b_p	c_p	b_n	c_n
Physical	0.0240	3.5673	0.0476	1.8698
Risk Neutral	0.0177	52.5002	0.0303	60.5921

Figure 10.2 presents the logarithm of the corresponding physical and risk-neutral densities.

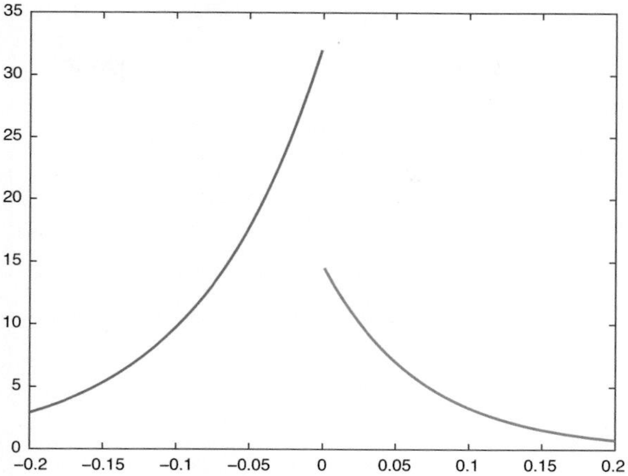

Figure 10.3 Ratio of risk-neutral to physical arrival rates for MSFT on October 7, 2013

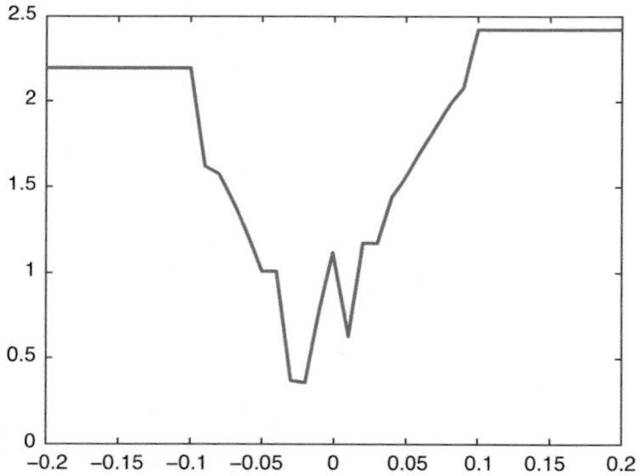

Figure 10.4 Measure change Z for MSFT on October 7, 2013

We observe that risk-neutral arrival rates are higher than their physical counterparts. Figure 10.3 presents the ratio at the various return levels. The ratios reflect a negative risk premium at all return levels on both sides, though they are large in absolute value for the downside returns.

Figure 10.4 presents the change of measure function that lifts tail events and discounts near money events.

Figures 10.5 and 10.6 present the values of calls and puts on Z with strikes above and below unity along with their upper bounds.

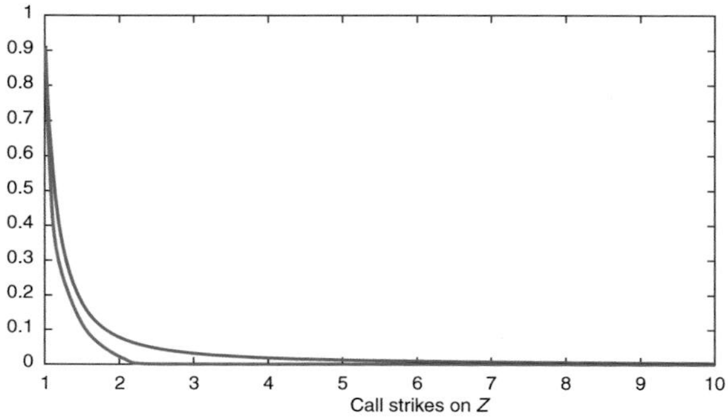

Figure 10.5 Calls on Z along with their upper bounds

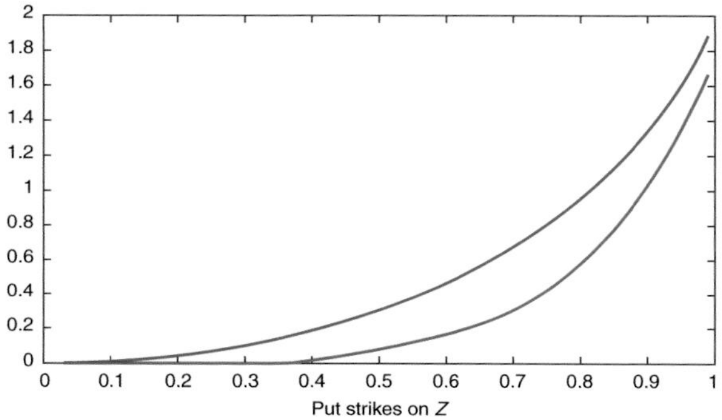

Figure 10.6 Puts on Z along with their upper bounds

11

Multivariate Static Hedge Designs Using Measure-Distorted Valuations

We now turn to the design of hedges in more than one dimension using traded securities. Two detailed examples are presented for dimensions 2 and 10 respectively. Practical higher-dimensional hedging exercises arise in insurance markets. With respect to hedge positioning for liability management at a specific maturity, an insurance-theoretic study is presented in Carr et al. (2016). Here the hedge and the targets are stylized constructions. Consider to begin with an exposure $y(x)$ in higher than one dimension for which one may seek hedges delivering $g(x)$ to access the residual

$$r(x) = y(x) + g(x).$$

If we implement a measure-distorted value-maximizing hedge, the abstract dual formulation of the problem is as follows:

$$\max_{g} \inf_{Z \in \mathcal{M}} E\left[Z\left(y + g\right)\right],$$

where Z is a change of measure from the base measure ν to $\widetilde{\nu}$ with

$$\widetilde{\nu}(dx) = Z(x)\nu(dx).$$

The restrictions on Z define the class of alternates $\mathcal{M}$.

Suppose now that there exists $g(x)$ at zero cost such that

$$a = \inf_{Z \in \mathcal{M}} E\left[Zg\right] > 0.$$

In this case there exists a zero cost position that is acceptable on its own. This may not be an arbitrage but just an acceptable or desirable outcome on the valuation operator being employed. Then for $\lambda > 0$ and a post-hedge position

$$r(x) = y(x) + \lambda g(x),$$

we have that

$$\begin{aligned}\inf_{Z \in \mathcal{M}} E\left[Z\left(y + \lambda g\right)\right] &\geq \inf_{Z \in \mathcal{M}} E\left[Zy\right] + \lambda \inf_{Z \in \mathcal{M}} E[Zg], \\ &= \inf_{Z \in \mathcal{M}} E\left[Zy\right] + \lambda a\end{aligned}$$

and with a large λ the result tends to infinite. One then has to introduce constraints or restructure $\mathcal{M}$ to avoid $a > 0$. We may use regularization or other constraints. Another alternative is distorted least squares. We report on both measure-distorted value-maximizing and distorted least-squares hedging.

11.1 A Two-Dimensional Example

First consider an example in two dimensions. The two underlying assets chosen are SPY and VIX. We begin by constructing $y(x)$ where x_1 is the instantaneous return on SPY and x_2 is the instantaneous return on VIX. The function $y(x)$ is built for a hedge to be constructed on December 31, 2018.

11.1.1 The Two-Dimensional Target Exposure

The function $y(x)$ is taken to be the exposure of a one-month bull spread on SPY as evaluated on December 31, 2018. For this purpose we employ a GPR summary of the volatility surface constructed for this date. The summary allows us to extract implied volatilities for arbitrary strikes and maturities on this date. The maturity is fixed at one month and the bull spread is constructed using two calls. The strikes for the bull spread were such that a put and call at these strikes had an absolute delta of 0.4. The spot level of the SPY was 248.8286; the two strikes were 245.9478 and 254.7767 respectively.

The function $y(x)$ was trained using GPR on a grid of exposure valuations based on shocks to SPY and VIX in the ranges of absolute values below 20% and 10% respectively in steps of 10 basis points. The position being valued was long a call at the lower strike and short a call at the upper strike.

Apart from independent shocks to implied volatilities we supposed that implied volatilities on SPY responded to shocks in the SPY return based on a regression of VIX returns, V, on SPY returns, X, estimated from time series data over the period January 3, 2007, to December 31, 2018. The result of this regression was

$$V = -4.65X + 7.6580X^2,$$

with t-statistics of -60.84 and 4.18 respectively. For a point x_1, x_2 on the SPY, VIX grid of possible shocks the implied volatilities for the two strikes were shocked by the magnitude

$$x_2 - 4.65x_1 + 7.658x_1^2.$$

The shocked implied volatilities for the two strikes along with the shocked spot level were used to revalue the bull spread and then, on subtracting the original bull spread value, the post-shock call spread exposure, $y(x_1, x_2)$, was calculated at the grid point x_1, x_2. The result was a matrix of size 401×201 for the call spread exposures. Taking every fifth value across the rows and columns, a smaller matrix of size 81×41 was obtained and a GPR was run on a vectorization of these outputs regressed on a vectorization of the associated (x_1, x_2) inputs on the grid. The output of this GPR is the function $y(x)$. Figure 11.1 presents a graph of the target exposure function to be hedged.

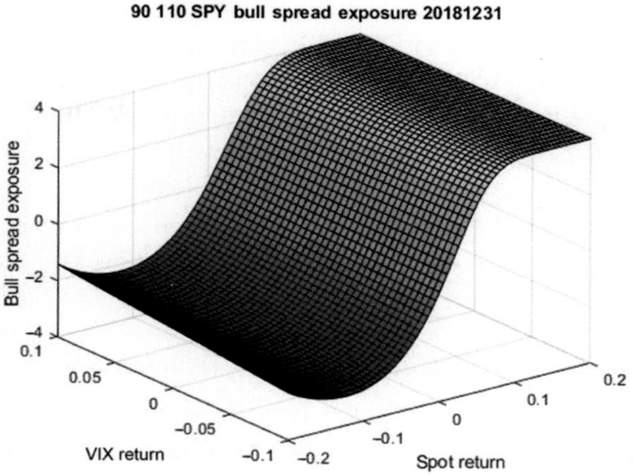

Figure 11.1 Exposure of one-month bull spread on SPY to movements in SPY and VIX

11.1.2 Hedging Asset Exposures

The next step is the construction of hedge functions $H(x)$ that evaluate the exposure of a unit position in a set of hedging assets. The hedging assets being employed are investment in SPY, using 21 out-of-the-money options on this underlier, and investment in the VIX, employing 10 out-of-the-money options on VIX. The total number of hedging assets is then 33. The spot exposures are clear. For the options we use the same structure of implied volatility movements as for the construction of $y(x)$ and create functions that take as inputs all information on the vectors of strikes, rates, dividend yields, the type of option, and their implied volatilities that are appropriately shifted to evaluate post-shock values less the pre-shock value to evaluate the hedge asset exposure function $H(x)$. This function outputs a vector of 33 exposures for all 33 hedging assets as a function of (x_1, x_2), the shock to the logarithm of the SPY and VIX levels.

11.1.3 Sample Space of Two-Dimensional Shocks

We next have to generate the sample space of joint moves on SPY and VIX on which the hedge is to be performed. For this we employ a joint law consistent with prespecified BG marginals on these asset returns. The BG marginal parameters were obtained from data on option prices as of December 31, 2018, on each underlier by specifying a Sato process for the physical densities across maturities depreferenced using the same five preference parameters across all maturities, for maturities between 0.06 and 0.16. The implied BG parameters for the two underliers were as follows:

	b_p	c_p	b_n	c_n
SPY	0.006873	1.075084	0.011451	0.557540
VIX	0.053585	1.147277	0.032252	1.8897756

The joint law was estimated by digital moment matching of tail probabilities in all quadrants generated at randomly selected pairs of return strikes with observed probabilities in the bivariate time series return data over the period January 3, 2007, to December 31, 2018. The joint law used was that of correlated multivariate Brownian motion with correlation ρ and drifts and variances as implied by the marginals, which are time changed by a single gamma process with unit drift and variance rate ν, plus independent BG marginals consistent with the prespecified BG marginals. The joint law was determined on estimating ρ, ν, and they were estimated at -0.899 and 1.8416 respectively. The estimation details are presented later in the 10-dimensional example.

11.1.4 Simulation of Measure-Distorted Value-Maximizing Hedges

The multivariate Lévy measure $k(z)$ for this joint law is known analytically, and the evaluation of measure-distorted valuations of post- and pre-hedge jump exposures $u(z)$ is conducted on evaluating integrals of the form

$$\int u(z)k(z)dz.$$

This integral is transformed into three components reflecting joint moves in both SPY and VIX returns implied by the multivariate VG model plus the two univariate components associated with independent moves in the two returns induced by the independent BG components. In each of the three components the integration is transformed into an expectations evaluation by simulating from a gamma process for the univariate components and from a gamma-distributed elliptical radius distribution for the multivariate VG components. We thus evaluate the integral by evaluating the expectation

$$E^g\left[u(z)\frac{k(z)}{g(z)}\right]$$

on simulating from the specific distribution $g(z)$. The measure change given by

$$m(z) = \frac{k(z)}{g(z)}$$

is needed explicitly to perform the expectation computation. We thus need to generate points z_i and the associated measure changes $m_i = m(z_i)$ for points z_i. The sample space consisted of $10,000$ points generated from the multivariate VG component using the gamma-distributed elliptical radius distribution and $5,000$ points each from the two independent BG components. The full sample space has $20,000$ points. Figure 11.2 presents a sample of shocks to the SPY and VIX returns. Figure 11.3 presents the measure change for the joint and independent components.

11.1.5 Explicit Hedge Constructions

One may evaluate the function $y(x)$ on the sample space to obtain a set of representative exposures to be hedged that are stored in a vector w of dimension $1 \times 20,000$. Similarly

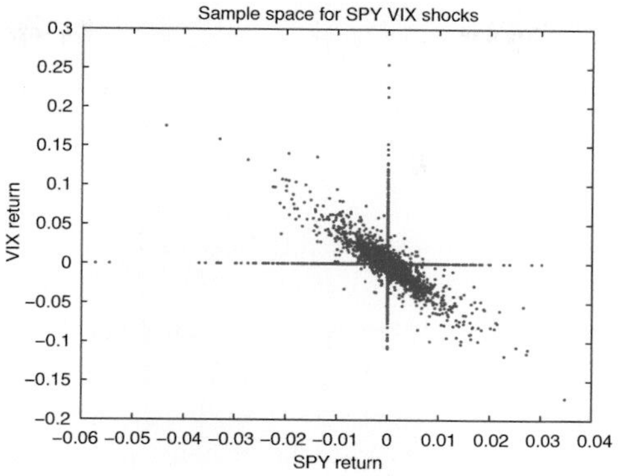

Figure 11.2 Sample of 20,000 shocks in the SPY and VIX returns

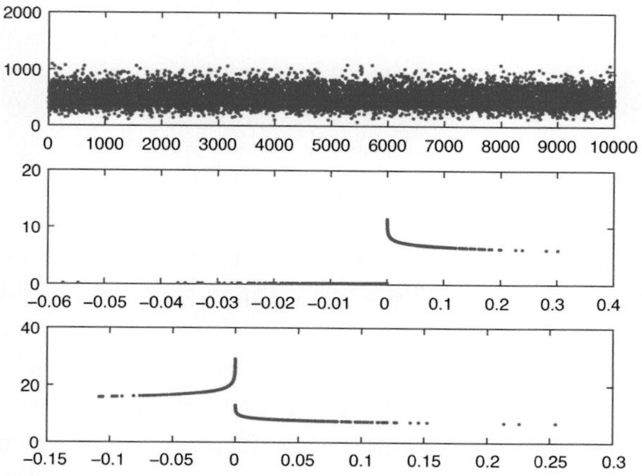

Figure 11.3 Sample measure changes for the joint moves in the top panel, and for the independent SPY and VIX moves in the second and third panels respectively

evaluating the hedge function produces a matrix of dimension $33 \times 20,000$ of responses delivered by the hedging assets. The post-hedge position for hedge positions a is

$$r = w + a'H.$$

The measure-distorted valuation-maximizing zero-cost hedge position a, solves by duality the problem of maximizing b for hedging asset prices V, subject to the constraints

$$b \leq \frac{1}{N} \sum_{i=1}^{N} Z_i m_i (w_i + (a'H)_i),$$

$$\frac{1}{N} \sum_{i=1}^{N} m_i \left(Z_i - \alpha_j \right)^+ \leq \Phi(\alpha_j), \ \alpha_j > 1.$$

$$\frac{1}{N}\sum_{i=1}^{N} m_i\,(\beta_k - Z_i)^+ \leq -\widetilde{\Phi}(\beta_k),\ \beta_k < 1,$$

$$a'V = 0,$$

$$Z_i \geq 0.$$

The vector Z is a measure change constrained to have call prices on strikes $\alpha > 1$ bounded above by $\Phi(a)$ and to have prices of puts with strikes $\beta < 1$ bounded above by $-\widetilde{\Phi}(\beta)$. We employed values of α_j from 1.01 to 500 in steps of 5 with a cutoff at $\Phi(\alpha) = 0.005$. Similarly we employed values for β of 0.01 to 0.99 in steps of 0.01 with a cutoff at $-\widetilde{\Phi}(\beta) = 0.005$. The measure-distortion parameters embedded in $\Phi, \widetilde{\Phi}$ were for b, c, γ at 1, 0.5, and 0.75.

This is an optimization problem of dimension 20,000 plus 33 on account of the values for Z_i that must be solved for. The dimension is reduced by postulating that the measure change Z is a smooth function of its arguments that may be interpolated from a smaller subset of representative values. We therefore performed a quantization of the sample space of 20,000 points down to 2,000 representative values. A GPR-based interpolator/extrapolator was employed to infer the values at the 20,000 points from the 2,000 points. We solve for $g = \log(Z)$ at the 2,000 representative points and employ a $20,000 \times 2,000$ matrix B obtained from a GPR kernel to obtain G at the 20,000 points by

$$G = Bg$$

and define $Z = \exp(G)$. The nonnegativity of Z is imposed by exponentiation. The dimension of the problem is reduced to 2,000 plus 33. The problem may be solved using a global optimizer like patternsearch in Matlab.

The direct problem without further constraints produced an unbounded solution, reflecting the fact that the hedging assets provide access to zero-cost portfolios with positive payoffs on the entire sample space or almost all of it that are highly acceptable on their own and may even constitute sample space arbitrages. The selection of hedges then requires the introduction of other constraints. Recognizing the desire on the part of hedges to move post-hedge positions closer toward zero and not to leverage up positions, we penalize a positive covariation between the hedged position and the exposure to be hedged. The covariation between hedging assets and the target exposure may be defined as

$$c_j = \frac{1}{N}\sum_{i=1}^{N} w_i H_{ji} m_i,$$

and we penalize the objective by maximizing

$$b - 10\sum_j a_j c_j.$$

In addition, we observe that the quadratic variation of the exposure to be hedged is

$$\text{qvar} = \frac{1}{N}\sum_{i=1}^{N} w_i^2 m_i.$$

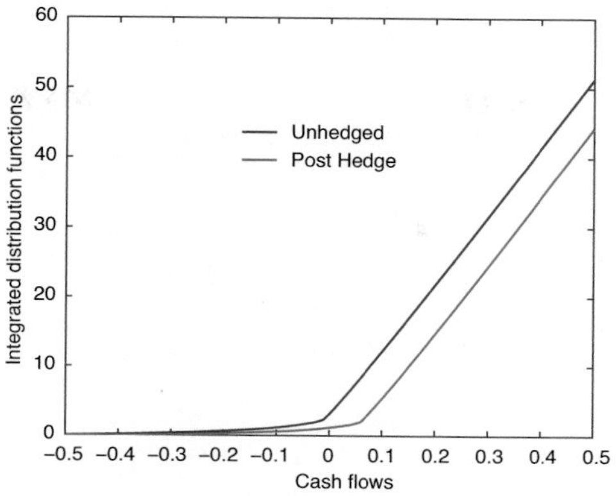

Figure 11.4 Integrated distribution functions for the unhedged and post-hedge cash flows associated with the covariation-penalized and quadratic variation-constrained hedge of a one-month bull spread on SPY

We introduce the further constraint that requires the quadratic variation of the hedge to be below that of the target or that

$$\frac{1}{N}\sum_{i=1}^{N}(a'H)_i^2 m_i \leq \text{qvar}.$$

The covariation-penalized and quadratic variation-constrained measure-distorted valuation maximization hedge had a finite solution. The resulting covariation was −1.3034.

Figure 11.4 presents the integrated distributed functions of the unhedged and post-hedge cash flows from this hedge. It is clear that on the sample space the post-hedge cash flow second-order stochastically dominates the unhedged counterpart.

Figure 11.5 presents the hedge positions being held in the underlying assets of both SPY and VIX.

11.1.6 Distorted Least-Squares Hedging

On any sample space one may evaluate the vector of target exposures w and construct a matrix of exposure outcomes of associated hedge instruments H. With positions a the post-hedge residual may then be defined as

$$r = w - a'H,$$

and least-squares hedging seeks to find a with a view toward minimizing the squared residuals. The chosen hedge will willingly trade reducing a large positive residual for an increase in the absolute value of some smaller negative residuals that are then more negative. But we do not wish to give up gains for the sake of increasing losses. Distorted least squares is sign sensitive and gives a low weight to high positive residuals and a high weight to the smaller negative ones that may prevent it from making such an adverse trade. We therefore

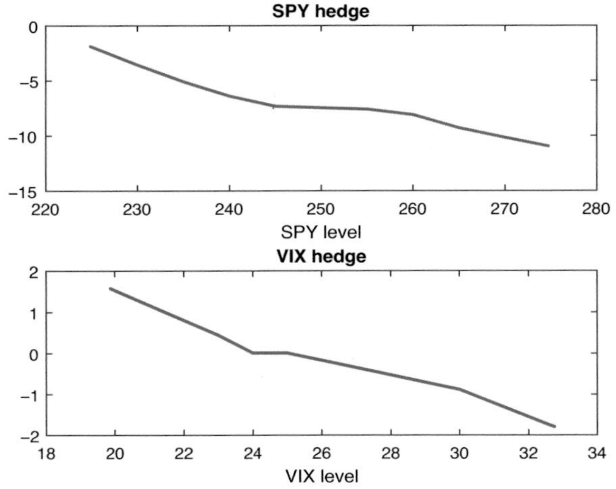

Figure 11.5 Hedge positions in SPY and VIX held by the covariation-penalized and quadratic variation-constrained measure-distorted value-maximizing hedge

consider selecting the hedge to minimize the distorted least-squares objective on our sample space.

Two solution methods were adopted. They follow the procedures of applying an ADAM optimizer with different methods for constructing the gradient. In the first we use the methods of simultaneous perturbation to evaluate the gradient in all directions by employing just two function evaluations. In the second we employ numerical differentiation built into the Matlab ADAM optimizer that approximates

$$f'(x) \approx \frac{\mathrm{Im}(f(x+ih))}{h},$$

where Im is the imaginary part of the complex number.

Figure 11.6 presents the integrated distribution functions for the hedge determined by simultaneous perturbation gradients input into an ADAM optimizer and the use of complex-valued gradient calculations embedded in the ADAM optimizer.

We observe that the use of more exact gradients slightly dominates the result delivered by simultaneous perturbation with both gradients employed in an ADAM optimizer. Figure 11.7 presents the hedges employed by DLSQ hedging for SPY and VIX for the two different gradient calculations. We note that these optimizations were conducted on their own sample spaces that were not identical. It is interesting that a large number of post-hedge cash flows are positive with both gradients.

11.2 A 10-Dimensional Example

The underlying hedging assets are nine sector ETFs, namely XLB, XLE, XLF, XLI, XLK, XLP, XLU, XLV, and XLY, and SPY, along with one-month maturity options on these. The

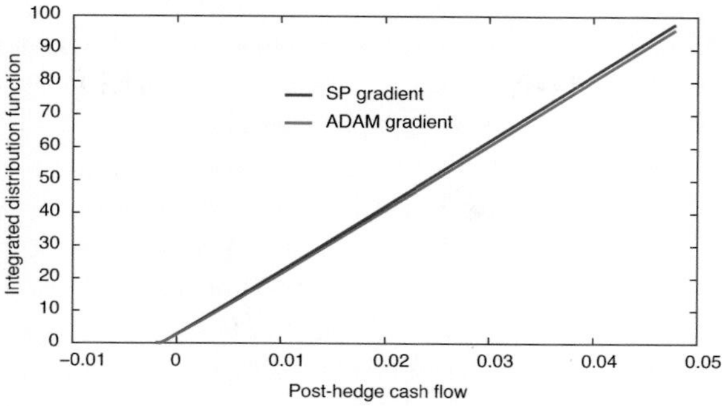

Figure 11.6 Integrated distribution functions for DLSQ hedges based on simultaneous perturbation gradients and complex-valued ADAM gradients

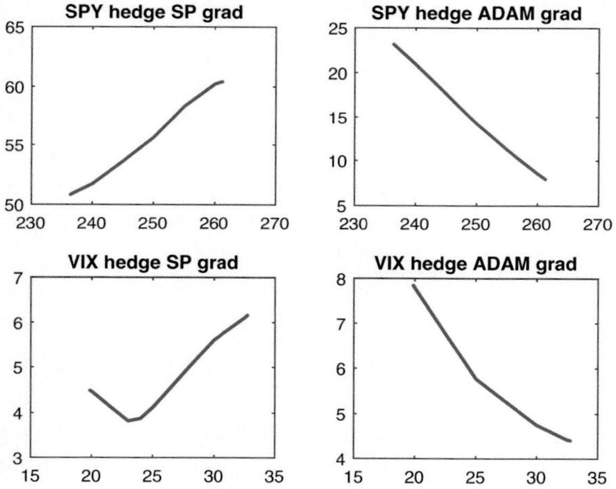

Figure 11.7 Distorted least-squares hedging of one-month bull spread on SPY. The hedge position is in SPY and VIX for two different gradient calculations.

hedge is constructed on December 31, 2018. For a 1,000 days of immediately prior returns on the 10 underliers we estimated parameters of the BG model and the result is presented in Table 11.1.

By way of a target exposure to be hedged we take the risk of an ETF with a relatively weak options market and consider hedging the risk by positions in the sector ETFs, SPY, and their options. The ETF exposure to be hedged here is that for the ticker for a specific ETF, namely BKLN. To build the target exposure function we performed a GPR of BKLN returns on the returns in the 10 underlying hedging assets and saved the GPR model created as the exposure function to be hedged. The result is the function $y(x)$ where x is now 10 dimensional. Alternatives for factor representations of risk exposures can include deep learning functions as formulated in Nakagawa et al. (2018).

Table 11.1 *Bilateral gamma parameters for the hedge assets*

	b_p	c_p	b_n	c_n
XLB	0.0079	1.4698	0.0117	0.9129
XLE	0.0087	1.4928	0.0114	1.0656
XLF	0.0105	1.0220	0.0154	0.6686
XLI	0.0076	1.2190	0.0118	0.7070
XLK	0.0065	1.3880	0.0114	0.7043
XLP	0.0040	1.8768	0.0065	0.9977
XLU	0.0044	2.2191	0.0071	1.2568
XLV	0.0050	1.7385	0.0079	0.9591
XLY	0.0068	1.3841	0.0111	0.7570
SPY	0.0069	1.0751	0.0115	0.5575

The hedge function is a function from the space of returns on the 10 assets to value exposures on the hedging assets. There were 95 hedging assets made up of the 10 underliers and one-month maturity out-of-the-money options on these. The number of options on these underliers were respectively 8, 7, 5, 8, 8, 5, 7, 9, 10, and 28. The option exposures were all evaluated at their implied volatilities and represent the difference between post-jump and original prices, taking account of movements in the underliers as given by the vector of return inputs for the 10 underlying assets. The result is a function $H(x)$ from where the output dimension is 95 for the 10-dimensional vector x.

We report on two hedges for the constructed target exposure. The first employs distorted least squares in the 95 hedging dimensions based on simulating daily returns from a joint law estimated for the vector of returns. The optimization is conducted using an ADAM optimizer with gradients evaluated using simultaneous perturbation. The second is based on maximizing a measure-distorted valuation with constraints penalizing positive covariations and constraining hedge quadratic variations to be bounded by the target quadratic variations. The measure distortion is applied to the joint Lévy measure associated with the estimated joint law for the vector of 10 returns.

11.2.1 The Joint Law and Its Estimation

The joint law considered is that of multivariate VG plus independent BG processes for the 10 components consistent with the already estimated marginal BG processes already estimated and reported on in Table 11.1. The drift and covariances of the multivariate variances are implied by the marginal BG processes, as are the scale coefficients of the independent BG processes. The speeds of the independent BG processes are those of the marginal BG processes less the reciprocal of the variance rate of the gamma process time changing the correlated multivariate Brownian motion with drift. It remains to estimate the correlation matrix C for the multivariate Brownian motion and the variance rate ν of the gamma time change.

For any candidate choice for C, ν one may simulate the joint law as a gamma time-changed multivariate Brownian motion with drift plus independent BG processes.

From the simulation one may then evaluate model tail probabilities $\widehat{T}_i$ evaluated at return points x_i, for $i = 1, \ldots, N$. These are given by the probability in the simulation for return vectors x satisfying

$$\widehat{T}_i = P\left(sign(x_{ik})(x_k - x_{ik}) \geq 0,\ k = 1, \ldots, 10\right).$$

These may be compared to comparable observed tail probabilities T_i in data. The parameters C, ν may be estimated to minimize the Anderson–Darling weighted least-squares objective

$$z = \sum_i \frac{\left(T_i - \widehat{T}_i\right)^2}{T_i(1 - T_i)}. \tag{11.1}$$

The optimization may be carried out once the points x_i and the tail probabilities T_i have been evaluated. For many points x_i that one may select the observed tail probability will be zero.

We randomly selected a number k, $2 \leq k \leq 10$, for which we randomly selected k assets from the 10, for which we randomly selected strike levels x_{ik} and evaluated the tail probability in the data for T_i and kept redrawing until we had 10,000 points with tail probabilities above 0.025. This set of 10,000 points constituted our points x_i with their associated tail probabilities T_i.

Given this choice of test tail probabilities the objective for the estimation is defined and we have to perform a search over all positive definite correlation matrices C and the value for ν the speed of the gamma process time changing our multivariate Brownian motion with known variances and drifts. There is a lower bound on ν given by $\underline{\nu}$ where

$$\underline{\nu} = \frac{1}{\min_{i=1,\ldots,10}(c_{pi} \wedge c_{ni})}.$$

The optimization of ν was constrained by this lower bound.

For the parametrization of C we follow Archakov and Hansen (2018). The proposed parametrization takes as parameters the lower diagonal g of the matrix logarithm of the correlation matrix C. Let $G(g, w)$ be the symmetric matrix with lower diagonal g and diagonal w. Given g there exists a unique vector $w^*(g)$ such that

$$\exp(G(g, w^*(g)))$$

is a correlation matrix. The vector $w^*(g)$ solves the equation

$$\begin{aligned} w &= F(w), \\ F(w) &= w - \log(\text{diag}(\exp(G(g, w)))). \end{aligned}$$

The vector w^* may be found on iterating F that is shown to be a contraction mapping.

With this parametrization of C and the lower bound on ν one may optimize over C, ν to minimize the least-squares objective (11.1). We further introduced a regularization term to minimize

$$z = \sum_i \frac{\left(T_i - \widehat{T}_i\right)^2}{T_i(1 - T_i)} + \lambda g' g \tag{11.2}$$

where the regularization penalty λ was taken at 10. The estimated value for ν was 1.8143 and the correlations ranged between 0.7566 and 0.8726.

11.2.2 Distorted Least-Squares Hedge

The joint law was simulated to construct 10,000 draws from the multivariate VG process plus the independent BG processes. The result was a matrix of size 10 X 10,000 giving the 10 joint returns for each of the 10,000 drawings. The function $y(x)$ was evaluated at the points to form the target vector w for the hedge. The hedge function was evaluated at these points to form the matrix H of size 95 by 10,000 that gave the responses of the 95 hedge assets to each of the joint return outcomes. For position a in the 95 hedge assets the post-hedge residual is

$$r = w - aH.$$

The distorted least-squares objective seeks to minimize over choices for a the criterion

$$\sum_i \omega_i \left(r_i - (aH)_i\right)^2,$$

$$\omega_i = \Psi\left(\frac{\sum_j \mathbf{1}_{r_j \le r_i}}{10000}\right) - \Psi\left(\frac{\sum_j \mathbf{1}_{r_j < r_i}}{10000}\right).$$

The distortion Ψ was minmaxvar at stress level 0.75. We employed an ADAM optimizer with gradients evaluated using simultaneous perturbation to find an optimal hedge position. The hedge was not constrained and its cost was 3.71.

As a diagnostic for the hedge we present a graph of the integrated distribution function for w less the integrated distribution function for the optimized r in Figure 11.8. The uniform positivity of this graph shows that all concave utilities prefer the hedged position on the sample space to the unhedged position. Equivalently the hedge second-order stochastically dominates the unhedged situation.

Additionally we present a graph of the cash flows held as functions of the 10 underliers as the hedge position in Figure 11.9.

11.2.3 Constrained Measure-Distorted Value Hedge

For the measure-distorted value-maximizing hedge the sample space consisted of 10,000 points sampled from a gamma-distributed elliptical radius simulation of a random variable

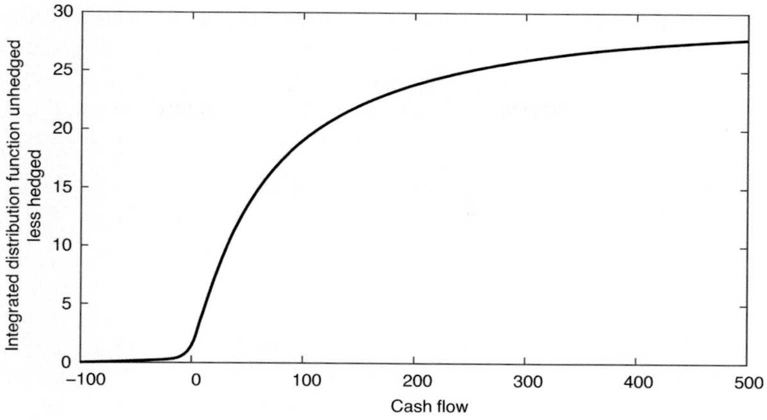

Figure 11.8 Integrated distribution of unhedged exposure less integrated distribution function for hedged exposure

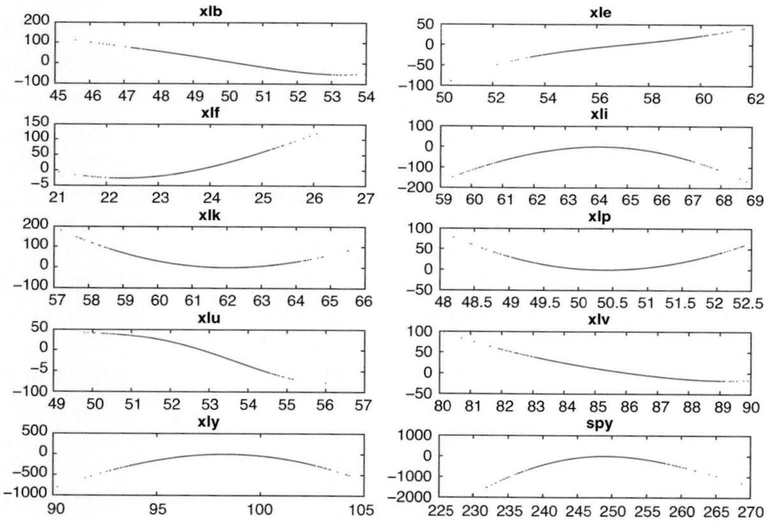

Figure 11.9 Hedge positions in the 10 underlying assets as evaluated by distorted least squares using ADAM optimization and simultaneous perturbation for gradient evaluation

with simulation density $g(x)$. The ellipse was given by the covariance structure of the multivariate Brownian motion being time changed. Also evaluated at each of these points was the measure change given by the ratio of the Lévy measure to the gamma-distributed elliptical radius density $g(x)$ to enable expectations under g reweighted by the measure change to be approximations of integrals with respect to the Lévy measure. Additionally 10 sets of 5, 000 points were simulated from a single gamma density with shape parameter

0.75 and unit scale along with the measure change ratio of the univariate independent BG Lévy measures to the gamma density. The total number of points was N equal to 60, 000. Denote the points by x_i and the associated measure change values by m_i.

The vector w now evaluates the target exposure at these 60, 000 points while the matrix H evaluates the hedge asset exposures at these points to produce H that is 95 by 60, 000.

The dual measure-distorted value-maximization problem seeks to find $Z_i \geq 0$ so as to maximize b subject to

$$b \leq \frac{1}{N} \sum_i Z_i m_i \left(w_i - (aH)_i \right),$$

$$\frac{1}{N} \sum_i (Z_i - u_k)^+ m_i \leq \Phi(u_k), \; u_k > 1,$$

$$\frac{1}{N} \sum_i \left(u_j - Z_i \right)^+ m_i \leq -\widetilde{\Phi}\left(u_j \right), \; u_j < 1.$$

The measure-distortion parameters embedded in $\Phi, \widetilde{\Phi}$ were $b, c, \gamma = 1, 0.5, 0.75$ respectively. The values u_k and u_j were truncated away from unity when the bounds were below 0.005. The dimension of Z at 60, 000 is quite large. We quantized the points down to 4, 096 representative points and sought g_l at these points. A kernel interpolator/extrapolator was used to define a matrix B of dimension 60, 000 by 4, 096 to construct

$$G = Bg, \qquad Z = \exp(G).$$

The covariation between the hedging asset and target was evaluated by

$$c_k = \frac{1}{N} \sum_i H_k(x_i) m_i w_i,$$

while the quadratic variation for the target was

$$q = \frac{1}{N} \sum_i m_i w_i^2.$$

The hedge sought was forced to be of zero value, satisfying

$$\sum_k a_k \pi_k = 0,$$

where π_k was the price of the hedging asset. The covariation was penalized at

$$10 \sum_k a_k c_k,$$

and the hedge quadratic variation was constrained by q.

11.2.4 Measure-Distorted Value-Maximization Results

Figure 11.10 presents the difference between unhedged and hedged distribution functions in this case. We observe a second-order stochastic dominance of the hedged position over

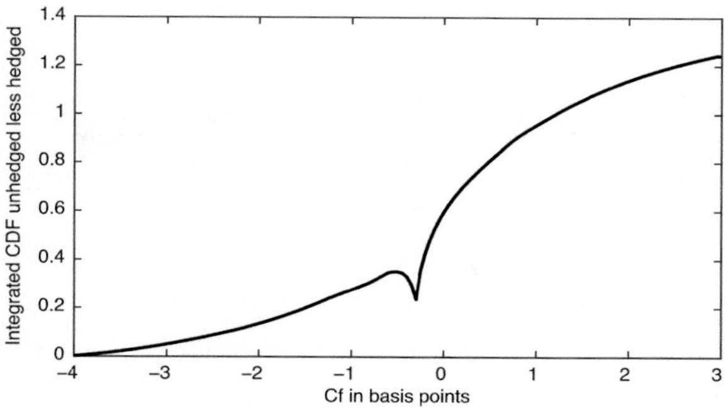

Figure 11.10 Integrated distribution function for unhedged less hedged exposures

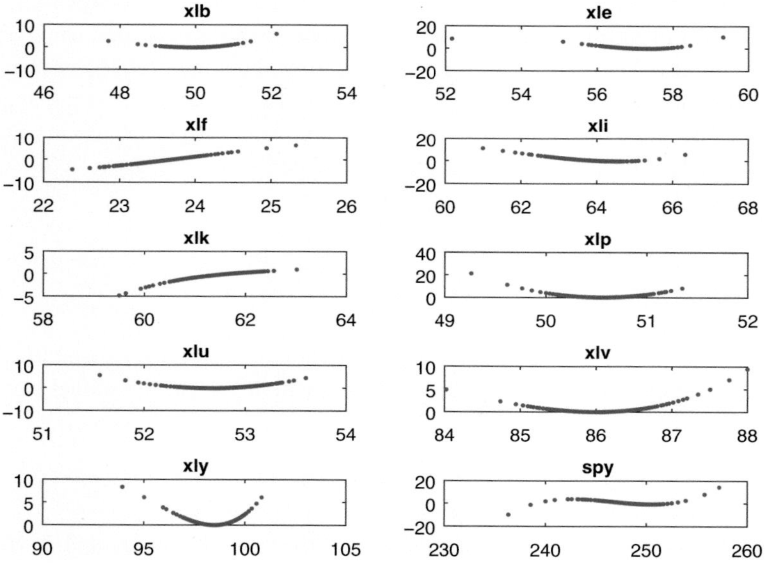

Figure 11.11 Positions taken in the 10 underliers by the measure-distorted value-maximizing hedge

the unhedged, but it is not as pronounced as the distorted least-squares position. However, this hedge was constrained to be zero cost as opposed to the unconstrained situation with distorted least squares.

Additionally we present in Figure 11.11 the positions held in the 10 assets and their options.

12

Static Portfolio Allocation Theory for Measure-Distorted Valuations

The portfolio allocation problem is traditionally studied in terms of joint distributions of multivariate outcomes at a horizon. Classic references include Markowitz (1991) and Haugen (2001). For a horizon-specific approach related to risks evaluated using probability distortions we cite Madan (2016b) and Madan and Schoutens (2016). A horizon-free approach focused on exposures is adopted in Madan (2018a), and this is the approach we follow here. The static allocation problem seeks a portfolio exposure $w(z, a)$, attained by investing a_i dollars in asset i that has exposure to log price returns z_i with an implied portfolio wealth exposure of

$$w(z, a) = \sum_{i=1}^{n} a_i \left(e^{z_i} - 1\right).$$

The portfolio choice is based on a knowledge of the multivariate arrival rate function $m(z)$ for the vector of possible returns z. In the case of the multivariate BG model one has that

$$m(z) = k_{\mathrm{mvg}}(z) + \sum_{i=1}^{n} k_i(z_i)\mathbf{1}_{(z_j=0, j\neq i)},$$

where $k_{\mathrm{mvg}}(z)$ is the Lévy compensator of the multivariate VG model and k_i is the Lévy compensators of the independent BG components. The compensating measure is given by

$$M(A) = \int_A m(z)dz. \tag{12.1}$$

For measure distortions G^+, G^- the measure-distorted value is given by

$$\mathcal{V}(a) = -\int_0^\infty G^+\left(M\left(w^-(z,a) > u\right)\right)du + \int_0^\infty G^-\left(M\left(w^+(z,a) > u\right)\right)du. \tag{12.2}$$

The measure-distorted valuation scales with the positions a and the measure-distorted value-maximization problem is a portfolio choice problem with the constraint $\mathbf{1}'a = 1$. Further constraints on long and short positions may also be considered. We refer to the solution of the problem of maximizing $\mathcal{V}(a)$ over possibilities for a as presented in equation 12.2 as the primal measure-distorted value-maximizing portfolio.

It is useful to note in passing that there is no riskless asset among the investment choices possible when dealing with exposures. This is because the only constant exposure is the identically zero exposure that may then be ignored.

The measure-distorted value-maximizing portfolio problem may also be presented in a risk–reward format on defining the variation as the reward where the variation $V(a)$ is given by

$$V(a) = -\int_0^\infty M\left(w^-(z,a) > u\right) du + \int_0^\infty M\left(w^+(z,a) > u\right) du.$$

The risk charge is then defined as the – positive by construction – difference between the variation and the conservative or distorted variation by

$$\mathrm{RC}(a) = V(a) - \mathcal{V}(a).$$

One may observe that

$$\mathrm{RC}(a) = \int_0^\infty \Gamma^+\left(M\left(w^-(z,a) > u\right)\right) du + \int_0^\infty \Gamma^-\left(M\left(w^+(z,a) > u\right)\right) du, \qquad (12.3)$$

where

$$\Gamma^+ = G^+ - I, \qquad \Gamma^- = I - G^-,$$

and $I(x) = x$ is the identity function.

For completeness we note that if we go back to probability distortions, then the risk charge is given by

$$\begin{aligned}\mathrm{RC}(a) &= \int_{-\infty}^0 (\Psi(F(x)) - F(x))dx + \int_0^\infty \left(\widehat{F}(x) - \widehat{\Psi}(\widehat{F}(x))\right) dx \\ &= \int_{-\infty}^0 (\Psi(F(x)) - F(x))dx + \int_0^\infty (1 - F(x) - (1 - \Psi(F(x))))\, dx \\ &= \int_{-\infty}^\infty (\Psi(F(x)) - F(x))dx. \qquad (12.4)\end{aligned}$$

A conservative expectation is then seen as an expectation shaved by the possibility of an excessive reliance on tail probabilities in the expectation computation. These are known with less confidence, and to the extent they feed into the expectation calculation they should be discounted. Hence the conservative or shaved expectation is the more realistic assessment of the situation.

The valuation presented in (12.2) is the lower conservative valuation. The upper valuation reverses the roles of G^- and G^+ and may be expressed as the variation $V(a)$ plus an upper risk charge that reverses the roles of Γ^+ and Γ^-. The two risk charges are in general different but for probability distortions the upper risk charge is given by

$$\begin{aligned}\mathrm{RC}^U &= \int_{-\infty}^0 (F(x) - \widehat{\Psi}(F(x)))dx + \int_0^\infty (\Psi(1 - F(x)) - (1 - F(x)))dx \\ &= \int_{-\infty}^\infty (\Psi(1 - F(x)) - (1 - F(x))dx \\ &= \mathrm{RC}.\end{aligned}$$

Three solutions by approximation are constructed for this problem and implementations are reported for two of them. The first approximation evaluates the compensating measure integrals (12.1) by simulation. The second approach formulates the problem by the duality characterization of measure-distorted values. This also applies to probability distortions presented in Madan and Schoutens (2016) and is presented here for completeness. The third method approximates the measure-distorted value-maximization problem by a probability distortion-maximizing problem at a small time step to be judiciously chosen. The three solutions are presented in separate sections. Implementations are reported next. This is followed by the construction of efficient frontiers and the strategies for alpha generation implied by these frontiers.

12.1 Measure Integrals by Simulation

We need to evaluate the measures

$$M\left(w^{-}(z,a) > u\right), M(w^{+}(z,a) > u)$$

that are then distorted by G^{+}, G^{-} and integrated over u to form the measure-distorted valuation $\mathcal{V}(a)$. The set of possibilities for z are those associated with the Lévy compensator for the multivariate VG model, or the univariate outcomes associated with independent BG process outcomes when working with the multivariate BG model consistent with prespecified marginal BG distributions. In each of these cases we choose a suitable density $g(z)$ and write

$$\int_{A} m(z)dz = E^{g}\left[\mathbf{1}_{A}\frac{m(z)}{g(z)}\right].$$

In the case of the multivariate VG model the density g is that for a gamma-distributed elliptical radius already introduced and employed in earlier chapters. For the independent BG components we employ just a gamma density with a small shape coefficient. We used the same shape parameter for all the components and the gamma distributed radius. The required measure change functions are known analytically and the outside expectation is given by an average over the relevant simulated outcomes. The relevant measures are distorted and integrated to build the measure-distorted valuation. The portfolio is then selected using the solution computed by a traditional constrained minimization optimizer.

12.2 Dual Formulation of Portfolio Problem

The dual approach recognizes that the measure-distorted valuation is the supremum over all alternate measures $\widetilde{\nu}(dz)$

$$\widetilde{\nu}(dz) = L(z)\nu(dz),$$

of the valuation

$$\int w(z,a)L(z)\nu(dz).$$

The measure change functions $L(z)$ satisfy the inequalities

$$\int (L(z) - u)^+ \nu(dz) \leq \Phi(u), \qquad u > 1,$$
$$\int (u - L(z))^+ \leq -\widetilde{\Phi}(u), \qquad u < 1,$$

where $\Phi(u)$, and $\widetilde{\Phi}(u)$ are the conjugate duals of the concave and convex distortions G^+, G^-. The duals are analytically known for the measure distortions with parameters $b,\ c,\ \gamma$ presented and worked with earlier.

Converting the integrals to expectations under a density as in the primal approach, one may write the problem as maximizing b subject to the constraint

$$b \leq E^g \left[L(z)w(z,a)\frac{m(z)}{g(z)} \right],$$

with the restrictions on L that

$$E^g \left[(L(z) - u)^+ \frac{m(z)}{g(z)} \right] \leq \Phi(u),\ u > 1,$$
$$E^g \left[(a - L(z))^+ \frac{m(z)}{g(z)} \right] \leq -\widetilde{\Phi}(u),\ u < 1,$$
$$L(z) \geq 0.$$

In the case of probability distortions with density $f(z)$ and distortion Ψ one may maximize b subject to

$$b \leq E^f \left[L(z)w(z,a) \right],$$

with the restriction on L that

$$E^f \left[(L(z) - u)^+ \right] \leq \Lambda(u),\ u > 0,$$
$$E^f [L(z)] = 1,$$
$$L(z) \geq 0,$$

where $\Lambda(u)$ is the conjugate dual of the distortion Ψ.

For an implementation one may simulate from the densities either g or f and approximate the expectations by sample averages, discretizing the choices of test points u_k for the constraints on L. The search over $L_j = L(z_j)$ for the sample outcomes z_j makes the dimension of the optimization problem large. Strategies at further approximations may be considered by reducing the points at which L_j is evaluated, from which the values at other points may be accessed by an appropriate multidimensional interpolation/extrapolation procedure. But one then has to ensure the positivity of the interpolated and extrapolated values. We leave the dual portfolio problem's implementation open as an agenda for future research.

12.3 Approximation by Probability Distortion

The multivariate Lévy process modulo the small jumps may be approximated by a compound Poisson process with a finite arrival rate for the jumps that have not been deleted or left out of the approximation. This finite arrival rate will generally be large as the jumps being accounted for are expected to occur quite frequently. If we now consider a time step of the order of the reciprocal of this aggregate arrival rate, then this time step is small. If we are indifferent to the size of this small time step, then at a small enough time step we may have a single expected jump with a known multivariate probability distribution for the expected single jump. As an approximation we may consider portfolio selection by maximizing the distorted expectation for this multivariate probability distribution. Our third approach considers such a probability selection.

12.4 Implementation of Portfolio Allocation Problems

We report here on the optimal portfolios constructed from the maximization of the measure-distorted value using simulated measure integrals and probability distortion approximations. The underlying assets were the nine sector ETFs and SPY with the joint law as estimated and reported in Section 12.2 on hedging 10-dimensional risks.

For both approaches realizations were simulated using gamma-distributed elliptical radii for the multivariate VG part and gamma distributions for the independent BG components with shape parameters at 0.075 in both cases. There were $10,000$ readings from the multivariate VG component and $5,000$ readings in each of the 10 independent BG components. The total sample space consisted of $60,000$ points and the associated levels for the measure change function $m(z_i)$.

12.4.1 Simulated Measure Integral Results

The measure-distortion parameters were $b = 1$, $c = 0.5$ and there were three settings for γ at 0.25, 0.75, and 1.5. Table 12.1 presents the portfolio compositions at the three stress levels.

We observe that as the stress level rises, the proportion invested in XLP, XLU, and SPY rises and the short position in XLE is increased. The investment in XLI is decreased with the rise in risk aversion.

12.4.2 Probability Distortion Approximation

The first step here is the construction of the probability vector for the $60,000$ points drawn from the multivariate VG and the independent BG components. For this we first evaluated the expected measure change as an approximation of the aggregate arrival rate in the multivariate VG component λ_{mvg}, and λ_j, with $j = 1, \ldots, 10$ for the 10 independent components. The total arrival rate was

Table 12.1 *Simulated measure integrals*

	Stress Levels		
	0.25	0.75	1.5
XLB	0.0830	0.0274	0.0397
XLE	0.0549	−0.1218	−0.1396
XLF	0.1266	0.0888	0.0773
XLI	0.0844	0.0348	0.0487
XLK	0.0975	0.1046	0.0980
XLP	0.1141	0.1926	0.2382
XLU	0.1109	0.1758	0.2180
XLV	0.1227	0.2309	0.1836
XLY	0.1063	0.1513	0.0905
SPY	0.0996	0.1155	0.1456

Table 12.2 *Annualized mean returns*

XLB	−32.01
XLE	36.68
XLF	−0.66
XLI	−14.83
XLK	−29.24
XLP	27.13
XLU	−5.41
XLV	9.82
XLY	−6.74
SPY	25.58

$$\lambda = \lambda_{\text{mvg}} + \sum_{j=1}^{10} \lambda_j.$$

This value was 395.94. Its reciprocal was the time step and this was 3.64 minutes. The probabilities of a joint move from the multivariate VG component is $\lambda_{\text{mvg}}/\lambda$ while that of a move in the jth independent component is λ_j/λ. The conditional probability of a particular joint move is given by $m(z_i)/\sum_i m(z_i)$ running across the possibilities of joint moves. Similar calculations yield conditional probabilities for the independent BG components. Multiplying conditional probabilities by their probabilities results in a probability vector of length 60,000 for all the points sampled.

Table 12.2 presents the annualized mean returns on the 10 investment opportunities. The annualized mean returns are based on 24 hours per day and 252 days per year.

Table 12.3 presents optimal portfolios maximizing distorted expectations at three stress levels for the distortion minmaxvar. The stress levels are 0.25, 0.75, and 1.5.

We observe that increased risk aversion invests higher proportions in XLU and SPY while decreasing investment in XLP and XLV. The short positions in XLF are also decreased.

Table 12.3 *Optimal portfolios maximizing distorted expectations for minmaxvar*

	Stress Levels		
	0.25	0.75	1.5
XLB	−0.0154	−0.0142	−0.0016
XLE	−0.0031	−0.0090	−0.0350
XLF	−0.1804	−0.1335	−0.1191
XLI	−0.0203	−0.0194	−0.0041
XLK	0.0004	0.0176	0.0720
XLP	0.7037	0.5384	0.4174
XLU	0.2750	0.3442	0.3622
XLV	0.2410	0.2218	0.1720
XLY	−0.0147	−0.0117	−0.0170
SPY	0.0137	0.0657	0.1533

Table 12.4 *Mean returns and risk charges*

XLB	−36.44	129.15
XLE	31.36	144.76
XLF	−7.14	149.85
XLI	−18.53	117.26
XLK	−32.30	105.83
XLP	25.72	71.09
XLU	−7.34	84.52
XLV	7.58	91.79
XLY	−9.95	109.72
SPY	22.78	94.89

12.5 Mean Risk Charge Efficient Frontiers

The conservative valuation of an investment exposure may be expressed as reward measured by variation less a risk charge as expressed in equation (12.4). The annualized continuously compounded returns and risk charges for our 10 underlying assets at the stress level of 0.75 are presented in Table 12.4.

In the absence of risk-free exposures as commented earlier, one may seek to construct the minimum risk charge portfolio. For our example of 10 assets the minimum risk charge portfolio had an annualized risk charge of 58.95. The composition of this portfolio is presented in Table 12.5. The annualized return on the minimum risk charge portfolio was 14.95.

In keeping with traditional risk reward analyses, one may also seek to construct the efficient risk–reward frontier by minimizing the risk charge for a given mean return target or maximizing the mean return for a given risk charge. At the relatively high stress level of 0.75 the risk charges are high and the slopes of the associated efficient frontier are thereby small. One may consider the portfolio problem of maximizing the mean less a multiple θ

Table 12.5 *Minimum risk charge portfolio*

XLB	−1.14
XLE	−1.08
XLF	−13.78
XLI	−1.62
XLK	2.50
XLP	53.48
XLU	34.58
XLV	21.60
XLY	−1.06
SPY	6.52

Table 12.6 *Efficient frontier portfolio*

XLB	−45.72
XLE	20.07
XLF	−19.24
XLI	−29.36
XLK	−60.59
XLP	101.19
XLU	2.85
XLV	40.17
XLY	−25.51
SPY	116.14

times the risk charge. The optimal portfolio on the frontier for $\theta = 0.013$ is shown at a risk charge of 105 and a mean return of 107.19%. Its composition is presented in Table 12.6.

Mimicking classical arguments, one may infer asset pricing equations associated with points on the efficient frontier. Suppose that a^* is the optimal portfolio and one offers an investor holding this portfolio the opportunity of investing t dollars in asset i. The portfolio-constrained objective as a function of t, $Z(t)$ is

$$Z(t) = a^{*\prime}\mu + t\mu_i - \theta \text{RC}\,(a^* + te_i) - \eta((a^* + te_i)' - 1).$$

The first-order condition with respect to t evaluated at $t = 0$ implies that

$$\mu_i = \theta \frac{\partial \text{RC}(a^*)}{\partial a_i} + \eta, \tag{12.5}$$

which delivers the associated theory for required returns and hence a new perspective on abnormal returns or alpha.

Table 12.7 presents the required returns and alpha values for the 10 assets. The value of μ_{zrg} may be estimated as the return on the portfolio with a zero risk charge gradient. This value is −0.2956.

Table 12.7 *Mean returns, required returns, and alpha*

	Mean	Req. Ret.	Alpha
XLB	−36.44	−36.41	−0.03
XLE	31.36	31.14	0.22
XLF	−7.14	−7.21	0.08
XLI	−18.53	−18.50	−0.03
XLK	−32.30	−32.15	−0.15
XLP	25.72	25.53	0.19
XLU	−7.34	−7.33	−0.01
XLV	7.58	8.22	−0.64
XLY	−9.95	−9.93	−0.02
SPY	22.78	22.71	0.06

The associated theory of required returns may also be deduced by considering the investment of a marginal t dollars in asset i that is withdrawn from the zero risk charge gradient portfolio to infer that the excess return over μ_{zrg} the zero risk charge gradient return must be the market price of risk θ times the risk exposure measured by the risk charge gradient.

One may further infer that return on the efficient portfolio a^* is

$$\mu^* = a^{*\prime}\mu = \theta a^{*\prime}\frac{\partial \text{RC}(a^*)}{\partial a^*} + \eta.$$

Since risk charges are homogeneous of order 1 we get

$$\mu^* = \theta \text{RC}(a^*) + \eta,$$

or that μ_{zrg} is the intercept of the tangent line to the frontier at the point $(\text{RC}(a^*) \cdot \mu^*)$. More exactly, we may write that

$$\mu_i = \mu_{\text{zrg}} + \theta\frac{\partial \text{RC}(a^*)}{\partial a_i}. \tag{12.6}$$

Figure 12.1 presents a graph of the efficient frontier along with the location of the 10 individual assets in the reward–risk charge frontier.

12.5.1 Negative Required Returns

We observe a number of negative required returns. They arise when the risk charge is negatively related to investment in the asset or when the risk charge gradient is not high enough. When a position in the asset lowers or insufficiently raises risk charges, it is beneficial and the investor is willing to pay and invest as the asset has insurance or relative risk-reducing features. We note that many of the assets with observed negative returns also had negative required returns, and the theory picks up on the relative risk-reducing aspects of these assets.

We may contrast with variance as a risk measure by looking at the variance of

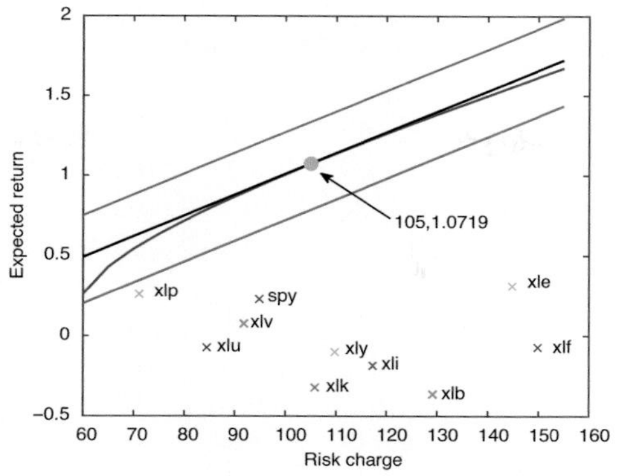

Figure 12.1 Graph of efficient frontier for the nine sector ETFs and SPY

$$X + hY,$$

and this is

$$\sigma_X^2 + h^2\sigma_Y^2 + 2h\sigma_{XY}.$$

To first order the introduction of Y reduces risk if $\sigma_{XY} < 0$ and required returns may be negative if $\beta < 0$. If we write

$$\mu = r + \beta\lambda,$$

where λ is the price of market risk, then a negative required return can arise if β is sufficiently negative. Classically this is not likely to occur in the equity market with respect to the market portfolio. In the conic exposure formulation there is no risk-free asset in the portfolio and η is the negative return on the portfolio with a zero risk charge gradient. For a positive return risk charge gradients must dominate the zero risk charge gradient return.

Theories should explain negative returns when they exist. The view of requiring all required returns to be positive is a mistaken consequence of assuming an investment horizon and the presence of a risk-free asset at this horizon. In fact, positions are held in assets with a wide range of stated and unstated maturities. Furthermore, all these assets are exposed to the risk of changes in valuations instantly and from an exposure viewpoint there can be no riskless asset. The reference asset is a zero-risk gradient asset with a negative reference return. Explaining observed negative returns is then both a theoretical and an empirical possibility.

12.5.2 Lowering the Stress Levels

We also present comparable graphs and results for two other stress levels of 0.025 and 0.015.

Table 12.8 presents the required returns and alpha on each of the 10 assets for the stress level 0.025.

Table 12.8 *Mean return, required return, and alpha*

XLB	−36.44	−36.40	−0.04
XLE	31.36	31.29	0.07
XLF	−7.14	−7.27	0.12
XLI	−18.53	−18.50	−0.04
XLK	−32.30	−32.26	−0.04
XLP	25.72	25.71	0.01
XLU	−7.34	−6.82	−0.52
XLV	7.58	7.55	0.03
XLY	−9.95	−9.96	0.00
SPY	22.78	22.63	0.14

Table 12.9 *Efficient portfolio*

XLB	−60.60
XLE	19.77
XLF	−14.37
XLI	−23.81
XLK	−76.64
XLP	151.35
XLU	0.00
XLV	31.92
XLY	−14.30
SPY	86.67

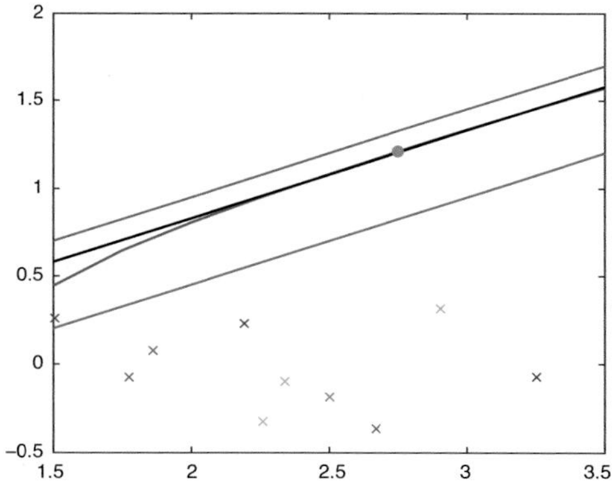

Figure 12.2 Frontier at stress level 2.5%

The value of μ_{zrg} was −0.1671. The return on the minimum risk charge portfolio at this stress level was 18.10. The efficient frontier graph is presented in Figure 12.2.

The efficient portfolio for $\theta = 0.5$ is shown in Table 12.9.

Table 12.10 *Minimum risk charge portfolio*

XLB	−1.14
XLE	−0.16
XLF	−17.89
XLI	−2.63
XLK	0.02
XLP	58.93
XLU	34.17
XLV	27.99
XLY	−3.29
SPY	3.98

Table 12.11 *Asset risk charges*

XLB	2.6699
XLE	2.9039
XLF	3.2516
XLI	2.5013
XLK	2.2595
XLP	1.5053
XLU	1.7738
XLV	1.8606
XLY	2.3393
SPY	2.1926

The minimum risk charge was 1.3677. The minimum risk charge portfolio is given in Table 12.10 and the asset risk charges are given in Table 12.11.

Reducing the stress level further to 0.015, one may employ a value of θ equal to unity. The graph is shown in Figure 12.3.

The efficient frontier portfolio is

Efficient Frontier Portfolio	
XLB	−33.08
XLE	5.82
XLF	−17.98
XLI	−14.29
XLK	−34.78
XLP	126.63
XLU	1.78
XLV	29.39
XLY	−9.51
SPY	46.02

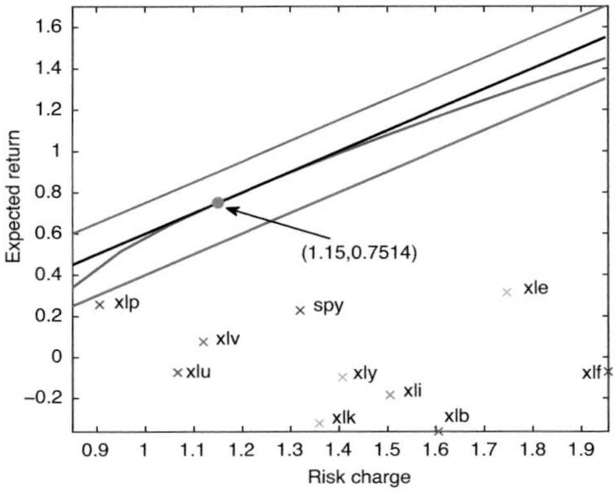

Figure 12.3 Efficient frontier at stress 1.5%

The minimum risk charge was 0.8226 and the return on the minimum risk charge portfolio was 15.02. The minimum risk charge portfolio is

Minimum Risk Charge Portfolio	
XLB	−1.14
XLE	−0.16
XLF	−17.89
XLI	−2.63
XLK	0.02
XLP	58.93
XLU	34.17
XLV	27.99
XLY	−3.29
SPY	3.98

The risk charges on the asset are as follows.

Risk Charges	
XLB	1.6054
XLE	1.7466
XLF	1.9585
XLI	1.5046
XLK	1.3590
XLP	0.9058
XLU	1.0666
XLV	1.1193
XLY	1.4068
SPY	1.3200

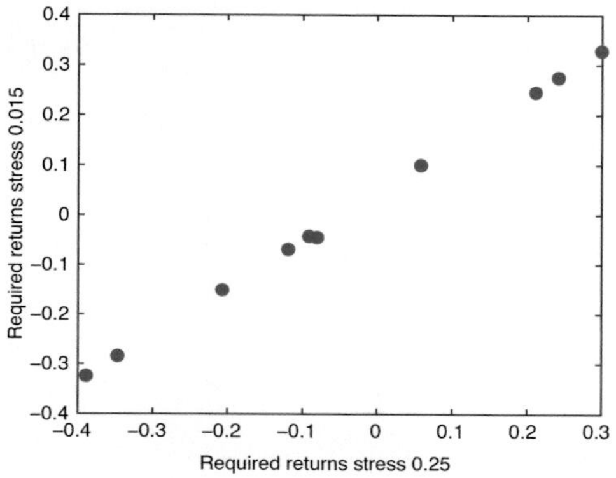

Figure 12.4 Required returns as evaluated from different points on the efficient frontier

The required returns and alpha are as follows. The value for μ_{zrg} was -0.427. The alpha results are

Mean Returns, Required Return, and Alpha			
XLB	−36.44	−35.22	−1.21
XLE	31.36	30.01	1.35
XLF	−7.14	−7.07	−0.08
XLI	−18.53	−17.93	−0.60
XLK	−32.30	−31.24	−1.06
XLP	25.72	24.69	1.02
XLU	−7.34	−7.29	−0.05
XLV	7.58	7.20	0.38
XLY	−9.95	−9.71	−0.25
SPY	22.78	21.78	1.00

12.6 Sensitivity of Required Returns to Choice of Points on Frontiers

Given the absence of a risk-free exposure, different market participants may well settle at different points on the efficient frontier, reflecting different zero-risk gradient returns and market prices of risk or θ. However, the risk gradients at different points on the frontier will also be different, but the effect on required returns may not be significant as the first-order optimality conditions are targeting observed mean returns.

With a view to assessing this sensitivity, we constructed required returns for the stress level 0.015 for two different market prices of risk or values of θ. These were 1.0 and 0.83. Figure 12.4 presents the required returns evaluated from the asset-pricing equations associated with the two corresponding separate points on the frontier.

We may observe that though there is some difference in the returns required on the assets in the portfolio, this is insignificant in comparison to the differences on required

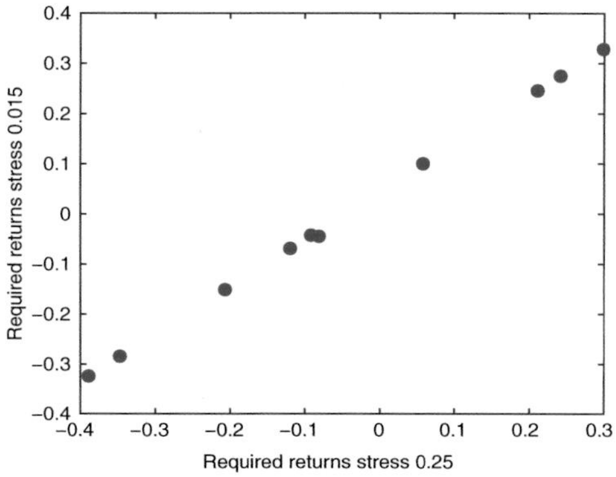

Figure 12.5 Efficient required returns on the individual assets implied by points on different frontiers associated different stress levels.

returns between assets. The predicted returns are robust to the selection of points on the frontier. Since different stress levels yielding different frontiers also target in their first-order conditions the same mean returns the required returns are also likely to be robust to the selection of stress levels as well as points on the frontier. They are essentially a property of the assets being combined to form efficient frontiers and not dependent on stress levels and the selection of points on the various frontiers. The result is a relatively robust collection of asset pricing models. Figure 12.5 presents required returns for the 10 assets using points on different frontiers associated with different stress levels.

12.7 Conic Alpha Construction Based on Arrival Rates

Let $r = (r_t, t = 1, \ldots, T)$ be the time series data for returns on a stock. Let X be the matrix of returns on the nine sector ETFs and SPY over the same period. With a view to capturing the dependence of r on the factors X, which may be nonlinear, we perform a GPR of r on X to generate the function

$$r_t = f(x_t) + \varepsilon_t.$$

Eleven GPRs were run, one of r on all of X and then 10 for r on each X variate separately as a univariate GPR.

We ignore ε_t as representing idiosyncratic or unpriced risks. On our sample space we have generated z_i with probability p_i and an efficient portfolio a^* with returns

$$r_i^* = a^* z_i.$$

We now construct returns for the asset in question as

$$\widehat{r_i} = f(z_i).$$

When z is from the multivariate VG component we use the GPR on all of X. The univariate GPRs were used when z comes from the independent bilateral components.

We now wish to evaluate the risk charge for

$$\mathrm{RC}(r_i^* + h\widehat{r}_i)$$

and evaluate the risk charge gradient as

$$\rho \approx \frac{\mathrm{RC}(r_i^* + h\widehat{r}_i) - \mathrm{RC}(r_i^* - h\widehat{r}_i)}{2h}.$$

The required mean return is given by

$$\mu = \theta\rho + \eta.$$

We report in the next section these constructions for stress levels 0.015, $\theta = 1$ and 0.025, $\theta = 0.5$ for assets on December 31, 2018.

12.7.1 Alpha Calculations for 191 Stocks

For the two stress levels and associated θ levels we evaluated the conic alpha calculation as just described for 191 stocks. Table 12.12 presents mean returns, required returns, and the conic alpha for a handful of stocks.

12.8 Fixed Income Asset Efficient Exposure Frontiers

Portfolio theory generally considers allocation strategies across a variety of equity assets. The static theory for investment over a specified time horizon often includes the existence of a risk-free asset with a unique risk-free return over the specific horizon. However, most investors hold a variety of fixed income securities with numerous maturities, all of which have an exposure to changes in values driven by movements in the term structures of interest rates. From an exposure viewpoint there are no risk-free assets among the many investable fixed income securities. These considerations motivate the construction of efficient risk exposures of fixed income assets in the absence of risk-free exposures.

For a sample construction we take the prices of 10 pure discount bonds with maturities of 1 to 10 years. The maturities are fixed at the start of the year 2017. The traded securities must have fixed maturity dates and cannot be of a constant maturity structure. Hence three months later the two-year pure discount bond has a maturity of a year and three quarters. A time series of prices for these pure discount bonds with a fixed maturity date may be constructed from daily data on the term structure of continuously compounded spot interest rates. The associated daily continuously compounded returns may then be formed from the price series.

12.8.1 Joint Law Estimation

An efficient frontier construction requires the specification of the joint law for these returns. We estimate BG marginal distributions and then formulate a joint law consistent with these

Table 12.12 *Sample of alpha calculations based on efficient frontiers for nine sector ETFs and SPY.*

	Mean Return	Required Return Stress 0.025	Required Return Stress 0.015	Alpha Stress 0.025	Alpha Stress 0.015
aapl	−0.0568	−0.4294	−0.4142	0.3725	0.3574
adbe	0.3053	−0.1409	−0.1229	−0.4462	0.4282
amzn	0.3486	−0.2421	−0.2271	0.5907	0.5757
ba	0.2240	−0.2620	−0.1751	0.4860	0.3991
bac	0.0797	−0.2103	−0.1831	0.2900	0.2628
c	−0.0021	−0.1307	−0.0947	0.1286	0.0927
cbs	−0.2049	−0.1770	−0.1932	−0.0279	−0.0117
cmcsa	−0.0606	0.0691	0.1181	−0.1297	−0.1787
csco	0.0119	0.0075	0.0143	0.0044	−0.0025
cvs	−0.1414	0.1642	0.1970	−0.3056	−0.3384
cvx	0.0956	0.1361	0.0857	−0.0405	0.0100
dis	−0.3276	−0.0229	−0.0037	−0.3047	−0.3238
ebay	−0.0508	−0.2751	−0.2767	0.2243	0.2259
f	−0.2874	−0.1284	−0.1188	−0.1590	−0.1686
fdx	−0.0281	−0.1553	−0.0804	0.1272	0.0523
gs	−0.0657	−0.1683	−0.0898	0.1026	0.0240
hp	−0.1151	0.1244	0.1541	−0.2395	−0.2692
hpq	0.0093	−0.1628	−0.0489	0.1721	0.0582
ibm	−0.1223	−0.2197	−0.2112	0.0974	0.0889
intc	−0.0471	−0.3095	−0.2361	0.2624	0.1890
jnj	0.0481	0.0079	−0.0647	0.0402	0.1128
jpm	0.0376	−0.1277	−0.1096	0.1653	0.1472
ko	0.0520	0.0512	0.0149	0.0008	0.0371
mcd	0.1244	−0.0525	−0.1094	0.1770	0.2339
mmm	0.1061	−0.0573	−0.0245	0.1634	0.1306
mrk	0.0343	0.0473	0.0322	−0.0130	0.0021
ms	−0.0074	−0.1782	−0.0971	0.1708	0.0896
msft	−0.0884	−0.2707	−0.2492	0.1823	0.1608
nke	−0.0241	−0.2482	−0.2042	0.2241	0.1801
orcl	−0.0494	−0.0978	−0.0947	0.0484	0.0453
oxy	−0.0061	0.0490	0.0002	−0.0550	−0.0063
pfe	−0.1246	0.0422	−0.0106	−0.1668	−0.1141
pg	−0.1125	0.1035	0.0917	−0.2160	−0.2042
pnc	0.0957	−0.1531	−0.1593	0.2488	0.2550
rok	−0.0359	−0.2951	−0.2337	0.2592	0.1977
spx	0.0159	−0.0268	−0.0249	0.0427	0.0408
t	−0.1046	−0.0730	−0.1433	−0.0317	0.0387
txn	0.1223	−0.3832	−0.3353	0.5055	0.4576
ups	0.0635	−0.0710	−0.0522	0.1346	0.1158
usb	0.0103	−0.0994	−0.0896	0.1097	0.0999
vz	0.0477	0.0209	0.0047	0.0269	0.0430
wfc	−0.1461	−0.1413	−0.1375	−0.0048	−0.0085
wmt	−0.0721	0.1085	0.1007	−0.1806	−0.1728
xom	−0.0820	0.2522	0.1703	−0.3342	−0.2523
xrx	−0.1638	−0.0605	0.0182	−0.1034	−0.1821

Table 12.13 *Bilateral gamma marginals*

Maturity	b_p	c_p	b_n	c_n
1	0.0026	2.7415	0.0233	0.0932
2	0.0158	1.5807	0.0177	1.1576
3	0.0306	2.2725	0.0237	2.7632
4	0.0402	3.5207	0.0284	4.8410
5	0.0589	3.5765	0.0355	5.8005
6	0.0119	107.1200	0.0118	107.1200
7	0.0137	118.0500	0.0137	118.0500
8	0.0176	99.7300	0.0176	99.7300
9	0.0193	109.9400	0.0192	109.9400
10	0.0236	94.7300	0.0236	94.7300

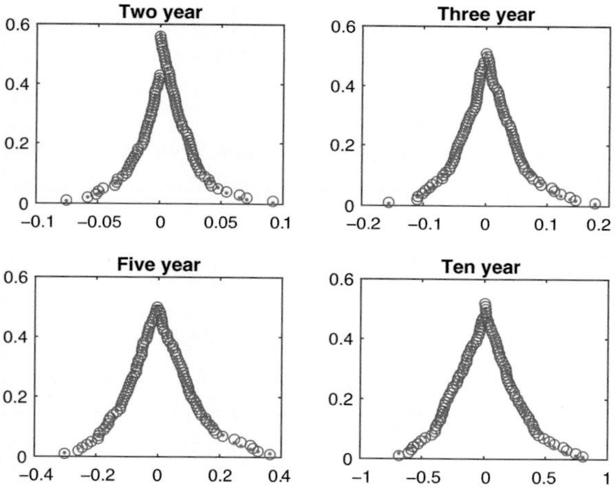

Figure 12.6 Sample fit of marginal bilateral gamma densities to tail probabilities. The data are represented by circles while the model tail probabilities are shown as dots.

marginals. The joint law requires the estimation of a 10×10 correlation matrix C for a multivariate Brownian motion time changed by a gamma process with variance rate ν to which we add independent BG shocks with a view to matching the already estimated marginal BG laws.

The parameters of the estimated 10 BG marginals are presented in Table 12.13.

We observe a symmetry with a low value for volatility of the gamma process for maturities 6 to 10, suggesting that these distributions are close to Gaussian. Figure 12.6 presents a sample of the fit for 4 of the 10 maturities.

For the estimation of correlations and the variance rate of the gamma process in the multivariate VG we matched randomly selected multidimensional tail probabilities. The search of correlation matrices was done using parameterization by the lower triangular part

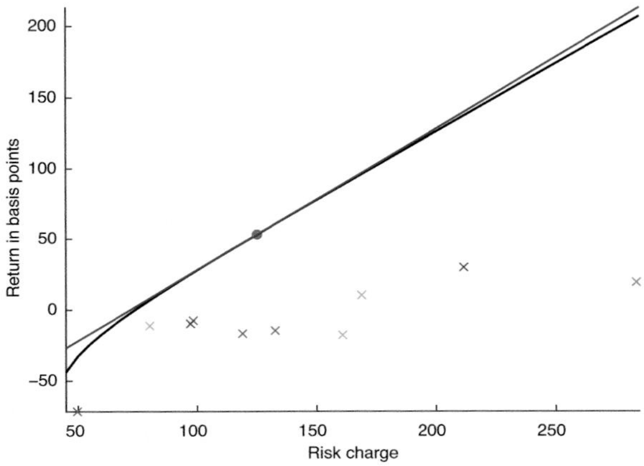

Figure 12.7 Fixed income efficient frontier. The crosses present the 10 mean returns and their associated risk charges. The point on the frontier has a slope of unity.

of the matrix logarithm of the correlation matrix. The value of ν was 10.7339 indicative of a high kurtosis in this component. The estimated correlation matrix for the subset of odd maturities was

$$\begin{pmatrix} 1 & 0.3744 & 0.4014 & 0.3061 & 0.3718 \\ 0.3744 & 1 & 0.5502 & 0.4347 & 0.5957 \\ 0.4014 & 0.5502 & 1 & 0.5707 & 0.7460 \\ 0.3061 & 0.4347 & 0.5707 & 1 & 0.6118 \\ 0.3718 & 0.5957 & 0.7460 & 0.6118 & 1 \end{pmatrix}.$$

12.8.2 Efficient Frontier Construction

We used the method of probability approximation at a small time step to simulate 10-dimensional shocks coming from the multivariate VG and the independent BG components. The total number of points simulated was 60,000 along with their probabilities. One may then construct mean returns and distorted expectations along with their associated risk charges. The stress level of 0.0375 was used in the distortion minmaxvar and permits a tangency to the frontier with a unit slope. The selected point on the frontier has this slope. Figure 12.7 presents a graph of the frontier constructed.

The minimum risk charge attainable was 42.78 with an associated mean return of −56.71 basis points. The zero-risk gradient mean return was −71.98 basis points.

Table 12.14 presents the 10 mean returns and risk charges for this stress level.

Table 12.15 presents the composition of the efficient portfolio shown in Figure 12.7.

Table 12.16 presents the mean returns, required returns, and alphas on the fixed income 10 securities.

Table 12.14 *Mean returns and risk charges*

Maturity	Mean	Risk Charge
1	−71.45	49.65
2	−7.57	98.27
3	10.13	168.78
4	29.69	211.57
5	19.09	283.75
6	−11.16	80.15
7	−9.82	97.15
8	−16.81	118.89
9	−14.79	132.66
10	−18.03	160.82

Table 12.15 *Efficient portfolio composition*

Maturity	Proportion
1	−0.9250
2	0.6610
3	0.0961
4	0.1201
5	−0.1210
6	1.0347
7	0.4486
8	−0.0139
9	−0.0005
10	−0.3001

12.8.3 Fixed Income Alpha Generation

Any efficient portfolio constructed from a set of assets with positions a^* and risk charge function $\mathcal{R}(a)$ has an equation for the asset drifts μ_i given by equation (12.6) as the zero-risk gradient expected return plus the market price of risk θ times the risk charge gradient. Such equations may be employed to determine required returns for assets responding to or depending on the returns reflected in the frontier. Essentially, if the residuals of return projections onto the space of frontier returns are viewed as idiosyncratic unpriced risks, then the projection function may be employed to evaluate risk exposure or the risk charge gradient.

As in the alpha generation section for equity returns, we may project, using GPR, other fixed income asset returns onto the frontier returns given by the returns for the traded pure discount bonds to determine the function

$$r = \phi(z),$$

Table 12.16 *Mean, required return, and alpha*

Maturity	Mean	Required Return	Alpha
1	−71.45	−72.07	0.6142
2	−6.57	−7.87	0.3089
3	10.13	9.88	0.2445
4	29.69	29.68	0.0149
5	10.09	18.89	0.1973
6	−11.16	−11.48	0.3151
7	−9.82	−10.11	0.2974
8	−16.81	−17.07	0.2574
9	−14.79	−14.18	−0.6070
10	−18.03	−18.35	0.3250

Table 12.17 *Mean returns, required returns, and alpha*

Maturity	Mean Returns	Required Return	Alpha
0.5	−34.88	−71.98	37.10
1.5	−68.26	−69.32	1.05
2.5	158.70	−45.85	204.55
3.5	406.40	22.65	383.74
4.5	347.48	57.48	290.00
5.5	345.17	71.38	273.80
6.5	328.57	76.62	251.95
7.5	263.16	73.41	189.75
8.5	144.38	66.10	78.29
9.5	−111.38	52.54	−163.92
10.5	−673.47	30.87	−704.34

where z is the vector of return on the traded pure discount bond frontier returns, and r is a candidate return on some other fixed income asset not included in the frontier.

For simulated frontier returns z_k with probability p_k we have efficient portfolio returns

$$r_k^* = \sum_i a_i^* z_{ki}.$$

The risk charge gradient for $\phi(z)$ may be evaluated by

$$\frac{\partial \text{RC}}{\partial r} \approx \frac{\text{RC}(r_k^* + \varepsilon\phi(z_k)) - \text{RC}(r_k^* - \varepsilon\phi(z_k))}{2\varepsilon},$$

and the required return then may be evaluated as

$$\rho = \mu_{\text{zrg}} + \theta \frac{\partial \text{RC}}{\partial r}.$$

Table 12.17 presents such required returns and alpha values for pure discount bond maturities as indicated that were not on the frontier.

13

Dynamic Valuation via Nonlinear Martingales and Associated Backward Stochastic Partial Integro-Differential Equations

Before branching into the details of our exercise a few general remarks may be helpful. Valuation is sometimes seen as a cost of replication when the latter is possible. Such valuation concepts are the central thesis of the Black and Scholes (1973) and Merton (1973) option pricing models that were later simplified by Cox et al. (1979). However, in the absence of replication, by just no arbitrage and the law of one price, value is a linear function (Ross, 1978) that translates to expectation under a change of measure. Value was later related in the absence of arbitrage to conditional expectations under a measure change and the martingale property (Delbaen and Schachermayer, 1994, 1998). The measure change encapsulates the risk premia embedded in valuation via risk charges for factor exposures (Ross, 1976). Hedges may no longer be present as replication is lost. Furthermore, whatever hedges accomplish with respect to utility and/or other risk assessments, they have no impact on value. The value of the target plus the hedge is the value of the target alone under a self-financed zero-cost hedge. Hedging problems are formulated and solved for a variety of personalized reasons and/or objectives, (Duffie and Richardson, 1991; Duffie et al., 1997; Cvitanić, 2000; Schweizer, 2005; Basak and Chabakauri, 2010), but value maximization is not a hedging concern. When replication is possible the value is tied to the hedge. In its absence, under the assumptions of the law of one price and no arbitrage, the two are divorced.

In two-price markets with prices reflecting the direction of trade, market valuations are no longer linear but concave for the lower price and convex for the upper price. Further with the scaling of value with outcomes, there are just the two prices, and hedges impact both the upper and lower valuations. The hedge and value are once again linked.

When tied to replication, values inherit the martingale property via the replication hedge. In its absence they attain the martingale property by virtue of being conditional expectations. In either case they are then connected to partial differential or integro-differential equations via Itô's lemma (Stroock, 2003). In the two-price framework, both prices are by construction nonlinear martingales with the hedged value higher for the lower price and lower for the upper price than its unhedged counterpart. They are then connected to nonlinear partial differential or integro-differential equations via the theory of backward stochastic differential equations (Peng, 1992, 2019; Barles et al., 1997; Rosazza, 2006; Royer, 2006).

Here we wish to consider the conservative valuation and hedging of a book of derivatives on a single underlier. The book is specified by writing down a sequence of spot slides due at a set of maturities, $t_i \leq T$, at which the functions $c_i(S(t_i))$ are due. The object is to work out the lower prudential valuation $V(S,t)$ post-dynamic hedge positions $a(S,t)$ in the stock that bring in the cumulated cash flow

$$\int_0^T \int_{-\infty}^{\infty} a(S,u)\,(e^x - 1)\,\mu(dx,du),$$

for an integer-valued random-valued measure μ, in the case of finite variation jump processes.

The present value of cash flows to date are then

$$-\sum_{t_t \leq t} c_i(S(t_i))e^{-rt_i} + \int_0^t \int_{-\infty}^{\infty} e^{-ru} a(S(u),u)\,(e^x - 1)\,\mu(dx,du),$$

with the final present value being

$$L(T) = -\sum_{t_t \leq T} c_i(S(t_i))e^{-rt_i} + \int_0^T \int_{-\infty}^{\infty} e^{-ru} a(S(u),u)\,(e^x - 1)\,\mu(dx,du).$$

The post-hedge valuation, in a Markovian context, is given by $V(S(t),t)$. By construction

$$V(S(T),T) = L(T).$$

The construction of the valuation and the hedge strategy at earlier dates $t < T$ as a function of the current spot price $S(t)$ and time t brings together results presented in a number of papers. First we recognize that conservative valuations are not additive and nor are the associated hedges; hence both are to be done at the level of a book of positions and not as the sum of results for each position separately conducted.

Given any proposed hedge policy, the associated valuation follows the $\mathcal{G}$-expectations approach introduced in Peng (2007) and implemented in Eberlein et al. (2014b). In this approach the valuation with time discounting occurring at a constant interest rate of r is defined as a nonlinear expectation that is a unique viscosity solution to a nonlinear equation of the form

$$V_t = \mathcal{G}(V) - rV \tag{13.1}$$

for the boundary conditions

$$V(S,0) = c_T(S),$$
$$V(S,t_i) = V(S,t_{i-}) + c_i(S),\ t_i > 0.$$

The result $V(S,t)$ is a nonlinear valuation at time t of the future cash flows yet to be realized if the spot price at time t is at the level S.

The definition of the operator $\mathcal{G}$ requires first the specification of the dynamics of motion for the underlying spot price $S(t)$ or its logarithm $X(t)$. We suppose that the local motion $X(t)$ is that of a BG process but allow for spatial inhomogeneity by letting the parameters depend on the level of $X(t)$. The compensator ν of the integer-valued random measure μ may then be written as

$$\nu(X, dx)dt = \left(\frac{c_p(X)}{x} \exp\left(-\frac{x}{b_p(X)} \right) \mathbf{1}_{x>0} + \frac{c_n(X)}{|x|} \exp\left(-\frac{|x|}{b_n(X)} \right) \right) dxdt.$$

The BG process is a special case when the parameters are constant with no dependence on the level of the spot price. For the existence of processes with such a functional dependence of parameters we refer the reader to Stroock (1975), Stroock and Varadhan (1979), Bass (1988), and Angelos (2013).

Define the spatially inhomogeneous measure

$$\nu(X, A) = \int_A \nu(X, dx).$$

The operator $\mathcal{G}$ is defined in terms of measure distortions G^+, G^- in the presence of a hedging strategy $a(X, t)$ by

$$\mathcal{G}(V) = -\int_0^\infty G^+ \left(\nu(X, [a(X,t)(e^x - 1) + V(Xe^x, t) - V(X,t)]^- > w\right) dw$$
$$+ \int_0^\infty G^- \left(\nu\left(X, [a(X,t)(e^x - 1) + V(Xe^x, t) - V(X,t)]^+ > w\right)\right) dw.$$

The hedge selection follows the development in Madan et al. (2017a) whereby

$$a(S,t) = \max_a \left(-\int_0^\infty G^+ \left(\nu(X, [a\,(e^x - 1) + V(X + x, t) - V(X,t)]^- > w\right) dw \right.$$
$$\left. + \int_0^\infty G^- \left(\nu\left(X, [a\,(e^x - 1) + V(X + x, t) - V(X,t)]^+ > w\right)\right) dw \right).$$

The task to be accomplished here is first the estimation of a specific spatially inhomogeneous BG dynamics for the logarithm of the underlying asset price. Next we take a book of options on the underlying asset for which we implement the conservative valuation and hedging of this book of derivatives.

13.1 Backward Stochastic Partial Integro-Differential Equations and Valuations

The general development follows Madan et al. (2017a), where detailed proofs are presented in the case of an underlying Lévy process with the jump measure compensating stock price jumps that lie in the interval $[-1, \infty)$. Here we reformulate without proofs for jumps in the logarithms of the price and an underlying Markov process, allowing compensators to depend on time and the level of Markov variates.

The lower valuation may be related to the solution of backward stochastic partial integro-differential equations. For this purpose we work on a filtered probability space associated with a pure jump Markov process with finite-variation jump compensators for the logarithm of the positive Markov variables X of the form

$$\nu(X(t), t, dx)dt,$$

with

$$X(t) = \int_{(0,t]\times\mathbb{R}^k\setminus\{0\}} ((\exp\circ I)(x) - \mathbf{1}_k)\,\nu(X(s), s, dx)$$
$$+ \int_{(0,t]\times\mathbb{R}^k\setminus\{0\}} ((\exp\circ I)(x) - \mathbf{1}_k)\,\widetilde{N}(ds\times dx),$$

where $\widetilde{N}(ds \times dx)$ is a compensated jump measure, $(\exp\circ I)(x) = (e^{x_1}, \ldots, e^{x_k})^{\mathrm{T}}$, and $\mathbf{1}_k$ is the k-dimensional vector with all unit entries.

Finite variation yields the result

$$\int_{(0,t]\times\mathbb{R}^k\setminus\{0\}} |x_i|\nu(X(s), s, dx) < \infty.$$

On this probability space let χ be a terminal random variable and let B_t be a lower prudential valuation for χ at time t associated with a driver function

$$g(t, z),$$

defined on $(0, T] \times \mathcal{L}^2(\nu(X(t), t, dx))$ that is positive, homogeneous, and convex in z. The backward stochastic partial integro-differential equation solves for (B_t, Z_t) for $t < T$, the equation

$$B_t + \int_t^T g(s, Z_s)ds + \int_{(t,T]\times\mathbb{R}^k\setminus\{0\}} Z_s(x)\widetilde{N}(ds\times dx) = \chi. \tag{13.2}$$

In this Markovian context the lower prudential valuation can be written as

$$B_t = V(t, X(t)),$$

where the function V satisfies

$$V(T, X(T)) = \chi,$$

and the function V satisfies the semilinear partial integro-differential equation

$$\dot{V} + \mathcal{K}V(t, x) - g(t, \mathcal{D}V_{t,x}) = 0,$$
$$\mathcal{K}V(t, x) = d_t^{\mathrm{T}}\nabla V + \int_{\mathbb{R}^k\setminus\{0\}} \left(\mathcal{D}V_{t,x} - \nabla V(t, x)^{\mathrm{T}} x(e^y - 1)\right) \nu(dy),$$
$$\mathcal{D}V_{t,x} = V(t, xe^y) - V(t, x),$$
$$d_t = \int_{\mathbb{R}^k\setminus\{0\}} x\,(e^y - 1)\,\nu(x, t, dy). \tag{13.3}$$

The lower prudential value is given by

$$B_t(\chi) = \inf_{\mathbb{Q}\in\mathcal{S}^g} E^{\mathbb{Q}}[\chi|\mathcal{F}_t],$$

where the Radon–Nikodým derivative ξ of $\mathbb{Q}$ is the stochastic exponential of M or

$$\xi = \mathcal{E}(M),$$

and

$$M = \int_{(0,T]\times\mathbb{R}^k\setminus\{0\}} H_s(y)\widetilde{N}(ds \times dx),$$

with

$$\int z(y)H_s(y)\nu(X(s), s, dy) \le g(s, z),\ z \in \mathcal{L}^2\left(\nu(X(s), s, dy)\right).$$

13.2 Nonlinear Valuations and BSPIDE

We now connect the nonlinear valuation equation (13.1) and backward stochastic partial differential equations. The equation (13.1) is a nonlinear martingale condition for discounted values whereby

$$e^{-rt}V(t, X(t) = \inf_{Q\in\mathcal{M}} E^Q\left[e^{-rT}\chi\right] = \mathfrak{E}[e^{-rT}X].$$

So the discussion of the previous section applies to

$$W(t, X(t)) = e^{-rt}V(t, X(t)).$$

We observe that

$$\dot{W} = -rW + e^{-rt}\dot{V}.$$

As time has been reversed in (13.1), we write the equation as

$$\dot{V} + \mathcal{G}V - rV = 0.$$

Now multiply by e^{-rt} to get

$$\dot{W} + \mathcal{G}W = 0. \tag{13.4}$$

For the moment we ignore the hedge and consider the unhedged valuation. The distorted variation is

$$\begin{aligned}\mathcal{G}W = &-\int_0^\infty G^+\left(\nu(X, [W(Xe^x, t) - W(X, t)]^- > w\right) dw\\ &+\int_0^\infty G^-\left(\nu\left(X, [W(Xe^x, t) - W(X, t)]^+ > w\right)\right) dw.\end{aligned}$$

The variation on the other hand is

$$\begin{aligned}\mathcal{V}(W) = &-\int_0^\infty \left(\nu(X, [W(Xe^x, t) - W(X, t)]^- > w\right) dw\\ &+\int_0^\infty {}^-\nu\left(X, [W(Xe^x, t) - W(X, t)]^+ > w\right) dw.\end{aligned}$$

Define

$$\Gamma^+ = G^+ - I, \qquad \Gamma^- = I - G^-,$$

and observe that

$$\mathcal{G}W = \mathcal{V}W - \int_0^\infty \Gamma^+ (\nu(X, [W(Xe^x, t) - W(X, t)]^- > w)\, dw$$
$$- \int_0^\infty \Gamma^- \left(\nu\left(X, [W(Xe^x, t) - W(X, t)]^+ > w\right)\right) dw.$$

Define

$$Z(y) = W(Xe^y, t) - W(X, t),$$

and let

$$g(t, z) = \int_0^\infty \Gamma^+ (\nu(X, [z(y)^- > w])\, dw + \int_0^\infty \Gamma^- \left(\nu(X, [z(y)^+ > w]\right) dw. \qquad (13.5)$$

We may then write

$$\mathcal{G}W = \mathcal{V}W - g(t, Z),$$

and our equation (13.4) becomes

$$\dot{W} + \mathcal{V}W - g(t, Z) = 0.$$

The semilinear equation (13.3) in W is

$$\dot{W} + \mathcal{K}W - g(t, \mathcal{D}W_{t,x}) = 0,$$

and observe that $\mathcal{K}W$ is just $\mathcal{V}W$ and hence the driver is identified as (13.5). Further the associated (13.2) holds with

$$e^{-rt}V(t, x) + \int_t^T g(s, Z_s)ds + \int_{(t,T]\times\mathbb{R}^k\setminus\{0\}} Z_s(y)\widetilde{N}(ds \times dy) = \chi.$$

We then have that

$$e^{-rt}V(t, X(t)) = \inf_{\mathbb{Q}\in\mathcal{S}^g} E^{\mathbb{Q}}[e^{-rT}\chi|\mathcal{F}_t].$$

Here $\mathcal{S}^g$ consists of all measures with densities ξ that are stochastic exponentials of M with representation

$$M(t) = \int_0^t H_s(y)\widetilde{N}(ds \times dx),$$

where for all $z \in \mathcal{L}^2\,(\nu(X(t), t, dy))$ it is the case that

$$\int_{\mathbb{R}^k\setminus\{0\}} H_s(y)z(y)\nu(X(t), t, dy) \le g(t, z).$$

13.3 Spatially Inhomogeneous Bilateral Gamma

We will estimate such a model for the underlier SPY. We may begin by partitioning the time series data into different SPY levels and estimate BG separately on these partitions.

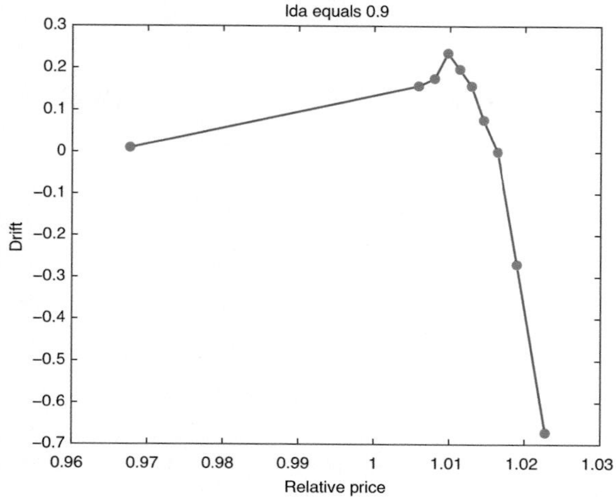

Figure 13.1 Spot drift as a function of the ratio of price to an exponentially weighted average of past prices

13.3.1 A Preliminary Investigation

The data have 3,020 observations on the level of SPY from January 2007 to December 2018. Return distributions cannot depend on the level of the underlying, but may depend on the relative level of the underlying. We consider relativizing to a number of exponentially weighted averages and select the best one.

For the last 1,000 days we constructed the ratio of SPY to an exponentially weighted average of past prices using $\lambda = 0.9$. The return was then split into deciles for the level of this ratio and the BG parameters were estimated separately for each decile. The drift on SPY was then estimated from the BG parameters. Figure 13.1 presents a graph of the drift against the level of ratio.

We observe the presence of momentum with the drift falling as the relative spot drops and rising as the spot rises. However, if it rises too far, then the drift is negative. Figure 13.2 presents a graph of the four parameters as a function of the relative price.

13.3.2 Markov Model for Ratio of Spot to an Average

Given that an exponentially weighted average of past prices is known at any time, the local dynamics is that of a spot-dependent BG structure with all four parameters having a possible functional form with the shapes presented in Figure 13.2.

We estimate a model where the BG parameters are a function of the ratio of the price to an exponentially weighted average of past prices. In a continuous time formulation one would define

$$Y(t) = \theta \int_{-\infty}^{t} e^{-\theta(t-u)} S(u) du.$$

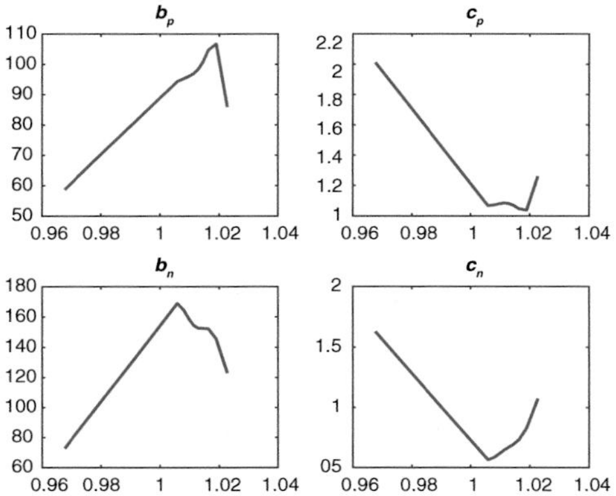

Figure 13.2 Dependence of bilateral gamma parameters on the ratio of price to an exponentially weighted average of past prices

We then see that

$$dY = \theta(S(t) - Y(t))dt = \theta\,(Z(t) - 1)\,Y(t)dt.$$

The compensator for X is

$$\nu\left(\frac{S}{Y}, dx\right) dt = \left(\frac{c_p(S/Y)}{x} \exp\left(-\frac{x}{b_p(S/Y)}\right) \mathbf{1}_{x>0} + \frac{c_n(S/Y)}{|x|} \exp\left(-\frac{|x|}{b_n(S/Y)}\right)\right) dxdt.$$

The process is jointly Markov in S, Y while the evolution of Y is deterministic with no martingale component. The valuation and hedge strategies must be of the form

$$V(S, Y, t), \qquad a(S, Y, t).$$

The solution of the hedge and valuation problem will have to be on a three-dimensional space-time grid. But what matters is the ratio of S to Y, and this lies in a narrow range.

Let $Z(t) = S(t)/Y(t)$ and let us consider the dynamics for this ratio:

$$dZ = Z(t)\frac{dS}{S} - Z(t)\frac{dY}{Y} = Z(t)\,((e^x - 1) * \mu(dx, du)) - \theta Z(t)\,(Z(t) - 1)\,dt.$$

Hence $Z(t), Y(t)$ is a Markov process in its own right and we may write

$$V(Z, Y, t), \qquad a(Z, Y, t),$$

and solve on the three-dimensional grid for Z, Y and time t.

The operator may be written as

$$\mathcal{G}(V) = -\int_0^\infty G^+\,(\nu(Z, [a(Z, Y, t)\,(e^x - 1) + V(Ze^x, Y, t) - V(Z, Y, t)]^- > w)\,dw$$
$$+ \int_0^\infty G^-\,\left(\nu\left(Z, [a(Z, Y, t)\,(e^x - 1) + V(Ze^x, Y, t) - V(Z, Y, t)]^+ > w\right)\right) dw.$$

We solve

$$V_t = \mathcal{G}(V) - rV - \theta Z(Z-1)V_Z + \theta(Z-1)YV_Y.$$

We specify

$$V(Z,Y,0) = C_T(YZ),$$
$$V(Z,Y,t_i) = V(Z,Y,t_{i-}) + c_i(Y(t_{i-})Z),\ t_i > 0.$$

The problem is now formulated as a two-dimensional Markov problem for which we determine the measure-distorted value-maximizing hedges and valuations after we have estimated the Z dependent BG evolution. We will leverage the negative drifts at high Z values into the hedge construction.

13.4 Dynamic Implementation of Hedging Problems

The implementation of the dynamic hedging problem is illustrated for a book of options on the underlier SPY. As the valuation is not linear, both the valuation and the hedge have to be constructed at the level of the whole book of open positions and not asset by asset. For the Markovian structure we take the four parameters of the BG motion to be dependent on Z the ratio of the spot price to an exponentially weighted average of past prices. We observe that the variable Z for geometric averaging with weight λ^n for the price n days back for $\lambda = 0.9$ lies between 0.96 and 1.03 in the data.

13.4.1 Maximum Likelihood Estimation of Markovian Model for Spot Average Ratio

From data on the ratio of the spot to this geometrically weighted average and the one-day forward return, the dependence of the BG parameters on the ratio is estimated by maximum likelihood. On a coarse grid between 0.85 and 1.05 in five steps of 0.05, the candidate value of each of the four BG parameters is taken as a free parameter to be estimated. The number of parameters to be estimated is then 20. The actual value of BG parameters at an arbitrary level for Z is constructed using a fixed Gaussian kernel smoother. The density for returns given the BG parameters is obtained by Fourier inversion of the characteristic function. Figure 13.3 presents the estimated smoothed dependence of each of the four parameters on the level of Z in the relevant range.

Figure 13.4 presents the dependence of the drift on the stock on the level of Z.

13.4.2 The Boundary Conditions or Payoffs

The claims to be hedged are functions of the stock spot price at maturities t_i, for $i = 1, \ldots, N$ with $t_N = T$ given by the functions $c_i(S)$. On the grid across Z, Y for the ratio Z and the average Y we write the terminal value function as

$$V(Z,Y,T) = c_N(YZ).$$

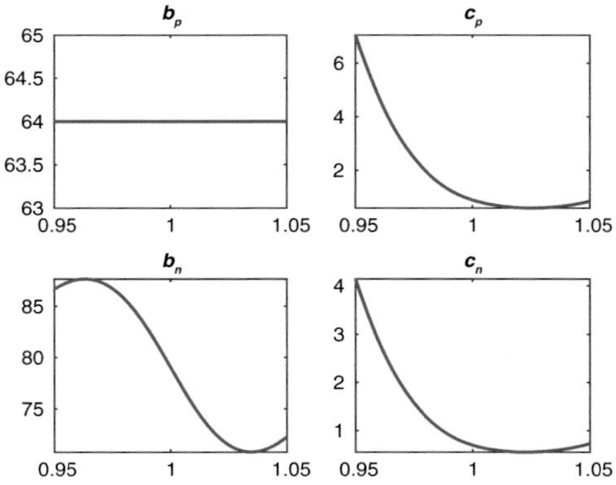

Figure 13.3 Dependence of bilateral gamma parameters on the level of the ratio Z of the spot price to the exponentially weighted average of past prices

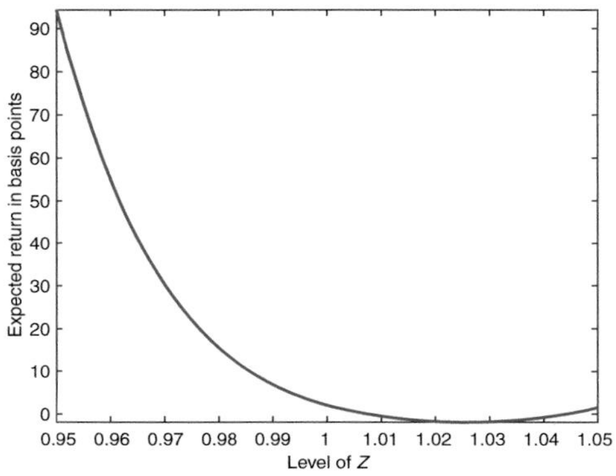

Figure 13.4 Dependence of the stock drift on the ratio Z of spot prices to the exponentially weighted average

For time t_i, solving backward in time, we write

$$V(Z, Y, t_i) = V(Z, Y, t_{i+}) + c_i(YZ).$$

The explicit functions $c_i(S)$ were obtained from information on dollar vega levels bucketed by strike and maturity at market close. Dividing the dollar vega by the option vega for the specific strike and maturity, we infer the number of options held at the specific strike and maturity. These option positions may then be transformed into payoff functions $c_i(S)$ using the payoff functions for each option. It is supposed that positions are for out-of-the-money options or for puts below the forward and calls above the forward.

The options hedged have maturities below a year. There were 29 maturities in all, ranging from 1 to 358 days. Figure 13.5 presents the cash flows to be hedged at a sample of the 29 maturities.

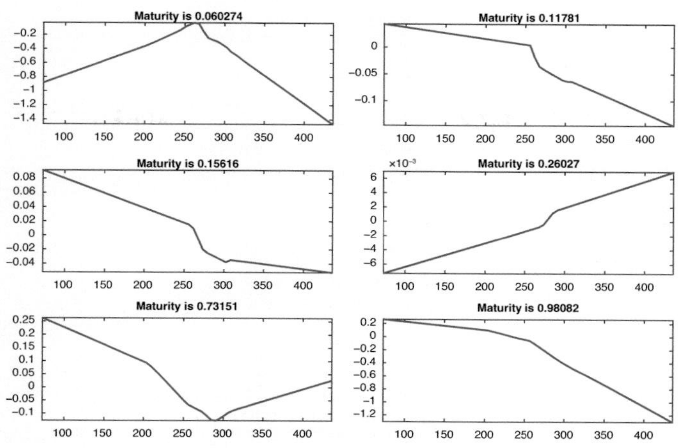

Figure 13.5 Cash flow functions to be hedged at a sample of maturities

13.4.3 Solution on a Three-Dimensional Space-Time Grid

We take a nonuniform grid of 50 points on the level of the average between half and twice the initial level. There is also a 40-point grid on the level of Z between 0.9 and 1.1. The year is partitioned into 20 time steps of 0.05. The integral of the Z compensator $\nu(Z, x)dx$ over sets A is accomplished by simulation using a gamma density with shape parameter 0.075. The potential jumps are taken to lie in the interval $\pm 30\%$ in steps of a basis point. The required measures for sets are distorted and integrated over tails to construct the measure-distorted value. Hedges are chosen to maximize the measure-distorted value. Both values and hedges are smoothed as functions of Z, Y using GPR at the end of each time step to iron out numerical disturbances.

Figure 13.6 presents the hedge positions at a sample of maturities as a function of the level of the ratio Z for three levels of the average Y, including the initial level and \$10 up and down.

Figure 13.7 presents the value of remaining open positions at a sample of future dates as a function of the ratio Z and levels of the average Y.

13.4.4 Regression Analysis of Hedge Positions

Across the space grid of Z, Y at selected levels for the remaining time the hedge position was regressed on the first and second derivatives of the lower prudential value function and the levels of Z, Y and their squares. Results of the regression are presented in Table 13.1 along with the R-square of the regression. Table 13.2 presents the associated t-statistics.

13.4.5 Remarks on Results

We note that as books are continuously refurbished, one never reaches a situation where the time to maturity of all claims held is below three months, for example. As near-maturity

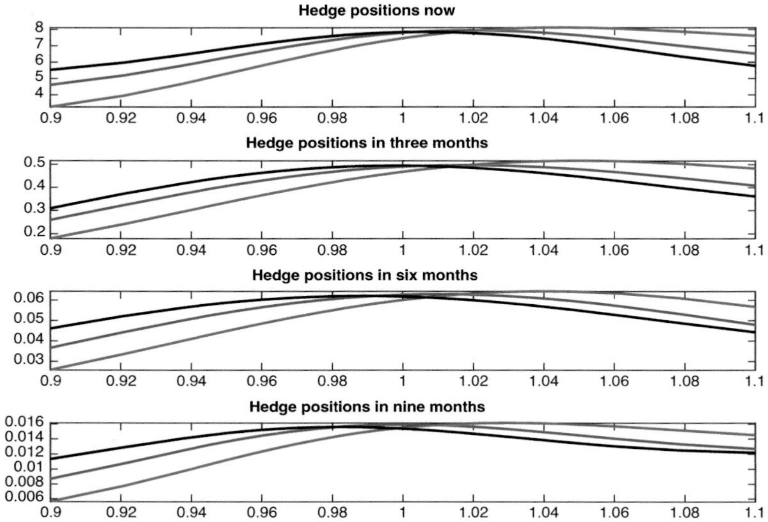

Figure 13.6 Hedge positions as a function of the level of the ratio Z and the level of the average Y at a sample of times to maturity. Shown in blue is the curve for Y at the initial level and in red and black for \$10 up and down for Y.

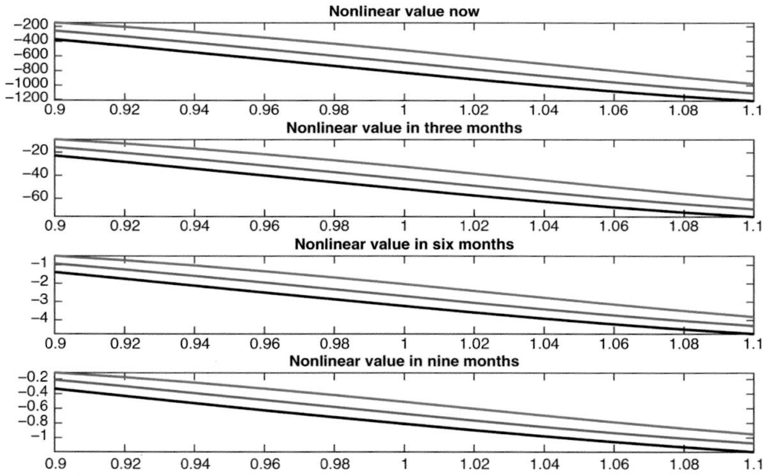

Figure 13.7 The lower prudential value of remaining open positions at a sample of maturities as a function of the ratio Z at different levels of Y. Shown in blue is the curve for Y at the initial level and in red and black for \$10 up and down for Y.

options expire, newer, longer-maturity options enter the book as hedges for what is now lying open at the new shorter maturities. Hence the only relevant rows are closer to the first row of the table, indicating positions slightly above the delta and responsive to the option gamma. The positions are also sensitive to both Z and Y and concave in both. The total size

Table 13.1 *Regression results and their* R-*squares*

Time Left	Delta	Gamma	Z	Y	Z^2	Y^2	RSQ
1	1.1293	29.9751	1723.6023	195.7961	−8.3502	−0.3197	0.9302
0.9	1.0964	36.8843	318.8334	48.4325	−1.5291	−0.0785	0.9243
0.8	1.4404	40.2417	95.7524	12.5620	−0.4641	−0.0205	0.9200
0.7	2.1472	66.0671	19.6692	3.4710	−0.0950	−0.0057	0.9080
0.6	2.0326	62.7417	8.7496	1.6764	−0.0425	−0.0027	0.9007
0.5	2.3538	73.7335	4.8765	0.8572	−0.0237	−0.0014	0.8791
0.4	5.6210	191.1055	2.6347	0.4515	−0.0129	−0.0007	0.8611
0.3	6.9205	197.6278	3.2125	0.4428	−0.0158	−0.0007	0.8342

Table 13.2 *The t-statistics associated with hedge position regression analysis*

Time Left	Delta	Gamma	Z	Y	Z^2	Y^2	RSQ
1	17.77	10.68	16.77	73.64	−16.24	−71.09	
0.9	15.46	11.71	11.67	68.52	−11.20	−65.67	
0.8	18.08	11.29	13.40	67.87	−12.99	−65.32	
0.7	22.23	15.23	9.77	66.50	−9.43	−64.38	
0.6	19.86	13.64	8.75	64.65	−8.50	−62.60	
0.5	19.28	13.47	8.65	58.71	−8.42	−56.99	
0.4	18.96	14.29	8.36	55.32	−8.17	−53.76	
0.3	21.73	13.70	9.45	50.24	−9.32	−49.09	

of positions falls as time passes on a fixed book, but as already noted the book is unlikely to remain fixed.

14
Dynamic Portfolio Theory

Dynamic portfolio theory recognizes the potential errors that may be induced into static portfolio theory by its narrow focus on short-term outcomes. Attention to longer-term interests may be relevant to the design of portfolios. No one is really managing a portfolio for some immediate result. The wealth accumulation being targeted is for some foreseeable future that will encompass a number of potential portfolio rebalancing instances. Hence one does not want to be profitable in the short term in ways that engender considerable costs in the longer term. Such actions could be penny wise and pound foolish. These considerations suggest that one take a longer term and hence more dynamic view of the portfolio management problem.

The dynamic portfolio problem is, however, a much harder problem to formulate, let alone solve. The static problem required the description of the joint law of returns across a number of investment opportunities. The non-Gaussian nature of marginal return distributions led us to construct multivariate laws consistent with these non-Gaussian marginals. We formulated as a candidate solution the use of the multivariate VG model enhanced by independent BG components. The resulting law for n assets has $4n + (n-1)n/2 + 1$ parameters that must now be allowed to move dynamically. The alternate would be to suppose the law of motion is a multivariate law of independent and identically distributed (i.i.d.) increments or a multivariate Lévy process. For such an i.i.d. formulation one has to ask what factors drive dynamic portfolio adjustments.

A large part of the literature on the subject follows Merton (1971) and considers the problem from the individual utility maximization perspective. From this perspective and locally Gaussian returns it is understood that portfolio allocations are driven by return covariations with marginal utilities of wealth. Movements in return distributions and marginal utilities then drive portfolio adjustments. If return distributions are constant, one may still have movements in covariations with marginal utilities that induce portfolio adjustments. For concave utility functions, marginal utilities must be varying but can in some cases have constant covariations, leading to constant portfolio proportions.

Another large strand of the literature formulates the problem from a terminal mean variance perspective. With locally Gaussian returns it is still covariations with marginal value functions that drive allocations. The value functions aggregate the effects on the variance of wealth as the nonlinear and nonconstant component. The mean variance objective is viewed as a quadratic approximation to a utility function. However, the approximation is only locally valid but the analysis conducted is global.

The perspective taken here is that of wealth management in a market economy. For funds broadly committed to wealth management for some horizon, the wealth manager seeks portfolios that will have a good realized market value at the distant horizon. It is market value that is important and not personal utilities of either the wealth manager or the many investors on whose behalf the strategies are being designed. As the portfolio will eventually be unwound or valued in the market, the focus of attention should be on market values. We have already noted that the maximization of market value is rendered problematic when the absence of arbitrage is combined with the law of one price, forcing market valuation functionals to be linear. This observation leads us to consider as an objective the maximization of a conservative market value that comes into existence once the law of one price is dropped. We take this conservative valuation operator to be concave and, in keeping with the size of the market being large relative to any portfolio being managed, we suppose that values scale with the size of invested funds. As a consequence the market valuation is an infimum of classical valuations taken over a set of test or scenario probabilities. The objective is thus a proper nonlinear expectation.

Hence the objective function we employ differs significantly from utility and mean variance. By the scaling property marginal valuations are constant at unity and cannot covary with anything. Portfolio adjustments are to be driven solely by variations in return distributions or investment opportunities. For i.i.d. returns there are no adjustments and dynamic portfolio rules agree with myopic ones. Our interest is not in the modeling of dissatisfaction arising at the margin with the accumulation of wealth or how marginal utility diminishes. Portfolio adjustments are then to be driven by changing risk reward possibilities.

One therefore has to model carefully and hopefully accurately the dynamic movements in risk–reward possibilities. The methods are illustrated in the context of investment in the nine sector ETFs for the US economy along with investment in SPY, the ETF tracking the S&P 500 index. The marginal distributions will be in the BG class and the joint law will be consistent with movements in these marginals. The movements are synthesized using the methods of GPR. The objective function will be a g-expectation related to a measure-distorted valuation operator. The solution will be constructed on a 10-dimensional space grid along with the 11th time dimension. Valuation functions at each time step are synthesized using support vector machine regressions (SVMR). Long-term and myopic policy functions will be synthesized using feedforward neural networks (NNET). Simulated wealth paths will be employed to compare the long-term and myopic solutions.

14.1 The Dynamic Law of Motion

We wish to construct for state variable s a dynamic portfolio position policy function $a(s,t)$ for the level of investment in n assets at time t when the state variables are at level s. The broad objective is to maximize the market value of the associated dynamic strategy. The final wealth is given by

$$W(T) = \int_0^T \int_{\mathbb{R}^n/\{0\}} a(s(t), t)^T (e^x - 1)\, \mu(dx, dt),$$

where the n-dimensional compensator of the random jump measure $\mu(dx, dt)$ is given by

$$\nu(s(t), x) dx dt.$$

We take ν to be in the class of multivariate VG laws enhanced by independent BG idiosyncratic shocks, with parameters that possibly depend on the state vector s.

What are the state variables? Numerous authors – Merton (1971), Aase (1984), Schroder and Skiadas (1999), to mention a few – have worked with general dynamics driven by multidimensional Brownian motion with drifts, volatilities, and covariances expressed as deterministic functions of the asset prices and time. However, one anticipates that one may have comparable multivariate return distributions being relevant at widely different levels for the levels of asset prices. The return distributions some 10 or 15 years apart may be quite comparable with asset prices at markedly different levels. The parameters of return distributions are likely to belong to some relatively compact set with price levels not being similarly constrained. The state variables driving return distributions should themselves belong to some compact set. If we consider the nine sector ETFs for the US economy and the ETF, SPY, we may consider as candidate state variables the relative levels s_i of the index levels S_i where

$$s_i = \frac{\alpha_i S_i}{\sum_i \alpha_i S_i}.$$

The coefficients α_i place all prices at a constant level at some average point. For example, if $\overline{S}_i$ is an average value of the ith stock over the recent past, then

$$\alpha_i = \frac{1}{\overline{S}_i},$$

making all stocks unity at the average index levels, and the state vector is at the center of the unit simplex with $s_i = 1/n$ for all i. Such state variables have been considered in the work on stochastic portfolio theory (see Fernholz, 2001, 2002, and Fernholz and Karatzas, 2005).

With a view to investigating such dependences, we estimated BG parameters every 5 days for 1,000 days using 252 prior returns each day on the 10 underliers: XL, XLE, XLF, XLI, XLK, XLP, XLU, XLV, XLY, and SPY. We then regressed the 200 parameters' estimates on the s_i' as they were observed on each day. Both least-squares regressions and GPR were performed. The GPR made a significant improvement in predictive power. Tables 14.1 and 14.2 report the root mean square error for least squares and GPR for all four parameters and the 10 underliers: b_p and b_n are in basis points.

Table 14.1 *Root mean square errors from least-squares regressions of bilateral gamma parameters on relative stock price vectors*

	b_p	c_p	b_n	c_n
xlb	8.3903	7.6590	8.3903	3.8324
xle	13.9489	5.5330	13.9489	5.0466
xlf	9.1862	22.1060	9.1862	9.0133
xli	5.3918	1.8763	5.3918	1.2153
xlk	8.5084	2.4489	8.5084	1.0682
xlp	5.7859	15.8956	5.7859	4.1146
xlu	9.2549	68.5703	9.2549	33.9736
xlv	6.5113	15.4450	6.5113	13.4144
xly	8.9914	63.4125	8.9914	23.6434
SPY	7.4677	1.4194	7.4677	0.8996

Table 14.2 *Root mean square errors from GPR of bilateral gamma parameters on relative stock price vectors*

	b_p	c_p	b_n	c_n
xlb	2.7948	3.5265	3.1210	1.5282
xle	5.0948	3.7003	4.3544	3.2563
xlf	3.2717	10.7324	4.2247	0.1359
xli	0.8783	1.2887	1.6236	0.3028
xlk	0.8783	1.2887	1.6236	0.3028
xlp	2.7491	0.0017	4.0589	0.0004
xlu	2.2967	13.4848	6.5371	2.1279
xlv	4.5302	0.0028	4.5096	0.0029
xly	2.4893	8.8528	3.5208	8.5993
SPY	1.0665	0.0428	2.0475	0.0434

14.2 Relativity Dynamics

For the dynamics of $s(t) = (s_1(t), \ldots, s_n(t))$ we note that as

$$s_i(t) = \frac{\alpha_i S_i(t)}{\sum_j \alpha_j S_j(t)},$$

it follows that

$$\begin{aligned} ds_i &= s_i(t)\,(e^{x_i} - 1)\,\mu_i(dx, dt) - s_i(t)\frac{\sum_j \alpha_j S_j(t)\,(e^{x_j} - 1)}{\sum_j \alpha_j S_j(t)}\mu_j(dx, dt) \\ &= s_i(t)\,(e^{x_i} - 1)\,\mu_i(dx, dt) - s_i(t)\sum_j s_j(t)\,(e^{x_j} - 1)\,\mu_j(dx, dt), \end{aligned}$$

and

$$\sum_i ds_i(t) = 0.$$

In terms of the compensated jump measure, $\widetilde{\mu}(dx, dt) = \mu(dx, dt) - \nu(s, x)dxdt$, we write

$$\begin{aligned} ds_i &= s_i(t)\left(\theta_i(s) - \sum_j s_j(t)\theta_j(s)\right) dt \\ &\quad + s_i(t)\,(e^{x_i} - 1)\,\widetilde{\mu}_i(dx, dt) - s_i(t)\sum_j s_j(t)\,(e^{x_j} - 1)\,\widetilde{\mu}_j(dx, dt), \\ \theta_i(s) &= (e^{x_i} - 1) * \nu(s, x) \\ &= \int_{\mathbb{R}^d\setminus\{0\}} \left(e^{\langle x, u_i\rangle} - 1\right) \nu(s, x)dx, \end{aligned}$$

where u_i is the vector with unity in the ith component and zeros elsewhere.

The state variables have a Markovian dynamics with the multivariate law depending on the state vector $s(t)$. In the special and simplified case when only the marginal scale parameters are modeled as functions of the state vector with marginal speeds, correlations, and the variance rate of time change for the multivariate Brownian motion taken as constant, the compensator for μ has the form

$$\begin{aligned} \nu_i(s, x) &= m_{\text{mvg}}(x; b_p(s(t)), c_p, b_n(s(t)), c_n, C, \nu) \\ &\quad + \sum_j k_j \left(x_j; b_{\text{p}j}(s(t)), c_{\text{p}j} - \frac{1}{\nu}, b_{\text{n}j}(s(t)), c_{\text{n}j} - \frac{1}{\nu}\right) \prod_{i \neq j} \mathbf{1}_{y_i = 0}. \end{aligned}$$

The functions $b_p(s(t)), b_n(s(t))$ are to be estimated from past data using GPR.

14.3 The Full Sample

The earlier reported results were for the 1,000 days ending on December 31, 2018, with estimations every 5 days with 200 observations in the GPR and LSQ. The analysis was repeated on a daily basis from January 3, 2008, to December 31, 2018, with 2,768 days in the GPR and LSQ. Tables 14.3 and 14.4 provide the root mean square error.

14.4 Portfolio Construction

We wish to construct $a(s, t)$ and $V(s, t)$, the policy and value functions. The dependence on t arises from the valuation of $W(T)$ at some fixed T. Alternatively we may work on a rolling horizon basis, implementing always the stationary policy $a(x) = \widetilde{a}(x, 1)$, regarding the horizon T as moving forward as we move forward in calendar time. The reported results take this approach. The intermediate values that are computed $a(x, t)$ are then never employed in any strategy. They are just intermediate calculations relevant only if the horizon is fixed, and one then recognizes that each time step brings us closer to the fixed horizon. In practice this is never the case as there is no such fixed horizon. The only purpose for considering the dynamic model is to incorporate the effects of looking forward

Table 14.3 *Root mean square errors for regressions by least squares of bilateral gamma parameters on stock price relativities using daily data*

	b_p	c_p	b_n	c_n
xlb	6.6530	50.2471	6.6530	4.7604
xle	7.7139	87.9947	7.7139	10.5912
xlf	18.6163	0.9441	18.6163	0.7348
xli	6.5272	1.2920	6.5272	0.7801
xlk	4.8761	3.2919	4.8761	2.5025
xlp	2.7676	1.9058	2.7676	0.9617
xlu	2.6963	1.9487	2.6963	1.0897
xlv	3.3542	1.6667	3.3542	0.9146
xly	7.6992	1.2856	7.6992	0.7950
SPY	5.2251	1.1757	5.2251	0.6194

Table 14.4 *Root mean square errors for Gaussian process regression of bilateral gamma parameters on stock price relativities using daily data*

	b_p	c_p	b_n	c_n
xle	0.1713	0.0887	0.1497	0.0146
xlf	0.4143	0.0024	0.4925	0.0015
xli	0.0714	0.0113	0.1196	0.0133
xlk	0.0714	0.0113	0.1196	0.0133
xlp	0.1253	0.0028	0.1813	0.0021
xlu	0.1382	0.0059	0.2506	0.0038
xlv	0.0868	0.0034	0.1494	0.0035
xly	0.2905	0.0043	0.3060	0.0023
SPY	0.1994	0.0015	0.3472	0.0012

in making current decisions, and we continuously look forward a similar number of time steps for the longer-term or dynamic perspective. It is unclear what mathematical problem the implemented strategy solves, and we leave this as undefined.

Whatever way we use the outputs, we have to determine the functions $a(s,t)$ and $V(s,t)$. In small space dimensions, like one or two, this is typically done on a grid across the space dimensions by time stepping backward from the horizon of T back to time 1. The valuation operator solves a partial integro-differential equation with known terminal value $V(s,T) = 0$ and $a(s,T) = 0$ as there is liquidation and no further investment involved.

Specifically, the value function satisfies the partial integro-differential equation

$$V_t + \sum_i V_{s_i} s_i \left(\theta_i(s) - \sum_j s_j \theta_j(s) \right)$$
$$+ \left[\begin{array}{l} -\int_0^\infty G^+ \left(\nu(s, \{x | (\widetilde{a}(s, ih)' (e^x - 1) + V(s(x), (i+1)h) - V(s, (i+1)h))^- > w\} \right) dw \\ +\int_0^\infty G^- \left(\nu(s, \{x | (\widetilde{a}(s, ih)' (e^x - 1) + V(s(x), (i+1)h) - V(s, (i+1)h))^+ > w\} \right) dw \end{array} \right]$$
$$= 0.$$

In higher dimensions, like the 10 dimensions we work with in our illustrative case, the use of a grid is not computationally feasible. Instead calculations are performed for a representative collection of points s_n that one is likely to visit. To assess these visitation points, we take the observed time series data on the observed values s_t. This was enhanced by simulation of the multivariate BG processes for all underliers at median values for the parameters. The result when pooled with the time series data is a large collection of points $s_k, k = 1, \ldots, K$. This cloud of points was quantized to 4,396 points at which the policy and value functions are computed. Adding 20 time steps at the time step of $h = 0.05$, we get a scattered set of points in space-time for the computation. The time points are ih for $i = 1, \ldots, 19$ at which we perform the computations at the 4,396 space points. For $i = 19$ the outcome does not include any valuation of future states, and we just take for the outcome

$$a(s)'(e^x - 1),$$

with compensating measure

$$\nu(s, x)dx.$$

At each space point s we seek to define

$$\widetilde{a}(s, 19h) = \arg\max_a \Big[-\int_0^\infty G^+(\nu(s, \{x \mid (a'(e^x - 1))^- > w\})\,dw \\ + \int_0^\infty G^-\left(\nu(s, \left\{x \mid (a'(e^x - 1))^+ > w\right\}\right)dw\Big].$$

After the first or myopic step, for the following steps define

$$\widetilde{a}(s, ih) = \\ \arg\max_a \Big[-\int_0^\infty G^+(\nu(s, \{x \mid (a'(e^x - 1) + V(s(x), (i+1)h) - V(s, (i+1)h))^- > w\})\,dw \\ + \int_0^\infty G^-\left(\nu(s, \left\{x \mid (a'(e^x - 1) + V(s(x), (i+1)h) - V(s, (i+1)h))^+ > w\right\}\right)dw\Big].$$

On discretization of the required partial integro-differential equation we write

$$V(s, ih) = V(s, (i+1)h) + h \sum_i V_{s_i} s_i \left(\theta_i(s) - \sum_j s_j \theta_j(s)\right) + \\ h\Big[-\int_0^\infty G^+(\nu(s, \{y \mid (\widetilde{a}(s, ih)'(e^x - 1) + V(s(x), (i+1)h) - V(s, (i+1)h))^- > w\})\,dw \\ + \int_0^\infty G^-\left(\nu(x, \left\{y \mid (\widetilde{a}(s, ih)'(e^x - 1) + V(s(x), (i+1)h) - V(s, (i+1)h))^+ > w\right\}\right)dw\Big].$$

One may then perform a support vector regression to build the functions $V(s, ih)$. The only policy to be continuously implemented is $a(s) = \widetilde{a}(s, 1)$, where the function $a(s)$ is built using a feedforward neural net.

14.5 Stationary Exposure Valuation

Given that one is just interested in the implementation of the initial policy and that there is no real specific horizon in mind, the difference between the myopic problem and the stationary dynamic one is that the latter takes account of changes in value. Thus beyond the immediate result of exposure to realized returns to a portfolio there is the change in the value of the state to be accounted for. Let $V(s)$ be this value function associated with a policy $a(s)$. This value must then satisfy the valuation equation for jump size x yielding the new state $s(x)$,

$$V(s) = \Big[- \int_0^\infty G^+ (v(s, \{y|(a(s)'(e^x - 1) + V(s(x)) - V(s))^- > w\}) \, dw \\ + \int_0^\infty G^- \left(v(s, \left\{y|(a(s)'(e^x - 1) + V(s(x)) - V(s))^+ > w\right\}\right) dw \Big].$$

In addition the function $a(s)$ satisfies

$$a(s) = \arg\max_a \Big[- \int_0^\infty G^+ (v(s, \{y|(a'(e^x - 1) + V(s(x)) - V(s))^- > w\}) \, dw \\ + \int_0^\infty G^- \left(v(s, \left\{y|(a'(e^x - 1) + V(s(x)) - V(s))^+ > w\right\}\right) dw \Big].$$

The corresponding myopic problem defines the value as

$$W(s) = \Big[- \int_0^\infty G^+ (v(s, \{y|(a(s)'(e^x - 1))^- > w\}) \, dw \\ + \int_0^\infty G^- \left(v(s, \left\{y|(a(s)'(e^x - 1))^+ > w\right\}\right) dw \Big],$$

with policy function $b(s)$ satisfying

$$b(s) = \arg\max_a \Big[- \int_0^\infty G^+ (v(s, \{y|(a'(e^x - 1))^- > w\}) \, dw \\ + \int_0^\infty G^- \left(v(s, \left\{y|(a'(e^x - 1))^+ > w\right\}\right) dw \Big].$$

For the long-term policy we iterate with $V_0(x) = 0$ that delivers $b(s)$ and define

$$V_{n+1}(s) = \Big[- \int_0^\infty G^+ (v(s, \{y|(a(s)'(e^x - 1) + V_n(s(x)) - V(s))^- > w\}) \, dw \\ + \int_0^\infty G^- \left(v(s, \left\{y|(a(s)'(e^x - 1) + V_n(s(x)) - V(s))^+ > w\right\}\right) dw \Big]$$

to build $a_n(s)$ as

$$a_n(s) = \arg\max_a \Big[- \int_0^\infty G^+ (v(s, \{y|(a'(e^x - 1) + V_n(s(x)) - V_n(s))^- > w\}) \, dw \\ + \int_0^\infty G^- \left(v(s, \left\{y|(a'(e^x - 1) + V_n(s(x)) - V_n(s))^+ > w\right\}\right) dw \Big].$$

Table 14.5 *Quantiles across 4,396 states for distances between portfolios at stated iterations of value function updates*

	Distance Quantiles			
quantile	0 to 5	5 to 10	10 to 15	15 to 20
1	0.1232	0.0005	0.0005	0.0005
5	0.1548	0.0008	0.0008	0.0008
10	0.1769	0.0010	0.0010	0.0010
25	0.2105	0.0014	0.0014	0.0014
50	0.2577	0.0022	0.0021	0.0021
75	0.3187	0.0035	0.0035	0.0035
90	0.3924	0.0057	0.0057	0.0058
95	0.4401	0.0076	0.0074	0.0075
99	0.5649	0.0154	0.0153	0.0137

The results presented in the following sections are for the stationary exposure policy presented here. From the construction we see that $V(x)$ is determined up to a constant and hence upon convergence the difference between V_{n+1} and V_n should be a constant.

14.6 Stationary Value and Policy Results

The recursion was implemented for 20 time steps at each of which the optimal portfolio was computed in the 10 assets at 4,396 points for the relativities x. Table 14.5 reports on the distance quantiles across the 4,396 points for states s at which they were computed between the myopic portfolio and the portfolio after 5 time steps, the portfolio after 5 time steps and the one after 10 time steps, the one between 10 and 15 time steps, and finally the one between 15 and 20 time steps. Table 14.5 essentially demonstrates that the long-term problem has converged after 5 time steps.

The value functions upon convergence may, as already noted, differ by just a constant. Table 14.6 presents quantiles for value differences modulo a constant, after the indicated number of time steps.

Given the lower bound of -0.5 for the short side, we add 0.5 to the positions to make them positive and then divide by sum across all 10 assets to form relativities. We then evaluate the Herfindahl concentration index for the myopic portfolios and the long-term portfolios for each of the 4,396 cases where these portfolios have been computed. Table 14.7 reports the quantiles for the concentration indices.

We observe that the long-term portfolios are marginally more diversified than their myopic counterparts. The long-term portfolios are quite stable in the levels of their concentration indices. The myopic concentration indices can be more spread out.

14.7 Building Neural Net Policy Functions and Simulating Trades

In order to simulate and compare the wealth performance of a myopic investor with one who chooses portfolios that are optimal from a long-term perspective we need to build the two policy functions. These are the myopic and long-term portfolio proportions expressed

Table 14.6 *Quantiles across 4,396 states of differences between value functions at the states between iterations less a constant*

	Value Differences				
			Time Steps		
quantile	19	15	10	5	1
1	-0.0013	-0.0021	-0.0010	-0.0009	-11.2389
5	-0.0011	-0.0018	-0.0008	-0.0006	-8.5261
10	-0.0009	-0.0016	-0.0007	-0.0003	-7.5650
25	-0.0006	-0.0014	-0.0005	0.0002	-6.9512
50	-0.0004	-0.0011	-0.0002	0.0007	-5.7470
75	-0.0002	-0.0009	0.0000	0.0010	-4.8065
90	0.0001	-0.0008	0.0003	0.0013	-3.5482
95	0.0002	-0.0007	0.0004	0.0014	-2.1094
99	0.0006	-0.0005	0.0006	0.0017	2.3918

Table 14.7 *Quantiles of concentration indices for portfolios*

Quantile	Myopic	Long-Term
1	0.1065	0.1083
5	0.1082	0.1092
10	0.1089	0.1092
25	0.1102	0.1092
50	0.1119	0.1092
75	0.1137	0.1093
90	0.1154	0.1094
95	0.1163	0.1099
99	0.1176	0.1152

as a function of the state variable given by the stock price relativities. This is a mapping from the 10-dimensional inputs of stock price relativities to the 10-dimensional output of portfolio proportions that have been computed at 4,396 quantized relativities. We employ feedforward neural nets to synthesize these functions and then simulate the wealth path for the two types of investors.

We now ask the following question given two policy functions produced by the net. How do we distinguish between the traders trading these policies in environments where the true law for which the policies were derived are used to generate the prices and the two traders face the same price history.

For state x we have the cash flows given by

$$a_i(x)' \left(\exp(z) - 1\right),$$

with both traders facing the same sequence of x_t, z_t through time. The wealth paths are

$$w_{it} = a_i(x_t)' \left(\exp(z_t) - 1\right).$$

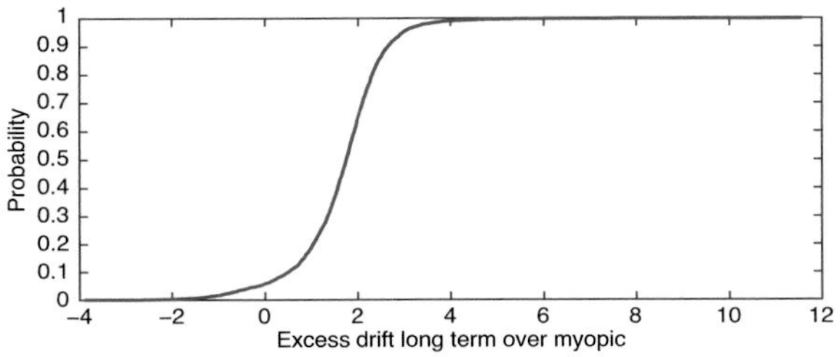

Figure 14.1 Excess drift for the long-term manager over the myopic

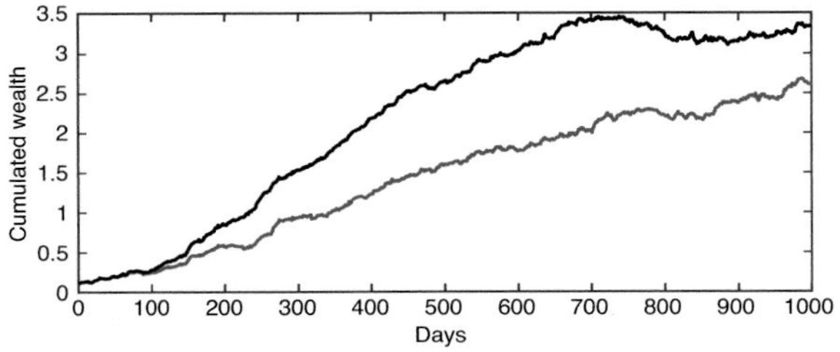

Figure 14.2 Wealth paths for the long-term and myopic managers on identical simulated price histories for the 10 assets

For trader 1 to outperform trader 2 in state x we must have

$$(a_1(x) - a_2(x))' (\exp(z) - 1) \geq 0.$$

For any holding period like a day, one may simulate from the joint law for z and evaluate given x:

$$\theta(x) = E_z \left[(a_1(x) - a_2(x))' (\exp(z) - 1) \right].$$

For trader 1 being the long-term manager with 20 time steps to maturity and trader 2 being the myopic trader with a unit time to maturity, we evaluated $\theta(x)$ for each point x at which the portfolio was evaluated. There were 4, 396 relativities for x at which the portfolios were computed. Figure 14.1 presents the distribution function for the excess drift for the long-term manager over the myopic one.

By and large in this specific case the long-term manager delivered by the net is superior to the comparable delivery for the myopic one with respect to trading the joint law both were trained on.

Figure 14.2 presents the cumulated wealth delivered by the long-term and myopic managers on identical simulated price histories for the two managers.

15

Enterprise Valuation Using Infinite and Finite Horizon Valuation of Terminal Liquidation

The material of this chapter comes from Madan (2019). Apart from the valuation of derivatives, other contingent claims, hedged liabilities, or portfolios of securities, using the principles of financial valuation, there is the fundamental question of applying valuation principles to the valuation of economic enterprise itself. Financial markets are regularly engaged in valuing the outstanding equity of corporate entities and thereby valuing their activities. Economic enterprises are engaged in producing economic value. Inputs are combined in a production process to make goods and services desired and purchased by consumers. Examples of industries include agriculture, automobiles, electronics, construction, and transportation to mention a few. In a perfect world everything could be synchronized with no surprises and the economy would function smoothly exactly fulfilling all demands with no delays, disruptions, or competitive destructions. Valuation would be much simpler and reduced to issues connected with discounting and assessing the time value of money. There is some uncertainty in these matters, but not as much as that faced by equity markets.

Unfortunately such levels of perfection are not possible and all the various processes are subject to a varied collection of risks. Agriculture must face the vagaries of weather, automobiles and electronics have to deal with variations in aggregate demand, construction has to cope with the effects of changes in interest rates, and transportation is affected by the movements in the costs of energy. Furthermore, all industries have to cope with the creative destruction of technological innovation. Industries like banking, insurance, and the capital markets come into existence to organize a better distribution of these necessary risks.

The financial sector is not needed in the perfect risk-free world, and it is not present in the classical models of the economy (Arrow and Debreu, 1954; Debreu, 1959) where the concept of equilibrium may itself be seen as a denial of risk, especially if one takes as one of the risks the possibility of a lack of clearance. Incorporating a utilitarian view, realized economic risks are converted to cash equivalents (Rothschild and Stiglitz, 1990; Pratt, 1992). Once real shortfalls are converted to cash, economic participants are relieved of loss exposure by using capital to cover the consequences of shortfalls. Participants in the economy may then act as if they do not have shortfall risks. Markets clear for them even if they do not in fact clear, for the associated risks have been transferred to the capital markets. Capital providers are absorbing risks and covering shortfalls as and when they occur. Consequently capital providers earn random returns. Hopefully sufficient capital is being held to maintain operations. But what is this capital and where does it come from? From a financial perspective this is equity capital that can be raised in the capital markets by any healthy enterprise.

Equity capital is classically defined as the value of assets less liabilities and is the primary concern of capital requirements as set out in the recommendations of the Basel Committee on Banking Supervision (2010). Equity capital is also viewed as equal to the current market value of the future random returns it generates. This follows from stock prices being modeled as claims to future dividend distributions (Williams, 1938). Alternate approaches emphasize stock prices as claims to future free cash flows (Copeland et al., 1994) or future abnormal earnings (Preinreich, 1938; Edwards and Bell, 1961; Ohlson, 1995).

This equity capital is, however, fickle as the prices employed in valuing assets and liabilities are subject to constant motion, especially for the liquid items on both sides of the balance sheet. The unstable nature of these prices has been noted and emphasized, for example, by Keynes (1936, 1937), Minsky (1982, 1986), and Keen (1995). These prices must necessarily change to afford returns that support the prices themselves, as noted, for example, in Madan and Schoutens (2017, 2019c), Madan and Wang (2017), and Madan et al. (2017b). A conservative or prudential equity capital determination would recognize that assets may fall in value while liabilities may have to be unwound at higher prices. We develop and implement here a theory for a prudential or conservative capital valuation that follows the principles of conservative valuation.

Finally the question arises as to whether the prudential capital provided by the capital markets is enough to render the enterprise viable given the risk exposures that must be undertaken. This brings us to issues of required regulatory capital, which is a subject different from equity capital determination. If the equity capital is insufficient, then the probabilities of facing operational losses beyond the capital cover available are too high to permit the enterprise to continue unchanged. Changes in the operations of the enterprise are then called for. Capital rules are based on evaluating such risks (Acerbi and Tasche, 2002; Rockafellar and Uryasev, 2002; Szegő, 2002; Gordy, 2003; Jorion, 2006).

Bringing all the considerations together, one has to first understand the enterprise and its risks. On the basis of this understanding one has to determine what the prudential capital level is that will be afforded, recognized, and delivered by the capital markets. Finally one has to assess whether the prudential equity capital is sufficient for the operational risk exposures inherent in the enterprise.

These are all difficult questions for which we develop some stylized answers. Actual enterprises like banks, insurance companies, or multinational corporations are quite complex entities with a rate of return process on invested wealth that is not easy to formulate, represent, and analytically specify. There is also a considerable paucity of data on such returns as balance sheets are generally available quarterly, maybe monthly in some cases.

For illustrative and tractable purposes we consider the holding of liquid assets like stocks, commodities, currencies, or bonds over some time horizon as the enterprise. Data on daily returns for such enterprises are widely available. Hence an analytical understanding of the enterprise and its risks is possible. One may expand the enterprises to include the holding of swap positions or more generally long and short positions in risky underliers. The resulting mark-to-market values may be either positive or negative. The associated return process is defined in continuous time by the stochastic logarithm (Jacod and Shiryaev, 2013) when the process is a positive one. However, signed value processes can be accommodated by extending the stochastic logarithm and defining the signed stochastic logarithm.

The signed value process or its signed stochastic logarithm is treated as arising in an efficient market where all price changes are surprises occurring at surprise times. The price and value processes are thus pure jump processes with no continuous deterministic or martingale components. Furthermore, recognizing the relatively large number of price movements possible in each day, the return process will be modeled as belonging to the class of limit laws. We work here with the BG model. Hence every enterprise is represented by the four parameters of the associated BG return process seen as the signed stochastic logarithm of the price process.

With enterprises represented as the signed stochastic exponential of a BG process, we seek to define the level of prudential capital for a dollar invested in such a process. Clearly the risk-neutral value of an invested dollar will be a dollar, but the prudential or conservative value will be lower depending on the risk exposures involved. The perspective of prudential equity capital is to take the long view and to let the enterprise run for some long period and then value a distant liquidation outcome (Ohlson, 1995; Philips, 1999; Zhang, 2000). This is the approach adopted here.

However, to come back to today from a distant future date is difficult to do in the context of a single large time step. We have little experience with how risks are or are to be evaluated over such single large time steps. One may learn how they are evaluated, with considerable experience in option markets, for example, over small time steps. A natural solution is to solve nonlinear integro-differential equations for the prudential or conservative capital value. In a Markovian context the solution may be implemented on a space-time grid.

Finally, for the required regulatory capital computation, we follow the traditional practice and references already cited of evaluating the 10-day value at risk or the expected shortfall for a coherent alternative.

The enterprise is thus represented by a BG process for the signed stochastic logarithm as the return process. Prudential capital is given by a nonlinear conditional expectation of the accumulated return to a distant liquidation date. Regulatory risk requirements are represented by the value at risk or the expected shortfall over the short term associated with the enterprise return process. If the equity capital is sufficient to cover regulatory capital requirements, the enterprise is economically acceptable. An analysis of economic acceptability follows in the next chapter.

15.1 Bilateral Gamma Enterprise Returns

All enterprises, financial enterprises in particular, have marked-to-market value processes of positions long or short in all markets, including the cash account. Denote this continuous time process by $V = (V(t), t > 0)$. Given that the positions taken may be long or short, the process is in principle signed and may be positive or negative. In understanding an enterprise or representing it as a mathematical object the interest is in the return, per dollar invested, that the process affords or makes possible.

When V is a positive process, one associates it with the return process $v = (v(t), t > 0)$ defined as the stochastic logarithm (Jacod and Shiryaev, 2013) by the equations

$$dV = V\_dv,$$
$$v = \mathcal{L}(V),$$

where $\mathcal{L}$ is the stochastic logarithm operator.

Equivalently we write

$$V = V(0)\mathcal{E}(v),$$

where $\mathcal{E}$ is the stochastic exponential operator.

We take the view that price moves in efficient markets must be surprises and hence they must occur at surprise times. As a consequence price processes are pure jump processes with no continuous deterministic or martingale components. This feature is inherited by V with the change in $V(t)$ between two time points being the sum of its jumps. For such processes with no Dirac measure components to the jump arrival rates, the process may change sign but never takes the value zero. Sign changes require a large jump in excess of $|V\_|$, and they occur finitely often.

For the return calculation when V is negative, we wish to have an increase in v reflected as a gain associated with an increase in V. We thus define the signed stochastic logarithm v by

$$dV = |V\_|dv, \qquad v = \widetilde{\mathcal{L}}(V),$$

where $\widetilde{\mathcal{L}}$ is the signed stochastic logarithm. For the inverse we write

$$V = V(0)\widetilde{\mathcal{E}}(v)$$

for the signed stochastic exponential.

For a negative $V(0)$ and $V\_$ we have that

$$dV = |V\_|dv = -V\_(dv) = V\_(-dv).$$

It follows that

$$V = V(0)\mathcal{E}(-v), \qquad |V| = |V(0)|\mathcal{E}(-v).$$

So on the positive side we take the usual stochastic logarithm while on the negative side we take the negative of the stochastic logarithm of the absolute value.

Let $X = (X(t), t > 0)$ be a pure jump process whereby

$$X(t) = \sum_{s \leq t} \Delta X(s).$$

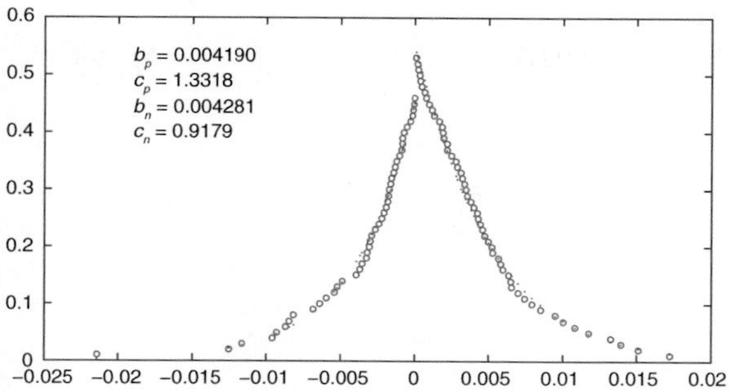

Figure 15.1 Observed tail probabilities are represented by circles. Model probabilities are given by dots.

Define the random measure $\mu_X(dx, dt)$ that is a Dirac measure if the process X jumps at time t by the magnitude $(x, x + dx)$. We may also write

$$X(t) = x * \mu_X = \int_0^t x\mu_X(dx, ds).$$

Define the process

$$v(t) = (e^x - 1) * \mu_X, \tag{15.1}$$

then

$$V(t) = V(0)\mathcal{E}(v) = V(0)\exp(X(t)).$$

We model the process $X(t)$, the data for which are given by log price relatives of the value or price process. In the signed case we take

$$x_t = \frac{\text{sign}(V(t-1)) + \text{sign}(V(t))}{2} \ln\left(\frac{|V(t)|}{|V(t-1)|}\right),$$

deleting all the zeros. We may further delete observations when $|V(t-1)| < \varepsilon$ to avoid a strip around zero that inflates returns.

The return process, like the price or value process, is pure jump. We take the return process to be the BG process or the difference of two independent gamma processes. Using 252 immediately preceding daily returns on 214 equity underliers for 2, 298 days, we obtained 491, 772 = 2, 298×214 sets of BG parameters representing equity return processes. Table 15.1 presents a table of 20 quantized representative points. The quantization was accomplished using vector quantization provided by Zavarehei as downloadable Matlab code in 2006.

Figure 15.1 presents a sample fit for the S&P 500 index.

For 35 hedge fund indices based on funds reporting daily returns of varying time lengths between 1, 142 and 3, 336 days, we estimated BG parameters on a year's worth of daily

Table 15.1 *Bilateral gamma parameters for equity underliers*

Case Number	b_p	c_p	b_n	c_n	Prop.
1	0.001586	103.9081	0.001674	103.1877	0.0120
2	0.003175	14:6479	0.003114	14.4926	0.0132
3	0.012837	1.4078	0.016275	1.0823	0.0643
4	0.005425	5.0741	0.005202	5.0482	0.0530
5	0.001750	66.3386	0.001790	62.8164	0.0133
6	0.011366	1.6209	0.011456	1.5476	0.0563
7	0.010716	1.7574	0.010801	1.6743	0.0570
8	0.013041	1.3736	0.012034	1.4099	0.1050
9	0.003954	9.7892	0.003849	9.6795	0.0232
10	0.010113	1.9138	0.009509	1.9222	0.1071
11	0.018172	1.0259	0.019889	0.9571	0.0743
12	0.009009	2.2237	0.008452	2.2348	0.0987
13	0.001667	86.9239	0.001661	84.8566	0.0068
14	0.007027	3.1510	0.006631	3.1517	0.0856
15	0.007998	2.6177	0.007489	2.6329	0.0928
16	0.001773	46.0522	0.001663	47.3736	0.0072
17	0.002392	24.8336	0.002321	24.8269	0.0111
18	0.006482	3.0159	0.011032	1.6304	0.0126
19	0.006182	3.9159	0.005879	3.8972	0.0691
20	0.004553	6.8867	0.004404	6.8365	0.0372

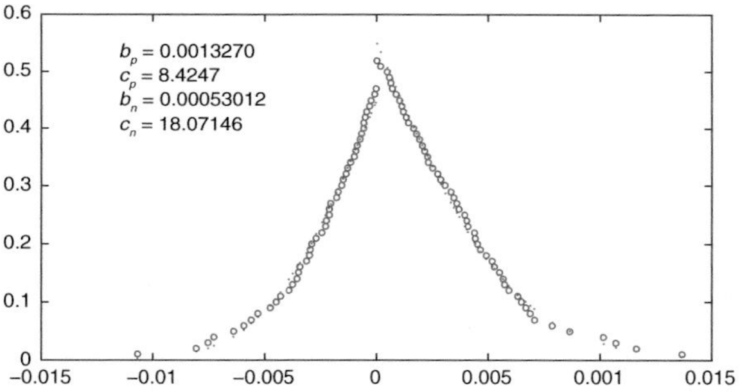

Figure 15.2 Observed tail probabilities are represented by circles. Model probabilities are given by dots.

returns. The data are from the HFRX series.[1] The total number of parameter sets was 97, 163. Table 15.2 presents a table of 20 quantized representative points. Figure 15.2 presents a sample fit of the BG model.

[1] HFR Indices source: Hedge Fund Research, Inc. `www.hedgefundresearch.com`. © 2017 Hedge Fund Research, Inc. All rights reserved.

Table 15.2 *Bilateral gamma parameter estimates for hedge funds*

Case Number	b_p	c_p	b_n	c_n	Prop.
1	0.001273	2.5475	0.000875	2.5331	0.0591
2	0.000424	22.1175	0.000397	21.4170	0.0128
3	0.003166	0.9168	0.001376	0.8643	0.0738
4	0.001725	1.7941	0.001140	1.7801	0.0634
5	0.000457	7.2298	0.001194	2.0124	0.0185
6	0.000920	2.3388	0.002095	0.6131	0.0523
7	0.000841	4.7944	0.000627	4.7859	0.0673
8	0.002042	1.2868	0.001090	1.2786	0.0695
9	0.001833	11.1076	0.000483	11.1303	0.0231
10	0.000686	3.4447	0.001497	0.9283	0.0527
11	0.001477	2.1262	0.000982	2.1176	0.0569
12	0.000711	6.7350	0.000557	6.7060	0.0450
13	0.001822	1.5236	0.001093	1.5142	0.0653
14	0.002263	1.0925	0.001062	1.0762	0.0712
15	0.003700	0.8426	0.002718	0.6089	0.0627
16	0.000270	60.9977	0.000255	61.0971	0.0209
17	0.001132	3.0381	0.000788	3.0290	0.0658
18	0.000998	3.7064	0.000727	3.6866	0.0691
19	0.000288	14.2818	0.000869	3.8143	0.0075
20	0.001317	1.7611	0.003205	0.4919	0.0432

Fixed income liabilities are often modeled as incurring an interest cost with the rate set one time step ahead and so is a known constant that varies as we move forward in time. However, if we treat the liability as a distant promise of a future dollar, its mark-to-market value is given by the debt price process $P = (P(t), t > 0)$. The associated return process may be obtained as the stochastic logarithm $Y = (Y(t), t > 0)$, and this process has a pure jump martingale component over any horizon. We take data on the yield curve and build the price process for maturities of 15, 20, 25, and 30 years and then construct Y as a BG process. The total number of BG parameter sets was 6, 768. Table 15.3 presents 20 quantized representative values. Figure 15.3 presents a sample fit.

15.2 Prudential Capital for Bilateral Gamma Returns

Let $X(t)$ be a pure jump return process. Starting with an initial investment of a dollar into the enterprise delivering the stochastic exponential V of v defined by (15.1) we accumulate at some distant date T the sum

$$V(T) = \exp(X(T)).$$

The risk-neutral value of this terminal payout or its equity value is unity as this is what was invested. Our interest, however, is in a prudential or conservative equity valuation.

Table 15.3 *Bilateral gamma parameters for bond price returns.*

Case Number	b_p	c_p	b_n	c_n	Prop.
1	0.003225	5.7703	0.003034	5.7455	0.0807
2	0.002123	10.9723	0.002018	10.9723	0.0421
3	0.000795	78.3815	0.000779	78.3815	0.0428
4	0.001331	23.9001	0.001281	23.9001	0.0225
5	0.004474	4.1343	0.004095	4.1093	0.0684
6	0.000699	100.8778	0.000690	100.8778	0.0228
7	0.001085	39.2250	0.001054	39.2250	0.0199
8	0.002601	8.0728	0.002450	8.0634	0.0554
9	0.001694	15.4534	0.001623	15.4534	0.0278
10	0.004518	2.2712	0.004170	2.2712	0.0709
11	0.003704	4.8822	0.003419	4.8688	0.0739
12	0.002269	9.4453	0.002143	9.4453	0.0455
13	0.001521	19.2267	0.001460	19.2267	0.0264
14	0.004817	3.3583	0.004349	3.3256	0.0652
15	0.000712	90.2116	0.000702	90.2116	0.0356
16	0.001222	30.3963	0.001181	30.3963	0.0211
17	0.000986	50.5162	0.000959	50.5162	0.0941
18	0.002806	6.8107	0.002637	6.7966	0.0721
19	0.002003	12.7585	0.001921	12.7585	0.0332
20	0.004514	1.3936	0.003944	1.3936	0.0795

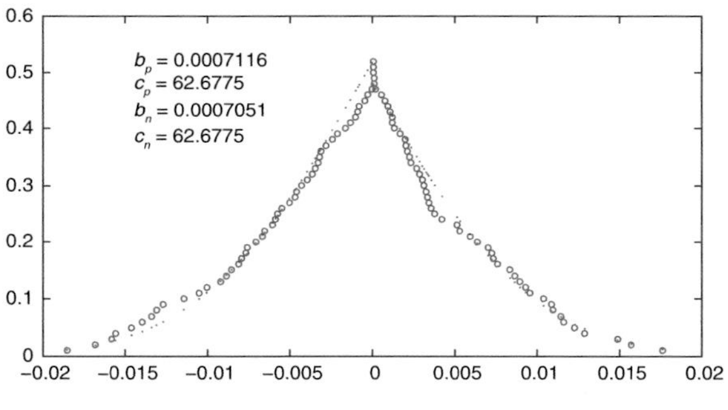

Figure 15.3 Observed tail probabilities are represented by circles. Model probabilities are given by dots.

A starting point for such considerations is to begin with a static one-period model. This is inadmissible from an equity standpoint as equity in capital markets will be taking a long view of evaluating the enterprise after it has been allowed to run for some time. However, our understanding of risk is built up over small time steps where we have access to some experience. We are unable to take the large time step directly.

In developing prudential capital evaluations for the large time step in continuous time, results are here briefly recalled first for a static one-period model. The one-period results are then applied recursively in a discrete time formulation. The relevant equations for the continuous time limit are presented next. Finally we conclude with solution details for the continuous time limit equations.

15.2.1 Lower and Upper Valuations in One Period

For the single time step we wish to establish a lower and upper prudential valuation for $V = \exp(X)$ where for notational convenience we drop the time dependence. For the lower and upper valuations we invoke the concept of acceptable risks from Chapter 6. Under the further assumptions of law invariance and comonotone additivity the two prices may be determined as distorted expectations. These are computed as expectations after the distortion of their distribution functions by a concave distribution function defined on the unit interval.

Specifically for a fixed concave distribution function $\Psi(u)$, with $0 \leq u \leq 1$, and any random outcome X with distribution function $F_X(x)$, the lower valuation $L(X)$ is given by

$$L(X) = \int_{-\infty}^{\infty} x d\Psi\left(F_X(x)\right). \tag{15.2}$$

The distorted expectation (15.2) may be related to Choquet expectations with respect to nonadditive probability on integration by parts to obtain

$$L(X) = -\int_{-\infty}^{0} \Psi\left(F_X(x)\right) dx + \int_{0}^{\infty} \left(1 - \Psi\left(F_X(x)\right)\right) dx. \tag{15.3}$$

The expression (15.3) may be written somewhat more symmetrically on defining the complementary convex distortion

$$\widehat{\Psi}(u) = 1 - \Psi(1-u)$$

and the complementary distribution function

$$\widehat{F}_X(x) = 1 - F_X(x)$$

as

$$L(X) = -\int_{-\infty}^{0} \Psi\left(F_X(x)\right) dx + \int_{0}^{\infty} \widehat{\Psi}\left(\widehat{F}_X(x)\right) dx. \tag{15.4}$$

Elementary computations between the supremum and infimum then lead to the result

$$U(X) = -\int_{-\infty}^{0} \widehat{\Psi}\left(F_X(x)\right) dx + \int_{0}^{\infty} \Psi\left(\widehat{F}_X(x)\right) dx. \tag{15.5}$$

We may now observe that as Ψ is above the identity and $\widehat{\Psi}$ is below the identity, that $L(X)$ and $U(X)$ must lie respectively below and above the expectation of X. The lower price lifts the tail probabilities of losses upward and reduces the tail probabilities of gains, thereby lowering the distorted expectation.

15.2.2 Lower and Upper Valuations in Discrete Time

For the development in discrete time we refer to Madan and Schoutens (2012) and Cohen and Elliott (2010). The essential idea is to come back in valuation from a distant liquidation date in a Markovian context by successively applying one step ahead distorted expectations at each point on a space-time grid. The results are related to the formulation of nonlinear conditional expectations as solutions of backward stochastic difference equations. In the interest of brevity attention will be focused on the lower prudential valuation with noted adjustments for the upper prudential valuation.

Consider the context of discrete time process with time step h. Let $X = (X(t), t \in \{0, h, 2h, \ldots\})$ be a finite state Markov chain with states e_i identified with the unit vectors of $\mathbb{R}^M$ for some large integer M. Let $\left(\Omega, \mathcal{F}, \{\mathcal{F}_t\}_{0\leq t\leq T}, P\right)$ be a filtered probability space, where $\mathcal{F}_t$ is the completion of the sigma algebra generated by the process X up to time t. In this context we follow Cohen and Elliott (2010) in defining dynamically consistent translation-invariant nonlinear expectation operators $\mathcal{E}(\cdot \mid \mathcal{F}_t)$. The operators are defined on the family of subsets $\left\{\mathbb{Q}_t \subset L^2(\mathcal{F}_T)\right\}$. For completeness we recall here this definition of an $\mathcal{F}_t$-consistent nonlinear expectation for $\{\mathbb{Q}_t\}$. This $\mathcal{F}_t$-consistent nonlinear expectation for $\{\mathbb{Q}_t\}$ is a system of operators

$$\mathcal{E}(\cdots \mid \mathcal{F}_t) : L^1(\mathcal{F}_T) \to L^1(\mathcal{F}_t),\ 0 \leq t \leq T$$

satisfying the following properties:

(1) (Monotonicity on $\mathbb{Q}_t$). For $Q, Q' \in \mathbb{Q}_t$, if $Q \geq Q'$ $\mathbb{P}$-a.s. componentwise, then

$$\mathcal{E}(Q \mid \mathcal{F}_t) \geq \mathcal{E}(Q' \mid \mathcal{F}_t)$$

$\mathbb{P}$-a.s. componentwise, with for each i,

$$e_i\mathcal{E}(Q \mid \mathcal{F}_t) = e_i\mathcal{E}(Q' \mid \mathcal{F}_t)$$

only if $e_iQ = e_iQ'$ $\mathbb{P}$-a.s.

(2) ($\mathcal{F}_t$-triviality). $\mathcal{E}(Q \mid \mathcal{F}_t) = Q$ $\mathbb{P}$-a.s. for any $\mathcal{F}_t$-measurable Q.

(3) (Recursivity). $\mathcal{E}(\mathcal{E}(Q \mid \mathcal{F}_t) \mid \mathcal{F}_s) = \mathcal{E}(Q \mid \mathcal{F}_s)$ $\mathbb{P}$-a.s. for any $s \leq t$.

(4) (Regularity). For any $A \in \mathcal{F}_t$, $\mathbf{1}_A\mathcal{E}(Q \mid \mathcal{F}_t) = \mathcal{E}(\mathbf{1}_AQ \mid \mathcal{F}_t)$ $\mathbb{P}$-a.s.

Furthermore the system of operators is dynamically translation invariant if for any $Q \in L^1(\mathcal{F}_T)$ and any $q \in L^2(\mathcal{F}_t)$,

$$\mathcal{E}(Q + q \mid \mathcal{F}_t) = \mathcal{E}(Q \mid \mathcal{F}_t) + q.$$

Such dynamically consistent translation-invariant nonlinear expectations may be constructed from solutions of backward stochastic difference and/or differential equations (El Karoui and Huang, 1997; Cohen and Elliott, 2010). These are equations to be solved simultaneously for processes Y, Z where Y_t is the nonlinear expectation and the pair (Y, Z) satisfy

$$Y_t - \sum_{t\leq u<T} F(\omega, u, Y_u, Z_u) + \sum_{t\leq u<T} Z_u M_{u+1} = Q$$

for a suitably chosen adapted map $F\colon \Omega \times \{0,\dots,T\} \times \mathbb{R}^K \times \mathbb{R}^{K\times N} \to \mathbb{R}$ called the driver and for Q an $\mathbb{R}$-valued $\mathcal{F}_T$-measurable terminal random variable. For all t, we have (Y_t, Z_t) are $\mathcal{F}_t$-measurable. Furthermore, for a translation-invariant nonlinear expectation, the driver F must be independent of Y and must satisfy the normalization condition $F(\omega, t, Y_t, 0) = 0$.

The drivers of the backward stochastic difference equations are the risk charges, and for our upper and lower price sequences at time step h we have drivers $F^{\mathrm{L}}, F^{\mathrm{U}}$ where

$$F^{\mathrm{L}}(\omega, u, Y_u, Z_u) = h \sup_{Q\in\mathcal{M}} E^Q\left[Z_u M_{u+1}\right],$$
$$F^{\mathrm{U}}(\omega, u, Y_u, Z_u) = h \inf_{Q\in\mathcal{M}} E^Q\left[Z_u M_{u+1}\right],$$

and the drivers are independent of Y. The process Z_t represents the residual risk in terms of a set of spanning martingale differences M_{u+1} and in our applications we solve for the nonlinear expectations Y_t without in general identifying either Z_t or the set of spanning martingale differences. We define risk charges directly for the risk defined, for example, as the zero mean random variable

$$Y_{t+h} - E_t^P\left[Y_{t+h}\right].$$

Dynamically consistent forward lower and upper price sequences for the time step h may be constructed as nonlinear expectations starting with

$$L(X(T)) = U(X(T)) = \exp(X(T)).$$

Thereafter we apply the recursions

$$U_{t-h}(X(t-h)) = E^P\left[U_t(X(t))\right] + h \sup_{Q\in\mathcal{M}} \left(E^Q\left[U_t(X(t)) - E^P\left[U_t(X(t))\right]\right]\right), \quad (15.6)$$

$$L_{t-h}(X(t-h)) = E^P\left[L_t(X(t))\right] + h \inf_{Q\in\mathcal{M}} \left(E^Q\left[L_t(X(t)) - E^P\left[L_t(S(t))\right]\right]\right). \quad (15.7)$$

Basing the drivers on one-step-ahead distorted expectations, we may write

$$F^{\mathrm{L}}(Z_u M_{u+1}) = -\int_0^\infty \Psi^{(\gamma)}\left(\Theta^L(x)\right)dx + \int_0^\infty \widehat{\Psi}^{(\gamma)}\left(\widehat{\Theta}^L(x)\right)dx, \quad (15.8)$$

$$F^{\mathrm{U}}(Z_u M_{u+1}) = -\int_0^\infty \widehat{\Psi}^{(\gamma)}\left(\Theta^U(x)\right)dx + \int_0^\infty \Psi^{(\gamma)}\left(\widehat{\Theta}^U(x)\right)dx, \quad (15.9)$$

$$\Theta^{\mathrm{L}}(x) = P\left(\left(L_t(X(t)) - E^P\left[L_t(S(t))\right]\right)^- \ge x\right),$$
$$\widehat{\Theta}^{\mathrm{L}}(x) = P\left(\left(L_t(X(t)) - E^P\left[L_t(S(t))\right]\right)^+ \ge x\right),$$
$$\Theta^{\mathrm{U}}(x) = P\left(\left(U_t(X(t)) - E^P\left[U_t(S(t))\right]\right)^- \ge x\right),$$
$$\widehat{\Theta}^{\mathrm{U}}(x) = P\left(\left(U_t(X(t)) - E^P\left[U_t(S(t))\right]\right)^+ \ge x\right).$$

15.2.3 Lower and Upper Valuations in Continuous Time

The distorted expectations are applied in discrete time to probability distortions of centered risks, but the centering expectations are then added back in the recursions for the lower and upper valuations one time step later, as given by equations (15.6) and (15.7). The continuous-time results follow on letting the time step go to zero (Madan et al., 2017a). As the time step tends to zero, probability elements converge to time step-scaled jump arrival rates. Madan et al. (2017a) study a careful dependence of the distortions employed on the time step to ensure the convergence of the distorted expectations to measure distorted expectations.

In the continuous time limit one introduces two measure distortions G^+ and G^- defined on the positive half line with G^+ increasing, concave, and bounded below by the identity while G^- is increasing, convex, and bounded above by the identity. In addition the following integrability conditions are assumed to hold:

$$\int_0^\infty \left(G^+(y) - y\right) \frac{dy}{y\sqrt{y}} < \infty, \qquad \int_0^\infty \left(y - G^-(y)\right) \frac{dy}{y\sqrt{y}} < \infty.$$

For the jump arrival rate function $k(x)$ for jump sizes x define the jump measure K by

$$K(A) = \int_A k(x)dx.$$

The lower prudential valuation function satisfies

$$L(X(T), T) = \exp\left(X(T)\right).$$

In the continuous time limit the lower prudential valuation function solves the nonlinear partial integro-differential equation

$$\begin{aligned} L_t &- \int_0^\infty G^+\left(K\left([L(X_t + x, t) - L(X_t, t)]^- > a\right) da\right. \\ &+ \int_0^\infty G^-\left(K\left([L\left(X_t + x, t\right) - L(X_t, t)]^+ > a\right)\right) da \\ &= 0. \end{aligned}$$

The corresponding equations for the upper prudential valuation are

$$U(X(T), T) = \exp\left(X(T)\right).$$

Furthermore

$$\begin{aligned} U_t &- \int_0^\infty G^-\left(K\left([U(X_t + x, t) - U(X_t, t)]^- > a\right) da\right. \\ &+ \int_0^\infty G^+\left(K\left([U\left(X_t + x, t\right) - U(X_t, t)]^+ > a\right)\right) da \\ &= 0. \end{aligned}$$

On making the time transformation $u = e^{-t}$, whereby time infinity gets mapped to zero and time zero is mapped to unity and setting,

$$\widetilde{L}(X,u) = L\,(X, -\ln u), \qquad \widetilde{U}(X,u) = U(X, -\ln u),$$

we get the boundary condition

$$\widetilde{L}(X,0) = \exp(X), \tag{15.10}$$
$$\widetilde{U}(X,0) = \exp(X), \tag{15.11}$$

and

$$\begin{aligned}\widetilde{L}_u =& \frac{1}{u}\Big[-\int_0^\infty G^+\left(K\left([\widetilde{L}(X_u+x,u)-\widetilde{L}(X_u,u))\right]^- > a\right)da \\ &+\int_0^\infty G^-\left(K\left(\left[\widetilde{L}\left(X_u+x,u\right)-\widetilde{L}(X_u,u)\right)\right]^+ > a\right)\right)da\Big]\end{aligned} \tag{15.12}$$

$$\begin{aligned}\widetilde{U}_u =& \frac{1}{u}\Big[-\int_0^\infty G^-\left(K\left([\widetilde{U}(X_u+x,u)-\widetilde{U}(X_u,u)\right)]^- > a\right)da \\ &+\int_0^\infty G^+\left(K\left(\left[\widetilde{U}\left(X_u+x,u\right)-\widetilde{U}(X_u,u)\right)\right]^+ > a\right)\right)da\Big].\end{aligned} \tag{15.13}$$

These equations are solved for

$$\widetilde{L}(X,1), \widetilde{W}(X,1), \widetilde{U}(X,1).$$

The computation of the expected value $\widetilde{W}(X,1)$ follows on eliminating the measure distortions. To determine the prudential value shave and prudential upper value add-on for an initial value of unity, we define X_0 by

$$\widetilde{W}(X_0,1) = 1,$$

and set the prudential shave equal to

$$1 - \widetilde{L}(X_0,1),$$

and the prudential add-on equal to

$$\widetilde{U}(X_0,1) - 1.$$

15.2.4 The Specific Measure Distortions

Here we follow Madan et al. (2017a) and Eberlein et al. (2014a) and model the measure distortions via the risk charge gaps defined by

$$\Gamma^+(x) = G^+(x) - x, \qquad \Gamma^-(x) = x - G^-(x).$$

They are both increasing, concave, and positive functions. For large levels of x associated with large tail measures or events near zero, we enforce no reweighting and therefore require that $G^{+\prime}$ falls to unity and $G^{-\prime}$ rises to unity. Such considerations suggest that the eventual gaps are constants that on rescaling may be taken to be unity.

Consider a candidate increasing concave function $K(x)$ that is zero at zero and unity at infinity and define for y, with $0 \leq y \leq 1$,

$$\Psi(y) = K\left(-\frac{\ln(1-y)}{c}\right).$$

Eberlein et al. (2014a) observe that the concavity of Ψ is ensured by

$$-\frac{K''}{K'} \geq c,$$

or a minimal concavity for K of c. As a result we have that

$$K(x) = \Psi\left(1 - e^{-cx}\right)$$

for some concave probability distortion Ψ. The measure distortions then take the form

$$G^{+}(x) = x + \alpha\Psi_{+}(1 - e^{-cx}),$$
$$G^{-}(x) = x - \frac{\beta}{c}\Psi_{-}(1 - e^{-cx})$$

for a choice of probability distortions Ψ_{+}, Ψ_{-}.

Since

$$\Gamma^{-\prime}(0) = 1 - \beta\Psi_{-}'(0) \geq 0,$$

$\Psi_{-}'(0)$ is bounded above. There is no such restriction for $\Psi_{+}'(0)$. We consider discounting gains at minimal concavity c and employ the maxvar distortion with stress γ for Ψ_{+} to formulate measure distortions

$$\Gamma^{+}(x) = x + \alpha\left(1 - e^{-cx}\right)^{\frac{1}{1+\gamma}}$$
$$\Gamma^{-}(x) = x - \frac{\beta}{c}\left(1 - e^{-cx}\right)$$

with parameters α, β, c, γ. Enforcing limiting gaps on the two sides to be identical α is set equal to β/c. It remains to select values for β, c, γ.

15.2.5 Details for the Lower and Upper Prudential Values As Solutions of Nonlinear Partial Integro-Differential Equations

The solutions of equations (15.12) and (15.13) are implemented as follows. The boundary condition is taken at $u = 0$. The transformed time steps are in u starting at $u = 0.01$ in steps of 0.01 and finishing at $u = 1$. The grid for the stock price ranges from 0.33 in steps of 1,069 points with a spacing of 0.0025 and finishing at 3.0.

The up and down standard deviations are

$$\sigma_p = b_p\sqrt{c_p}, \qquad \sigma_n = b_n\sqrt{c_n}.$$

The negative jump sizes go from $-20\sigma_n$ to $-0.001\sigma_n$ in steps of $0.01\sigma_n$ while the positive ones start at $0.001\sigma_p$ and go to $20\sigma_p$ in steps of $0.01\sigma_p$. The total number of possible jump sizes is $4,000$ split equally for positive and negative jumps. At each jump point the BG Lévy measure is evaluated as $k(x)dx$ for the jump size x and the result is stored in a vector k_i for jump y_i for $i = 1, \ldots, 4000$.

We solve for three value functions on the space-time grid stored in three matrices of size 1069×100. The three value functions represent the lower prudential value, the expectation, and the upper prudential value. The boundary condition sets the initial condition at the stock price for all three valuations. For the solutions we work forward in time u and set the change in value at each grid point equal to $h/u\Delta V$. Here h is the time step in u and ΔV is to be computed as a measure-distorted integral. There is no distortion when working with the expectation. Further details pertain to the computation of ΔV. Attention may be focused on the lower prudential valuation, the computations for the others being similar.

The first step at any grid point is to evaluate the $4,000$ post-jump stock prices. The value function at these prices is extracted by interpolation of the prior computed value function. Subtracting the value at the grid point under consideration gives the vector for the change in values. The changes in value possibilities are then sorted in increasing order, identifying the indices associated with the sort. Evaluating the Lévy measure vector at these indices and summing from the left for negative points and summing to the right for positive points delivers the tail measures that are to be distorted. Passing these tail measures through the distortion functions delivers measure-distorted tail measures. Summing these from the left for negative points and to the right for the positive ones delivers the required value.

The value functions can get noisy after a time step due to the interpolations involved in ascertaining the change in value, and over time steps the noise can accumulate. We employ GPR at each time step to smooth out the solution before proceeding to the next time step.

The level of the spot price S^* when at unit time the expectation is unity is then identified. The prudential shaves and add-ons are computed relative to unity for the ratio of the lower and upper values relative to the expectation at the point S^*.

15.3 Regulatory Risk Capital for Enterprises with Bilateral Gamma Returns

Equity capital in financial markets must take a long view in judging the value of the enterprise, prudentially or otherwise. The regulatory view, on the other hand, is explicitly concerned with the ability of capital in place to withstand operational losses in the short term. Hence the focus on value at risk and expected shortfall is over shorter periods like 10 to 20 days. The short period risk may be evaluated in a single time step and the expected shortfall is in fact a static distorted expectation. The distortion being used is quite an extremal one.

A special distortion links the expected shortfall to a distorted expectation. The expected shortfall distortion is

$$\Psi_{ES}(u; q) = \frac{u \wedge q}{q}. \tag{15.14}$$

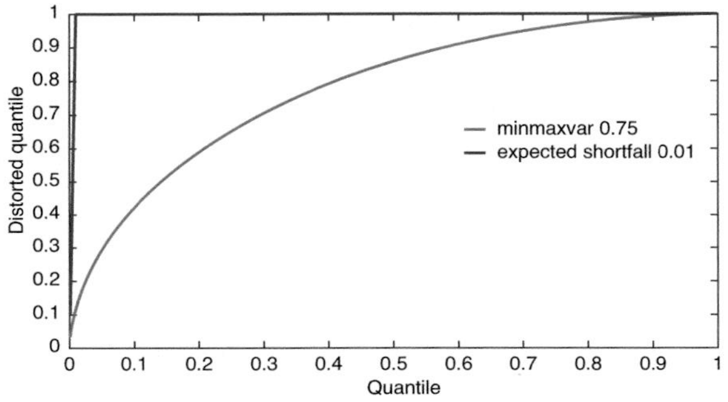

Figure 15.4 Shortfall versus minmaxvar distortions

The distorted expectation via equation (15.14) is then

$$L_t = \frac{1}{q} \int_{-\infty}^{F_{W_T}^{-1}(q)} w dF_{W_T}(w),$$

or the expected shortfall at the quantile $q < 1$. For $T = t + 10$ days the distorted expectation under the distortion (15.14) is precisely the expected shortfall. Figure 15.4 shows the minmaxvar distortion at stress level 0.75 contrasted with the expected shortfall distortion at loss probability level 99%. The concavity embedded in the expected shortfall distortion can then be seen to be quite extreme. However, relative to the long view for equity capital developed more fully in §15.2.3, the risk horizon is very limited by focusing attention on just the 10-day interval.

For BG return processes one may simulate 10-day returns from these distributions using their representations as the difference of gamma-distributed random variables. The final outcomes are then the realized values for $\exp(X(10))$. One may then compute the associated values for the value at risk (VAR) and the expected shortfall (CVAR). Table 15.4 presents VAR and CVAR levels associated with BG parameters reported in the rows of Table 15.1.

15.4 Calibration of Measure-Distortion Parameters

For measure distortions we may estimate at any maturity from data on option prices the risk-neutral tails $L^-(x)$ for $x < 0$ and $L^+(x)$ for $x > 0$. They are just integrals of Lévy densities to the left and right respectively. Given physical tails $K^-(x)$, for $x < 0$ and $K^+(x)$ for $x > 0$ we seek measure-distortion parameters accomplishing the dominations

$$\Gamma^-\left(K^-(x)\right) \leq L^-(x) \leq \Gamma^+\left(K^-(x)\right), \text{ for } x < 0,$$
$$\Gamma^-\left(K^+(x)\right) \leq L^+(x) \leq \Gamma^+\left(K^+(x)\right), \text{ for } x > 0.$$

The parameter c of Γ^- is sought to accomplish the two left inequalities while given c the parameter γ is sought to accomplish the two right inequalities. The parameter b is set to match the distorted physical derivatives to the risk-neutral derivatives for low values of tail integrals.

Table 15.4 *VAR and CVAR at 99.9% for a sample of bilateral gamma parameters*

Number	VAR	CVAR
1	0.2241	0.2433
2	0.1060	0.1221
3	0.1582	0.1819
4	0.1582	0.1226
5	0.1041	0.1234
6	0.1343	0.1541
7	0.1321	0.1531
8	0.1368	0.1570
9	0.1068	0.1237
10	0.1218	0.1426
11	0.1897	0.2172
12	0.1173	0.1357
13	0.1149	0.1346
14	0.1086	0.1258
15	0.1120	0.1304
16	0.0890	0.1050
17	0.0991	0.1149
18	0.1184	0.1398
19	0.1056	0.1212
20	0.1037	0.1197

Consider points $x_d^* < 0 < x_u^*$ such that

$$K^-(x_d^*) = \varepsilon = K^+(x_u^*).$$

We seek to equate

$$L^+(x_u^*) = \Gamma^-(K^+(x_u^*)) = K^+(x_u^*) - bK^+(x_u^*),$$
$$L^-(x_d^*) = \Gamma^-(K^-(x_d^*)) = K^-(x_d^*) - bK^-(x_d^*),$$

and hence

$$b = 1 - \frac{L^+(x_u^*)}{K^+(x_u^*)}, \qquad b = 1 - \frac{L^-(x_d^*)}{K^-(x_d^*)}.$$

The parameter b is selected to be positive and the larger of these two values. The value of ε employed was 0.001. The parameters c, γ are minimal levels accomplishing the required inequalities.

15.4.1 Estimation of Physical and Risk-Neutral Tail Measures

For the risk-neutral tail measures we estimate the Sato process based on the VG to options prices across strikes and maturities on a particular day. The Sato process named VGSSD was introduced in Carr et al. (2007). The risk-neutral distribution at each maturity is from the class of VG distributions, and one then identifies from the parameters the associated Lévy measure for any specific maturity. Tail integrals of the Lévy density yield the functions for the required tail probabilities.

For the physical Lévy measure associated with a particular maturity, we first estimate the VG model on daily returns using 252 immediately prior daily returns. For a maturity of H days we follow Eberlein and Madan (2010) and Madan and Schoutens (2020) to combine scaling and the running of an i.i.d. process to build the longer horizon returns. Just scaling leaves higher moments like skewness and kurtosis unchanged while they do fall in data. Running an i.i.d. process, however, forces skewness and excess kurtosis to fall too fast like the reciprocal of $\sqrt{H}$ and H respectively. Combining scaling with the running of an i.i.d. process produces intermediate and more reasonable effects on higher moments. We follow the recommendation of Eberlein and Madan (2010) and split the daily VG law X into

$$X \stackrel{(d)}{=} (cX) + X^{(c)},$$

where $X^{(c)}$ is independent of cX, with a density and simulation procedures identifiable from the VG characteristic function of X. For the longer horizon return X_H we run cX as a Lévy process to time H and scale $X^{(c)}$ by H^{η}. Hence we take

$$X_H \stackrel{(d)}{=} (cX)_H + H^{\eta}X^{(c)}.$$

The simulated data on X_H for $c = 0.5$ and $\eta = 0.5$ are then fit by a VG law to yield the Lévy density for the horizon H. Its tail integrals deliver the functions $K^{-}(x)$, $x < 0$ and $K^{+}(x)$, with $x > 0$. The distortion parameters can then be solved for each day. The underlier for these exercises was the S&P 500 index.

15.4.2 A Sample Calibration

First we fit the VG Sato process, VGSSD, to SPX option data on December 2, 2016, to get the risk-neutral parameters:

$$\sigma = 0.1898, \quad \nu = 0.9786, \quad \theta = -0.2039, \quad \gamma = 0.5296.$$

The fit of the Sato process to market option prices is presented in Figure 15.5.

We also fit the VG model by digital moment estimation to time series data for 252 days ending December 2, 2016, to get the statistical parameters

$$\sigma = 0.0106, \qquad \nu = 0.9029, \qquad \theta = 0.0001159.$$

Figure 15.6 presents the observed and fitted tail probabilities for the fit on December 2, 2016.

We then combined scaling and shaving to generate VG parameters for one month as

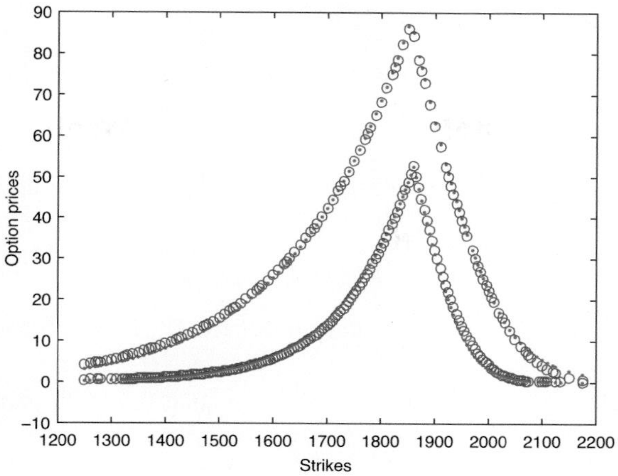

Figure 15.5 Fit of VGSSD to SPX option prices on December 2, 2016

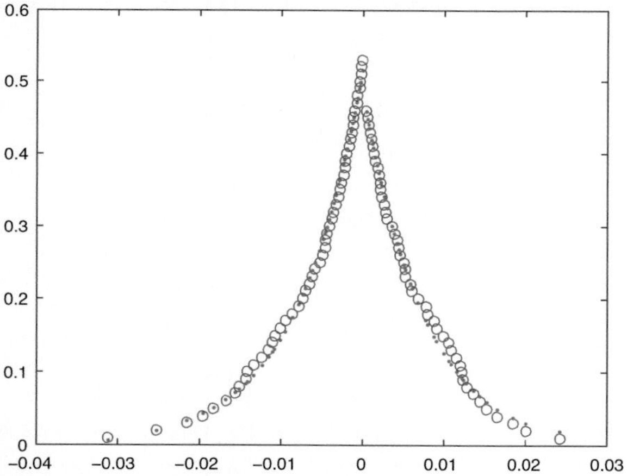

Figure 15.6 Variance gamma fit to tail probabilities. The circles are the observed tail probabilities while fitted probabilities are represented by dots.

	Quarter	Half Year
σ	0.0686	0.0970
ν	0.6119	0.6728
θ	−0.0073	−0.0071.

The risk-neutral VG parameters for the quarter and half year are as follows:

	One Month VG
σ	0.0463
ν	0.6580
θ	0.00031.

The measure distortions to be calibrated are:

$$G^+(x) = x + a(1 - e^{-cx})^{\frac{1}{1+\gamma}},$$
$$G^-(x) = x - \frac{b}{c}(1 - e^{-cx}).$$

For the maturity of a month we evaluate tail arrival rates for jump sizes in log space between ± 0.5. This gives us curves

$$K^+(x), K^-(x), L^+(x), L^-(x),$$

where x is the jump size between -0.5 and -0.001 for K^-, L^- and between 0.001 and 0.5 for K^+, L^+.

Next we solve for x_u^*, x_d^* such that $K^+(x_u^*), K^-(x_d^*)$ equals 0.04. The parameter b is defined by

$$b = \max\left(1 - \frac{L^+(x_u^*)}{K^+(x_u^*)}, 1 - \frac{L^-(x_d^*)}{K^-(x_d^*)}\right).$$

The value of b was calibrated at 0.8596. We then find c to enforce the inequalities

$$L^+(x) \geq \Gamma^-(K^+(x)),$$
$$L^-(x) \geq \Gamma^-(K^-(x)).$$

We take a balanced measure distortion and set

$$a = \frac{b}{c}.$$

We then search over γ to enforce the inequalities

$$L^-(x) \leq \Gamma^+(K^-(x)),$$
$$L^+(x) \leq \Gamma^+(K^+(x)).$$

Typically we have the left physical tail below the risk neutral that has to be lifted by calibrating γ. On the right we have the physical tail dominating the risk neutral, and it has to be lowered by estimating c. The distortion for this date was as follows:

a	0.6671
b	0.8596
c	1.2887
γ	1.7536.

Figures 15.7 and 15.8 present the risk neutral, physical, and distorted physical tails on the left and the right respectively.

Given estimates of the law of motion for the return process with BG parameters estimated by digital moment matching of tail probabilities and parameters for the measure distortion as calibrated to options on the SPX and the time series data on the index, we have eight inputs $b_p, c_p, b_n, c_n, a, b, c, \gamma$ that may be employed to solve the nonlinear partial integro-differential equations (15.12,15.13) subject to the boundary conditions (15.10,15.11) to

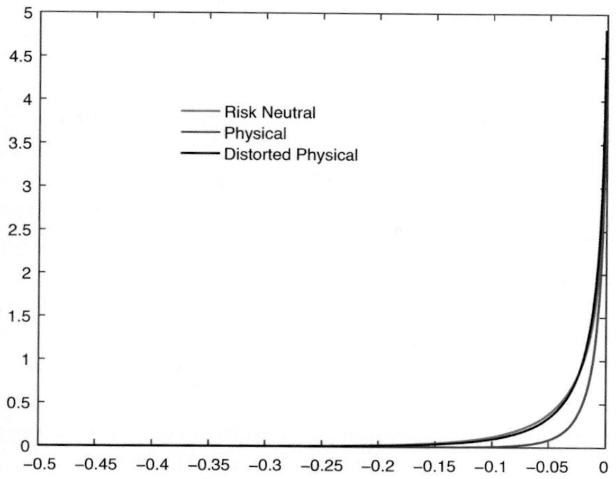

Figure 15.7 Risk-neutral, physical, and distorted physical left tails

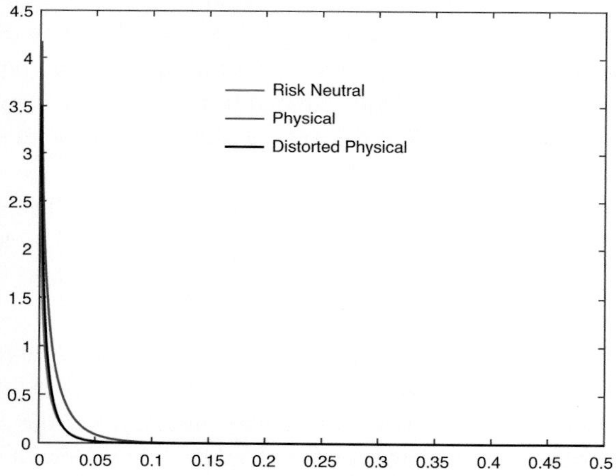

Figure 15.8 Risk-neutral, physical, and distorted physical right tails

determine the shave, and add-on consistent with notions of dynamic risk acceptability embedded in the nonlinear partial integro-differential equations.

15.4.3 Representative Measure Distortions Calibrated from S&P 500 Option Markets

The measure-distortion parameters were calibrated daily from January 4, 2010, to October 13, 2016, using options on the S&P 500 index and the time series data for the index for 252 days prior to the calibration date. The measure-distortion parameters were then quantized to five representative values. Table 15.5 presents these representative values and the proportion of data points they represent.

Table 15.5 *Quantized measure-distortion parameter values from daily calibrations to S&P 500 index options 2010 through 2016*

Case number	a	b	c	γ	Prop.
1	1.5408	0.9876	0.6410	0.1487	0.2302
2	3.3966	0.9637	0.2837	0.5145	0.1535
3	0.257	0.9788	3.8079	1.2324	0.0666
4	2.3594	0.9549	0.4047	1.3801	0.1591
5	3.2634	0.9851	0.3019	0.2087	0.3905

15.5 Results for Equity Enterprises

Representative parameter values for equity underliers presented in Table 15.1 and representative measure distortion parameters presented in Table 15.5 may be combined to form 100 cases for which the detailed algorithm presented in §15.4.2 may be applied to evaluate both the add-on and shave associated with a prudential or conservative valuation of the enterprise as an asset held or a liability promised. However, the long view for equity capital is actually appropriate for an enterprise held as an asset and the shave of a dollar delivers the level of prudential capital. For completeness, both the liability add-ons, Table 15.6, and asset shaves, Table 15.7, are presented for these 100 cases in two tables of 20 rows and 5 columns. The column numbers refer to the rows of Table 15.5 while the row numbers are those for Table 15.1.

The value that could be lost is unity less the VAR or CVAR. The prudential capital provided by capital markets under various measure distortions is unity less the prudential shave. Table 15.8 presents these values for all five measure distortions and two levels for VAR and CVAR quantiles of 99% and 99.9%.

We observe that the prudential capital is sufficient for regulatory requirements at even the 99.9% level. The calculations are, however, on very different bases. The prudential approach takes the long view to assess how much conservatively the enterprise may be worth. On the other hand, the regulatory approach takes a short view of often just 10 days and asks how much can be lost at a low probability and then applies the typical multiple of three. As was observed, the two approaches may be related to distorted expectations, but the distortions are very different. The CVAR distortion is very extreme compared to what would arise as relevant for the prudential capital computation. However, the former is applied in one step to the 10-day return while the latter accumulates risk charges by integration over long periods on solving a differential equation in time. The rank correlations between the five prudential capital levels and the 99.9% CVAR-based regulatory capital requirements are respectively 55.19, 43.91, 52.18, 37.59, and 53.68.

15.6 Results for Treasury Bond Investments

Apart from equity investments one may also invest in safer securities like Treasury bonds. Investments in long-term Treasury bonds are not exactly risk free as they are exposed to

Table 15.6 *Add-ons for bilateral gamma enterprises using different measure distortions*

Row	Column Number				
Number	1	2	3	4	5
1	0.0760	0.1487	0.0203	0.1254	0.1377
2	0.0831	0.1525	0.0226	0.1466	0.1370
3	0.1118	0.1849	0.0622	0.2545	0.1479
4	0.0920	0.1599	0.0350	0.1745	0.1386
5	0.0724	0.1406	0.0194	0.1215	0.1295
6	0.1056	0.1756	0.0546	0.2347	0.1415
7	0.1045	0.1740	0.0527	0.2287	0.1411
8	0.1072	0.1797	0.0584	0.2499	0.1424
9	0.0889	0.1603	0.0300	0.1607	0.1412
10	0.1022	0.1713	0.0501	0.2217	0.1396
11	0.1298	0.2164	0.0781	0.3213	0.1673
12	0.0992	0.1668	0.0466	0.2096	0.1376
13	0.0731	0.1433	0.0191	0.1222	0.1322
14	0.0936	0.1592	0.0398	0.1875	0.1345
15	0.0964	0.1628	0.0432	0.1983	0.1359
16	0.0645	0.1238	0.0179	0.1094	0.1134
17	0.0744	0.1393	0.0223	0.1281	0.1260
18	0.0935	0.1500	0.0423	0.1775	0.1277
19	0.0922	0.1583	0.0373	0.1798	0.1354
20	0.0886	0.1565	0.0304	0.1636	0.1373

movements in interest rates and thereby the prices of the underlying asset values. Using daily data on the yield curve from January 4, 2010, to June 15, 2017, time series of prices for Treasury bonds maturing in 15, 20, 25, and 30 years from January 4, 2010, were constructed. The BG model was estimated on the stochastic logarithm of these price series. The four maturities on each of $1,692$ days gave a total number of $6,768$ BG parameter sets. The parameter sets were quantized into 20 points and Figure 15.3 presents this sample of BG enterprise parameters for Treasury investment.

The prudential capital and regulatory capital levels for such enterprises are presented in Table 15.9.

The relative safety may be observed on noting that the market available prudential capital is consistently substantially well above the required regulatory capital.

15.7 Results for Hedge Fund Enterprises

For the 35 HFRX hedge fund indices with BG parameters summarized in Table 15.2, prudential and regulatory capital were evaluated and are reported in Table 15.10. We observe sufficient levels of prudential capital and low levels of regulatory capital. Parameter set 9 benefits from a strong drift and shows negative levels of regulatory capital.

Table 15.7 *Bilateral gamma shaves using different measure distortions*

Row Number	Column Number 1	2	3	4	5
1	0.0675	0.1286	0.0179	0.1108	0.1206
2	0.0804	0.1323	0.0254	0.1242	0.1207
3	0.1030	0.1622	0.0606	0.2162	0.1314
4	0.0842	0.1369	0.0334	0.1457	0.1216
5	0.0677	0.1239	0.0190	0.1088	0.1155
6	0.0955	0.1509	0.0509	0.1867	0.1240
7	0.0947	0.1490	0.0493	0.1853	0.1238
8	0.0963	0.0620	0.0530	0.1890	0.1234
9	0.0819	0.1379	0.0288	0.1371	0.1248
10	0.0924	0.1436	0.0470	0.1742	0.1218
11	0.1156	0.1798	0.0718	0.2441	0.1310
12	0.0899	0.1406	0.0435	0.1667	0.1203
13	0.0683	0.1308	0.0187	0.1078	0.1172
14	0.0855	0.1356	0.0376	0.1539	0.1181
15	0.0877	0.1378	0.0404	0.1594	0.1190
16	0.0606	0.1098	0.0174	0.1164	0.1018
17	0.0686	0.1214	0.0207	0.1120	0.1121
18	0.0879	0.1428	0.0469	0.1787	0.1184
19	0.0844	0.1353	0.0353	0.1495	0.1190
20	0.0801	0.1353	0.0304	0.1387	0.1208

Table 15.8 *Prudential and regulatory capital for bilateral gamma enterprises*

Row	Column Number					Multiple of Three			
Number	1	2	3	4	5	VAR 0.99	CVAR 0.99	VAR 0.999	CVAR 0.999
1	0.9325	0.9323	0.9181	0.9317	0.9314	0.6722	0.7299	0.7997	0.8482
2	0.8714	0.8761	0.8621	0.8692	0.8786	0.3180	0.3664	0.4243	0.4585
3	0.9821	0.9810	0.9712	0.9813	0.9793	0.4746	0.5456	0.6380	0.6929
4	0.8892	0.8912	0.8629	0.8922	0.8880	0.3179	0.3677	0.4289	0.4708
5	0.8794	0.8845	0.8752	0.8828	0.8879	0.3124	0.3702	0.4471	0.4905
6	0.9196	0.9045	0.9076	0.9145	0.9121	0.4028	0.4622	0.5323	0.5923
7	0.8677	0.8491	0.8564	0.8644	0.8572	0.3964	0.4594	0.5421	0.5899
8	0.9746	0.9491	0.9530	0.9624	0.9531	0.4105	0.4709	0.5456	0.5922
9	0.8758	0.8133	0.8258	0.8461	0.8213	0.3204	0.3712	0.4310	0.4696
10	0.8793	0.8760	0.8782	0.8819	0.8816	0.3653	0.4279	0.5107	0.5550
11	0.8970	0.9053	0.8844	0.9123	0.9156	0.5692	0.6516	0.7585	0.8240
12	0.8378	0.8510	0.8202	0.8622	0.8647	0.3518	0.4070	0.4739	0.5191
13	0.9394	0.9507	0.9282	0.9596	0.9647	0.3447	0.4038	0.4800	0.5209
14	0.7838	0.8147	0.7559	0.8406	0.8505	0.3258	0.3773	0.4384	0.4843
15	0.8686	0.8762	0.8690	0.8810	0.8810	0.3360	0.3911	0.4606	0.5041
16	0.9158	0.9037	0.9101	0.9394	0.9199	0.2669	0.3149	0.3757	0.4119
17	0.8631	0.9380	0.8594	0.8902	0.8647	0.2974	0.3447	0.4061	0.4488
18	0.9666	0.9470	0.9565	0.9826	0.9696	0.3553	0.4194	0.5078	0.5573
19	0.8543	0.8110	0.8333	0.8836	0.8613	0.3169	0.3636	0.4225	0.4558
20	0.8784	0.8766	0.8797	0.8982	0.8792	0.3111	0.3591	0.4210	0.4630

Table 15.9 *Prudential capital and regulatory capital for treasury bond investments*

Row	Column Number					Multiple of Three			
Number	1	2	3	4	5	VAR 0.99	CVAR 0.99	VAR 0.999	CVAR 0.999
1	0.9466	0.9378	0.9581	0.9592	0.9645	0.1924	0.2240	0.2595	0.2880
2	0.9117	0.8992	0.9264	0.9280	0.9347	0.1723	0.2032	0.2409	0.2685
3	0.9796	0.9745	0.9864	0.9872	0.9899	0.1739	0.2029	0.2379	0.2602
4	0.9067	0.8894	0.9298	0.9324	0.9519	0.1582	0.1854	0.2172	0.2424
5	0.9216	0.9115	0.9340	0.9341	0.9427	0.2134	0.2511	0.2972	0.3293
6	0.9537	0.9698	0.9551	0.9401	0.9499	0.1877	0.2178	0.2532	0.2791
7	0.9194	0.9432	0.9268	0.9043	0.9164	0.1672	0.1953	0.2298	0.2531
8	0.9842	0.9919	0.9781	0.9743	0.9815	0.1806	0.2109	0.2494	0.2721
9	0.9220	0.9513	0.9098	0.8927	0.9139	0.1656	0.1933	0.2289	0.2523
10	0.9256	0.9471	0.9363	0.9160	0.9252	0.1761	0.2044	0.2381	0.2626
11	0.9678	0.9636	0.9438	0.9701	0.9536	0.1965	0.2295	0.2679	0.2981
12	0.9374	0.9336	0.9073	0.9438	0.9198	0.1704	0.1997	0.2345	0.2614
13	0.9912	0.9894	0.9778	0.9919	0.9845	0.1640	0.1917	0.2264	0.2475
14	0.9480	0.9405	0.9018	0.9517	0.9222	0.2009	0.2375	0.2844	0.3145
15	0.9442	0.9387	0.9186	0.9477	0.9271	0.1768	0.2050	0.2394	0.2636
16	0.9616	0.9501	0.9534	0.9620	0.9665	0.1637	0.1917	0.2264	0.2472
17	0.9315	0.9158	0.9208	0.9315	0.9476	0.1685	0.1984	0.2375	0.2623
18	0.9882	0.9821	0.9838	0.9887	0.9818	0.1783	0.2098	0.2481	0.2716
19	0.9352	0.9149	0.9211	0.9377	0.9355	0.1797	0.2087	0.2447	0.2696
20	0.9370	0.9206	0.9285	0.9369	0.9567	0.1294	0.1540	0.1838	0.2063

Table 15.10 *Prudential and regulatory capital capital requirements for hedge fund enterprises as reflected in HFRX indices*

Row	Column Number					Multiple of Three			
Number	1	2	3	4	5	VAR 0.99	CVAR 0.99	VAR 0.999	CVAR 0.999
2	0.9808	0.9814	0.9583	0.9816	0.9809	0.0328	0.0412	0.0518	0.0587
3	0.9945	0.9950	0.9919	0.9939	0.9948	0.0116	0.0201	0.0304	0.0371
4	0.9780	0.9771	0.9626	0.9775	0.9786	0.0256	0.0337	0.0428	0.0503
5	0.9833	0.9783	0.9604	0.9843	0.9832	0.0220	0.0302	0.0401	0.0469
6	0.9870	0.9900	0.9906	0.9880	0.9879	0.0263	0.0358	0.0476	0.0566
7	0.9787	0.9830	0.9841	0.9840	0.9806	0.0189	0.0260	0.0347	0.0411
8	0.9973	0.9936	0.9947	0.9939	0.9970	0.0158	0.0229	0.0320	0.0376
9	0.9803	0.9752	0.9782	0.9799	0.9789	−0.3350	−0.3146	−0.2896	−0.2734
10	0.9805	0.9862	0.9873	0.9863	0.9829	0.0161	0.0240	0.0340	0.0403
11	0.9869	0.9883	0.9873	0.9837	0.9899	0.0226	0.0309	0.0414	0.0489
12	0.9813	0.9809	0.9804	0.9753	0.9820	0.0194	0.0267	0.0364	0.0416
13	0.9923	0.9955	0.9942	0.9893	0.9959	0.0200	0.0280	0.0383	0.0458
14	0.9759	0.9799	0.9770	0.9646	0.9795	0.0108	0.0176	0.0260	0.0312
15	0.9841	0.9830	0.9831	0.9801	0.9848	0.0391	0.0519	0.0690	0.0802
16	0.9867	0.9871	0.9888	0.9898	0.9870	0.0374	0.0463	0.0575	0.0661
17	0.9796	0.9827	0.9809	0.9810	0.9785	0.0195	0.0267	0.0361	0.0419
18	0.9935	0.9938	0.9965	0.9972	0.9912	0.0202	0.0273	0.0361	0.0424
19	0.9754	0.9780	0.9805	0.9834	0.9662	0.0234	0.0311	0.0412	0.0469
20	0.9825	0.9852	0.9829	0.9824	0.9839	0.0483	0.0620	0.0799	0.0926

Table 15.11 *Prudential and regulatory capital for short positions in equity underliers*

Row Number	Column Number 1	2	3	4	5	Multiple of Three VAR .99	CVAR .99	VAR .999	CVAR .999
1	1.0760	1.0724	1.0889	1.0731	1.0744	0.2929	0.3769	0.4850	0.5530
2	1.1487	1.1406	1.1603	1.1433	1.1393	0.4463	0.5095	0.5949	0.6464
3	1.0203	1.0194	1.0300	1.0191	1.0223	0.5621	0.6616	0.7871	0.8852
4	1.1254	1.1215	1.1607	1.1222	1.1281	0.4471	0.5153	0.5977	0.6508
5	1.1377	1.1295	1.1412	1.1322	1.1260	0.6078	0.6887	0.7915	0.8630
6	1.0831	1.1056	1.1022	1.0936	1.0935	0.5140	0.6087	0.7363	0.8116
7	1.1525	1.1756	1.1713	1.1592	1.1500	0.5049	0.5873	0.6947	0.7649
8	1.0226	1.0546	1.0501	1.0398	1.0423	0.5516	0.6478	0.7738	0.8691
9	1.1466	1.2347	1.2217	1.1875	1.1775	0.4568	0.5232	0.6087	0.6651
10	1.1370	1.1415	1.1396	1.1345	1.1277	0.5069	0.5905	0.6875	0.7838
11	1.1118	1.1045	1.1298	1.0964	1.0922	0.6567	0.7792	0.9389	1.0646
12	1.1849	1.1740	1.2164	1.1628	1.1583	0.4847	0.5612	0.6606	0.7246
13	1.0622	1.0527	1.0781	1.0432	1.0373	0.6641	0.7497	0.8599	0.9303
14	1.2545	1.2287	1.3213	1.1983	1.1798	0.4533	0.5268	0.6216	0.6849
15	1.1479	1.1411	1.1673	1.1359	1.1354	0.4738	0.5489	0.6450	0.7091
16	1.0920	1.1072	1.0992	1.0645	1.0886	0.4928	0.5571	0.6393	0.6918
17	1.1599	1.1797	1.1668	1.1238	1.1565	0.4519	0.5167	0.5958	0.6500
18	1.0350	1.0584	1.0466	1.0179	1.0304	0.4661	0.5323	0.6222	0.6800
19	1.1745	1.2499	1.2096	1.1094	1.1636	0.4461	0.5140	0.6009	0.6637
20	1.1386	1.1424	1.1376	1.1134	1.1373	0.4373	0.5031	0.5819	0.6419

15.8 Short Position Capital Requirements

One may undertake the enterprise to go short an existing enterprise. The enterprise accumulates to a terminal date T the sum of total returns of $\exp(X(T))$, and hence the short position is up to pay out this sum. The negative of the lower prudential value for the short position is the upper valuation that is unity plus the add-on evaluated by the solution of the nonlinear partial integro-differential equation.

From a regulatory perspective, one may determine the upper quantile for a short exposure to $\exp(X(t))$ for $t = 10$ days less the unit dollar received on going short for the amount that may be lost.

Table 15.11 presents the prudential capital values of short positions using the BG parameter values of Table 15.1 and the measure distortions of Table 15.5 along with the 10-day VAR and CVAR at the 99% and 99.9% levels. The corresponding values for the Treasury bond market using the BG parameter values of Table 15.3 are presented in Table 15.12.

Unlike the case for long positions, short positions in Treasury bonds require greater capital given the risk exposures being undertaken. The asymmetry carries over to equity underliers.

15.9 Equity versus Leveraged Equity

A position in equity may be financed by issuing or selling bonds. The initial value of the financed position with a fraction δ being financed is $V_0 = (1 - \delta)S_0$ for the stock price of S_0.

Table 15.12 *Prudential and regulatory capital for short positions in Treasury bond markets*

Row Number	Column Number 1	2	3	4	5	Multiple of Three VAR 0.99	CVAR 0.99	VAR 0.999	CVAR 0.999
2	1.0980	1.1139	1.0798	1.0779	1.0699	0.2602	0.2936	0.3371	0.3695
3	1.0212	1.0268	1.0140	1.0131	1.0102	0.2686	0.3027	0.3447	0.3751
4	1.1048	1.1284	1.0765	1.0733	1.0616	0.2436	0.2747	0.3138	0.3372
5	1.0856	1.0978	1.0720	1.0706	1.0643	0.3423	0.3905	0.4524	0.4897
6	1.0486	1.0311	1.0488	1.0640	1.0528	0.2545	0.2884	0.3292	0.3562
7	1.0869	1.0602	1.0809	1.1085	1.0922	0.2567	0.2910	0.3289	0.3581
8	1.0162	1.0081	1.0226	1.0254	1.0192	0.2757	0.3123	0.3574	0.3866
9	1.0861	1.0513	1.1002	1.1265	1.0966	0.2477	0.2807	0.3195	0.3456
10	1.0776	1.0558	1.0672	1.0921	1.0811	0.2431	0.2790	0.3233	0.3534
11	1.0332	1.0378	1.0598	1.0308	1.0487	0.3045	0.3464	0.4019	0.4378
12	1.0639	1.0712	1.1029	1.0595	1.0877	0.2603	0.2974	0.3445	0.3784
13	1.0087	1.0107	1.0232	1.0080	1.0159	0.2512	0.2840	0.3238	0.3549
14	1.0550	1.0636	1.1129	1.0509	1.0855	0.3350	0.3829	0.4461	0.4854
15	1.0591	1.0653	1.0891	1.0551	1.0787	0.2469	0.2823	0.3259	0.3552
16	1.0399	1.0526	1.0489	1.0395	1.0351	0.2570	0.2893	0.3307	0.3577
17	1.0738	1.0928	1.0868	1.0737	1.0578	0.2679	0.3027	0.3472	0.3741
18	1.0120	1.0186	1.0167	1.0115	1.0193	0.2703	0.3074	0.3529	0.3889
19	1.0682	1.0951	1.0874	1.0670	1.0788	0.2617	0.3004	0.3504	0.3848
20	1.0671	1.0821	1.0772	1.0673	1.0462	0.1929	0.2213	0.2579	0.2838

Table 15.13 *Leveraged and unleveraged bilateral gamma parameters for investment in S&P 500 index. The investment is financed by issuing 30-year Treasury bonds and a leverage of 50%.*

	b_p	c_p	b_n	c_n
Unleveraged	0.003074	3.0574	0.002530	3.05740
Leveraged	0.007541	2.4161	0.006628	2.41611

The number of bonds sold for an initial bond price of P_0 is $\delta S_0/P_0$. The subsequent value process reflects the movements in the bond price as a short position to generate the value process

$$V_t = S_t - \frac{\delta S_0}{P_0} P_t.$$

For a 50% leveraged position in the S&P 500 index starting on May 10, 2013, with $S_0 = 1,633.70$, one may obtain the BG parameters for investing in the index on a full equity and leveraged basis. The unleveraged and leveraged parameters were estimated as shown in Table 15.13.

Using the measure-distortion parameters of the first row of Table 15.5, the prudential capital for an unleveraged position is 0.9647, and this drops to 0.9231 for a unit dollar leveraged position. The reduction in value of 4.312% reflects the additional risk of a leveraged position. The one-year or 252-day period starting on May 10, 2013, was a particularly good period for the S&P 500 index, and regulatory capital for a 99.9% level based on the CVAR was just 0.1673 unleveraged that rises to 0.4054 on a 50% leveraged basis.

16

Economic Acceptability

Risk acceptability has been modeled as a convex cone of random variables that contains the set of nonnegative random variables. The nonnegative random variables are certainly acceptable to all as they have no loss exposure. The next question then is, "How is this cone defined in modern capitalist economies?" There is also a distinction between the acceptability of random outcomes and the acceptability of exposures. For outcomes realized in the future, theoretically one may consider adding cash to an unacceptable position, until with sufficient cash, it becomes acceptable. When it comes to the analysis of exposures, however, the addition of cash positions leaves exposures unchanged as cash has zero exposure and the other exposures are still there and unchanged. To reduce risk exposures, one must alter the risk positions themselves to manipulate the resulting exposure to an acceptable level.

Acceptable exposures are then exposures that market participants want, and hence the participants are willing to pay to access such exposures. They are then exposures with a positive market value. If the value is ascertained using a single valuation measure, then the set of acceptable risks is, as has been argued, a very large convex cone that would be an extraordinarily generous definition of acceptable risks. For a proper convex cone it must be the case that the infimum of a number of valuations is positive.

In modern capitalist economies it is not governments or regulatory bodies that define acceptable risks or viable enterprises. This is left to markets and more particularly to markets for equity capital that leave it to individual participants to decide what they want to pay to access a possible risk exposure. Regulators may make their calculations and ask for sufficient equity capital, and it is then the interaction between regulatory bodies and capital markets that ends up defining the economically acceptable enterprises. Here we report on a study of such interactions.

Economic enterprises are then seen as investing dollars with a view to accumulating returns. Equity markets, as argued in the previous chapter, take a long view of letting the enterprises flourish to some distant date at which accumulated returns are liquidated. Taking $X(t)$ to be the process for the logarithm of the invested value at date t, the liquidation value of an initial single dollar at the distant horizon H is $\exp(X(H))$. The current risk-neutral value would be unity. However, a conservative valuation of equity in a two-price economy is given by a nonlinear valuation operator at

$$L = \mathcal{E}\,(\exp(X(H))) = \inf_{Q \in \mathcal{M}^H} E^Q[\exp(X(H))], \tag{16.1}$$

for some set of test probabilities $\mathcal{M}^H$ supporting the prudential lower valuation L. By making the time change $u = e^{-t}$ one may set the boundary condition for liquidation at infinity or $u = 0$ with the current time set to unity or $u = 1$, as was done in Chapter 15.

In addressing the interplay between equity markets and regulatory concerns we wish to adapt the measure distortions employed in developing prudential equity valuations to market information. For regulatory concerns we work with multiples of 10- or 20-day VAR and CVAR. For both calculations one needs access to the law of exposure to changes in valuations. For this we consider the estimation of BG models on time series data using 252 and 1,000 days of immediately prior returns. Additionally we implement the technology of depreferencing option prices to extract option-implied physical distributions. We report on the relative distances of these distributions from the risk-neutral and time series-based distributions.

The strategy for adapting measure distortions to market data is then developed. This strategy is separately executed for the time series-based and option-implied physical distributions. Prudential equity valuations computed by dynamic measure-distorted valuations of time-reversed liquidation at infinity are then compared with regulatory capital requirements for SPY and some stocks from the financial sector.

16.1 Interplay between Equity Markets and Regulators

Regulatory capital requirements are set in terms of the ability of the enterprise to employ equity capital to absorb extreme losses that may occur by or at a nearby horizon h. If a coherent risk measure $\mathcal{R}$ is employed for this purpose with a regulatory capital at a multiple Λ of the risk measure, then the regulatory capital is

$$R = \Lambda\mathcal{R}(\exp(X(h))) = \Lambda \sup_{Q \in \mathcal{M}^h} E^Q\,[-\exp(X(h))]\,. \tag{16.2}$$

But as regulators seek sufficient equity capital to meet the regulatory capital requirements, the economic acceptability of an enterprise requires that L defined in equation (16.1) exceeds R defined in (16.2). The enterprise is then economically acceptable if

$$E^Q[\exp(X(H))] \geq \Lambda E^{Q'}\,[-\exp(X(h))],\ Q \in \mathcal{M}^H, Q' \in \mathcal{M}^h.$$

Denote by $\mathcal{A}$ the set of economically acceptable enterprises.

We may now write the economic acceptability condition as all enterprises for which

$$E^Q[\exp(X(H))] + \Lambda E^{Q'}\,[\exp(X(h))] \geq 0, Q \in \mathcal{M}^H, Q' \in \mathcal{M}^h. \tag{16.3}$$

Define by $\mathcal{N}^H$ the set of measures Q'' equivalent to P on $\mathcal{F}_H$ such that

$$E_h^{Q''}[\exp(X(H))] = \exp(X(h)),$$
$$Q''(A) = Q'(A) \text{ for all } A \in \mathcal{F}_h,\ Q' \in \mathcal{M}^h.$$

Economic acceptability then requires that

$$E^{\widetilde{Q}}[\exp(X(H))] \geq 0,$$
$$\widetilde{Q} = \frac{Q + \Lambda Q''}{1 + \Lambda},\ Q \in \mathcal{M}^H, Q'' \in \mathcal{N}^H, \tag{16.4}$$

with regulatory concerns receiving the weighting $\Lambda/(1 + \Lambda)$.

16.2 Candidate Physical Laws of Motion

The candidate physical laws were estimated in the BG class of return distributions each day using 1,000 days and 252 days of immediately prior returns computed as log price relatives between prices at the end of the day and the previous day. The estimations were conducted daily from January 2, 2013, to December 31, 2018, for a total of 1, 508 days. This gives us two sets of statistical parameters based on the time series of prices.

Additionally, from data on option prices at market close for maturities between five days and two months, we fit a GPR model to implied volatilities as a function of moneyness relative to the forward and maturity. The output of this regression was used to construct 41 option prices for the maturity of five days at moneyness levels ranging from negative to positive 10% in steps of 50 basis points. The BG risk-neutral model was fit to this data to extract a BG risk-neutral five-day density. We also fit a BG candidate for the option-implied physical density that was perturbed by a five-parameter preference model representing gamma-distributed risk aversions for participants on the long and short side of the market along with a parameter for the proportion short.

The risk-neutral BG parameters may be denoted by xrn_t. The option-implied physical parameters we denote by xoip_t. The two time series-based parameters may be denoted by xts_1 and xts_2. We thus have four sets of BG parameters that have each 1, 508 observations based on the daily estimations. To evaluate the status of the option-implied physical densities relative to the others, we compute six distances between rn to oip, ts1, and ts2, between oip and ts1, ts2 and between ts1 and ts2. Six distances are computed for each of 1, 508 days.

The distance measures used are Hellinger, Kolmogorov–Smirnov, Kullback–Leibler, and Wasserstein. Tables 16.1 to 16.4 present quantiles of the six distances across the 1, 508 cases for each of the four types of distances. It is clear that the option-implied physical density is closer to the time series physical densities than it is to the risk-neutral densities. Hence we take it as a forward-looking candidate for the physical density.

Though it is reasonable to consider the quantiles of distances constructed as a one-dimensional summary of the differences between the parameters, one cannot consider quantiles of parameter sets as they are tied together as four-dimensional vectors and representative values should be formulated in four dimensions. We may quantize them into, say,

Table 16.1 *Hellinger distance*

quantile	rn2oip	rn2ts1	rn2ts2	oip2ts1	oiptts2	ts12ts2
1	0.2173	0.9051	0.8561	0.0321	0.0446	0.0149
5	0.9197	0.9090	0.8735	0.0458	0.0676	0.0176
10	0.9408	0.9109	0.8867	0.0560	0.0777	0.0219
25	0.9553	0.9290	0.9177	0.0756	0.1007	0.0275
50	0.9674	0.9432	0.9400	0.1150	0.1334	0.0342
75	0.9777	0.9447	0.9468	0.1568	0.1693	0.0436
90	0.9852	0.9461	0.9522	0.1876	0.2115	0.0699
95	0.9897	0.9471	0.9541	0.2099	0.2378	0.0850
99	0.9981	0.9485	0.9555	0.6944	0.6802	0.0924

Table 16.2 *Kolmogorov–Smirnov distance*

quantile	rn2oip	rn2ts1	rn2ts2	oip2ts1	oiptts2	ts12ts2
1	0.2100	0.8839	0.8262	0.0099	0.0163	0.0058
5	0.9008	0.8883	0.8430	0.0174	0.0251	0.0070
10	0.9244	0.8901	0.8552	0.0214	0.0304	0.0094
25	0.9421	0.9085	0.8952	0.0288	0.0380	0.0117
50	0.9574	0.9255	0.9206	0.0408	0.0476	0.0147
75	0.9696	0.9278	0.9271	0.0540	0.0632	0.0241
90	0.9780	0.9298	0.9327	0.0749	0.0961	0.0401
95	0.9824	0.9306	0.9339	0.0889	0.1367	0.0502
99	0.9900	0.9331	0.9352	0.6754	0.6535	0.0630

eight groups and present the group centers and associated probabilities to get an appreciation of the differences in the respective magnitudes. We present in Tables 16.5 to 16.8 the levels of quantized parameters along with their proportions.

The risk-neutral speeds are different by a whole order of magnitude relative to the candidate physical ones that are comparable. The option-implied scales of down moves are higher than their time series counterparts.

16.3 Adapted Measure Distortions

The measure distortions we apply define

$$G^{+}(x) = x + a(1 - e^{-cx})^{\frac{1}{1+\gamma}},$$

$$G^{-}(x) = x - \frac{b}{c}(1 - e^{-cx}),$$

that have maximum risk charge gaps over and below the identity of $a, b/c$ for large levels of x or high arrival rates. Now $G^{-\prime}(0) = 1 - b \geq 0$, so the largest possible value for b is

Table 16.3 *Kullback–Leibler distance*

quantile	rn2oip	rn2ts1	rn2ts2	oip2ts1	oiptts2	ts12ts2
1	0.2171	2.2127	1.7692	0.0407	0.0778	0.0036
5	2.3819	2.2309	1.8939	0.0744	0.1619	0.0076
10	2.6795	2.2445	1.9914	0.1019	0.2305	0.0104
25	2.9851	2.4542	2.3088	0.1775	0.4578	0.0217
50	3.3560	2.6774	2.6068	0.5378	0.9018	0.0476
75	3.8016	2.7067	2.7069	1.3536	1.6640	0.1137
90	4.3384	2.7397	2.8145	2.2183	2.8278	0.2076
95	4.9057	2.7524	2.8577	2.8612	3.5974	0.2361
99	8.5747	2.7977	2.8965	5.2556	6.6989	0.3021

Table 16.4 *Wasserstein distance*

quantile	rn2oip	rn2ts1	rn2ts2	oip2ts1	oiptts2	ts12ts2
1	1.9116	45.8105	22.4524	12.2531	20.7824	3.4091
5	55.6524	46.5846	24.7839	26.2911	32.7910	6.2485
10	68.7651	47.0907	28.9269	32.6138	39.2234	7.0367
25	88.2772	51.2524	42.6208	45.0322	52.8340	10.9189
50	126.0672	59.5465	53.3478	93.7459	99.4287	16.7956
75	190.4511	64.8830	57.4908	173.1455	172.1560	19.6178
90	259.2582	73.5456	69.3665	239.8380	243.4602	21.8204
95	297.6999	76.0817	73.1779	289.5835	291.1358	22.9155
99	461.2873	82.5671	75.2014	440.5612	445.4952	26.8692

unity and we set $b = 1$. If we keep a symmetry for large risk charges on both sides, then we set $a = b/c$.

The maximum risk charge for high arrival rate events is $1/c$ with $b = 1$. For this to be, say, 2% on a $100 value one would need a value of $c = 0.5$. With this choice for c, it remains γ from market data.

The principle we have earlier invoked in the market calibration of distortion parameters was that of inducing a sufficient distortion such that risk-neutral measures are among the collection of distorted physical measures. These considerations lead us to compare physical and risk-neutral tails for the more common levels of up and down moves that we take to be plus or minus 2–5%. The tail measures can be explicitly computed from the BG physical and risk-neutral parameters.

However, these laws are generally not equivalent. For two Lévy measures K', K the equivalence of K' and K requires that $(1 - \sqrt{k})^2$ be integrable with respect to K (Jacod and Shiryaev, 2013, Chapter IV, Theorem 4.39 part [v]). This condition requires that risk neutral and physical speeds of upward and downward motion be identical. Estimates presented

Table 16.5 *Bilateral gamma risk-neutral parameter sets*

b_p	c_p	b_n	c_n	prop.
0.0082	35.2701	0.0117	87.9965	0.1658
0.0072	52.4408	0.0121	114.0794	0.1910
0.0061	730.9932	0.0056	985.4023	0.0232
0.0099	19.4380	0.0130	55.5541	0.2513
0.0063	134.0784	0.0151	171.5926	0.0723
0.0092	28.6801	0.0128	73.3989	0.1432
0.0072	525.2770	0.0078	630.2960	0.0279
0.0197	8.5789	0.0159	32.6441	0.1253

Table 16.6 *Bilateral gamma option-implied physical*

b_p	c_p	b_n	c_n	prop.
0.0051	2.5782	0.0124	1.5876	0.1916
0.0053	1.5352	0.0103	1.9179	0.1320
0.0054	2.5743	0.0103	3.1839	0.0676
0.0061	1.7398	0.0151	0.9356	0.2102
0.0059	1.1699	0.0112	1.4658	0.1088
0.0058	5.6233	0.0167	1.9098	0.0942
0.0435	0.8172	0.0107	2.8590	0.0696
0.0284	0.5595	0.0292	0.8531	0.1260

above show them to be different by an order of magnitude. The lack of integrability is associated with the behavior near zero that is essentially not observable. We ignore these events and force the risk-neutral and physical arrival rates of jumps of size 50 basis points to be the same on renormalizing by the total arrival rates of jumps greater than 50 basis points on each side. We then compute the maximum ratio of renormalized risk-neutral to physical arrival rates between 2 and 5% and determine γ such that G^+ of the physical arrival rate dominates the corresponding risk-neutral arrival rate. In cases where the risk-neutral arrival rates are below the corresponding physical ones, we let γ take a minimal value of 0.25. These calculations generate a sequence of daily estimates for γ to be employed for the measure-distorted valuations. Kernel estimation smoothing of values strictly above 0.25 delivers a sequence of γ values that are above the lower bound for each day.

Figure 16.1 presents the time series of calibrated gamma values across the 1,508 days for SPY. We observe that the option-implied tail arrival rates do not have to be lifted as much as the time series-based arrival rates. Also the ones based on 1,000 days require a smaller lift than the ones based on the relatively small sample of just 252 days.

16.4 Equity and Regulatory Capital Constructions

Given estimates of the statistical law of motion each day and the market-calibrated measure distortion for the day, one may solve on a space-time grid the nonlinear time-reversed

Table 16.7 *Bilateral gamma 1,000-day time series parameter sets*

b_p	c_p	b_n	c_n	prop.
0.0043	1.7905	0.0058	1.1021	0.0515
0.0059	1.2502	0.0085	0.7098	0.0465
0.0049	1.5480	0.0074	0.8515	0.1535
0.0075	1.0600	0.0117	0.5948	0.1782
0.0044	1.6879	0.0064	0.9674	0.1559
0.0078	1.1361	0.0124	0.6428	0.1965
0.0056	1.3877	0.0085	0.7730	0.1594
0.0092	1.0104	0.0148	0.5701	0.0584

Table 16.8 *Bilateral gamma 252-day time series*

b_p	c_p	b_n	c_n	prop.
0.0034	2.1224	0.0053	1.1297	0.1455
0.0011	12.9670	0.0026	4.5813	0.0589
0.0047	1.4494	0.0056	1.0291	0.1728
0.0062	1.2305	0.0087	0.7355	0.2426
0.0062	1.2599	0.0071	0.9456	0.1351
0.0020	4.6910	0.0041	1.9155	0.0450
0.0005	56.1942	0.0019	10.3951	0.0401
0.0060	1.0532	0.0089	0.5301	0.1599

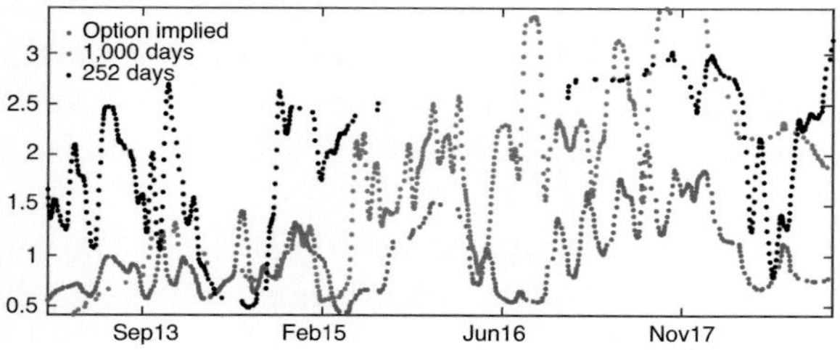

Figure 16.1 Market-calibrated measure-distortion stress parameter γ

prudential or conservative equity valuation equation for liquidation at infinity as a function of the current unobserved state on the space grid. One may also solve the linear equity valuation with no distortions on the same space-time grid. We then take the current value to be unity and determine the space point such that the equity value is unity and take for the conservative value the conservative solution at this space point. By construction this conservative equity value of a dollar will be below a dollar. This is the conservative equity

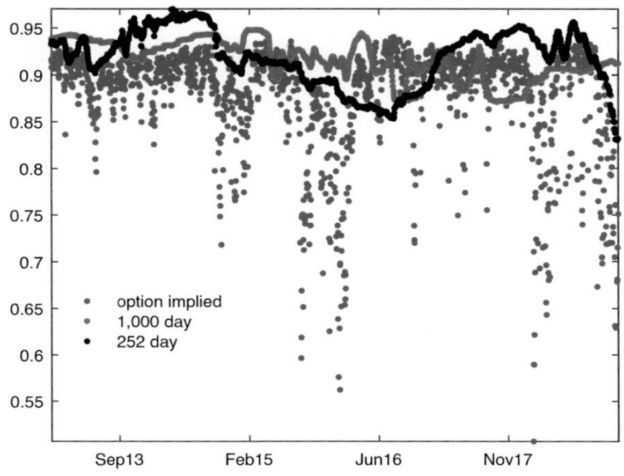

Figure 16.2 Market-calibrated measure-distortion stress parameter γ

value EC_t of day t. There are three such values depending on which of the three statistical models with their associated γ values one is using for the equity calculation.

Figure 16.2 presents the three sets of conservative equity values computed on the basis of the three different candidates for the law of motion. We observe that the 252-day estimates may not be that conservative on occasion as those based on 1,000 days. Both these estimates are not as conservative as those based on option-implied laws of motion.

For the regulatory capital calculation we simulate $100,000$ paths of the appropriate BG process for 20 days and evaluate the VAR and the CVAR at the 99.9% loss point. The regulatory capital is set at a multiple of three times the VAR and CVAR. Figures 16.3 and 16.4 present the results for calculations based on the two time series estimations of the physical law of motion, respectively. We see that on a multiple of three the equity capital is well above the required regulatory capital. A multiple of nine would bring them on to a comparable basis.

The option implied calculations are presented in Figure 16.5. The multiple was reduced to two in this case and the two capital calculations are at similar levels displaying period of sufficient and insufficient capital.

16.5 Financial Sector Capital during and after the Financial Crisis

With respect to possibly observing periods of both sufficient and insufficient equity capital relative to regulatory requirements we consider calculations for stocks in the financial sector beginning in January 2007 and through to the end of December 2018. The stocks selected for analysis were the larger banks with a sufficiently rich set of strikes and maturities trading in their option markets. Six US assets, along with one each from the UK and the Netherlands, will be reported on. For the US assets we report on the full period. For the

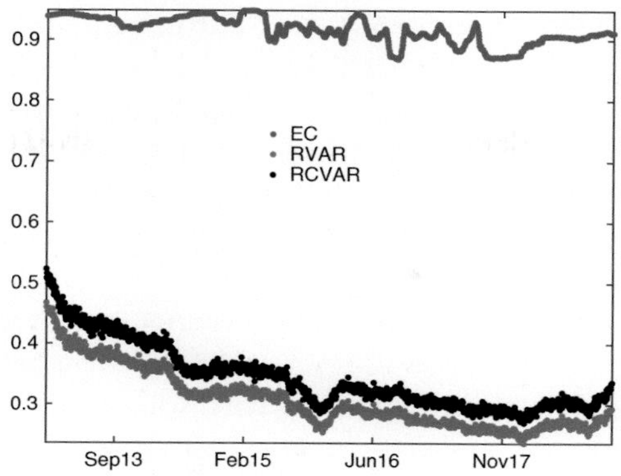

Figure 16.3 Equity and regulatory capital based on 1, 000-day time series estimations

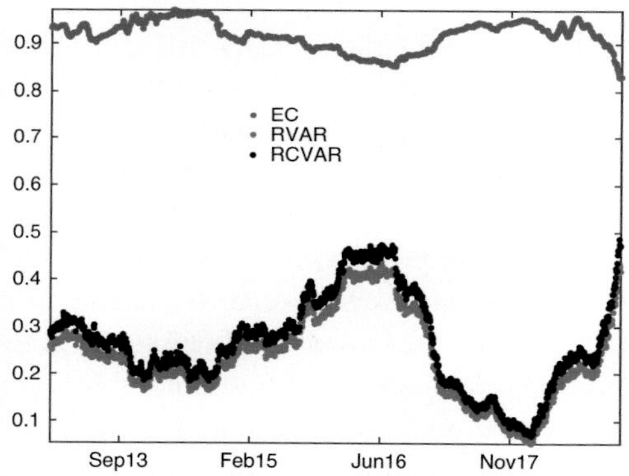

Figure 16.4 Equity and regulatory capital based on estimations using 252 days of time series data

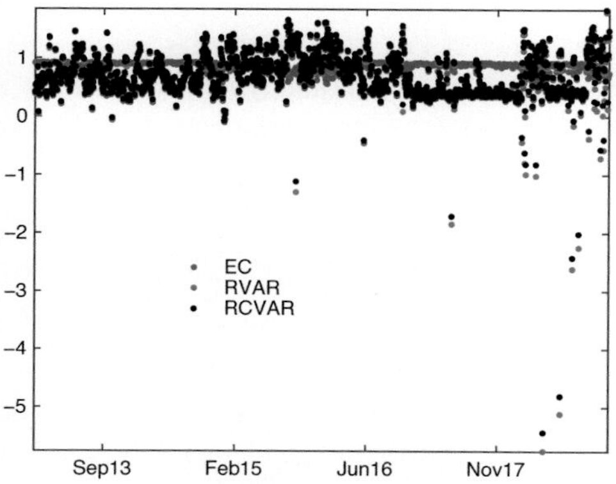

Figure 16.5 Equity and regulatory capital based on option-implied physical laws of motion and a regulatory multiple reduced to two

UK assets we report on the period from January 2010, and for the Netherlands assets we report from January 2017 onward. The US ticker symbols were JPM, C, WFC, MS, GS, and AIG. For the UK we report on RBS, and for the Netherlands we report on ING. A different strategy was employed for calibrating the market stress parameter, and this is presented first, followed by results on the equity and regulatory capital levels based on VAR and CVAR. We used a five-day period for the VAR and CVAR calculations at the 99% loss level with a multiple of just one half, so as not to be too stringent.

For this long period there were no real short maturity options trading with maturities under a month. The option maturities used for calibration were between a month and six months with calibration to a maturity of three months for strikes lying between plus and minus 20% of the forward.

16.5.1 Stress Calibration

Two models are fit to the three-month options: a BG risk-neutral model and a BG physical candidate with a five-parameter preference model for the depreferencing of the option prices. This results in two BG Lévy measures estimated on each day that we may denote by $k_P(x)$ and $k_Q(x)$ for the physical and risk-neutral counterparts.

For the measure-distortion parameters we take b to be unity. The other parameters c, γ are to be selected such that the distorted physical tail dominates the risk-neutral tail. For the tail risks we consider the tails at jump sizes between 10 and 20% in absolute value. The gap between the two tails may be computed on a sample of points at the estimated parameter values, and the maximum gap post distortion with b at unity is $(1/c)$. The value of $c = 0.5$ used above for SPY was too high with the maximum gap too low. We used $c = 0.25$ for all assets excepting C and AIG where the value of c was 0.1. Given b, c, the value of γ is determined to be the smallest distortion allowing the distorted physical to dominate the risk-neutral in the range of positive or negative moves with absolute value between 10 and 20%.

Figure 16.6 presents the time path of the calibrated γ values over the full period. At the peak of the crisis around the default of Lehman Brothers, the value of γ reaches levels like 3.5. There is another peak around 2011. The terminal values are well below unity.

16.5.2 Equity and Regulatory Capital Levels

For the five banks m, JPM, C, WFC, MS, and GS, and AIG, calibrated stress levels were employed to determine the conservative value of equity as a solution to the time-reversed nonlinear measure-distorted partial integro-differential equation. The candidate option implied physical process was distorted for the valuations. The same BG physical process estimated each day was simulated for five days to evaluate the VAR and CVAR at the 99% loss level. Regulatory capital was set at a multiple of one half of these values. The estimations were conducted every 10 days over the entire period.

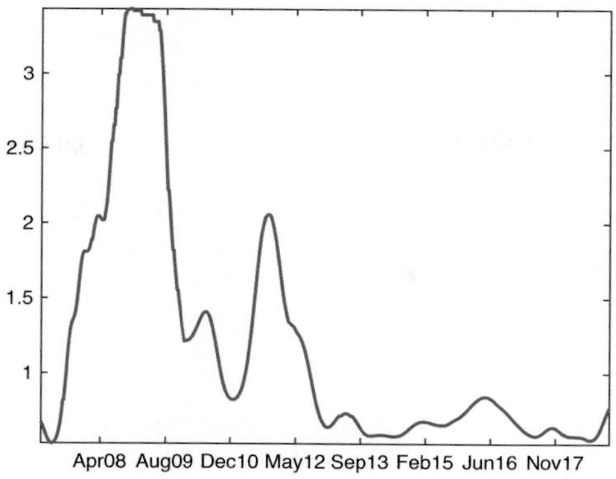

Figure 16.6 Distortion stress levels calibrated to option data on JPM using option-depreferencing technology for the option-implied physical return process.

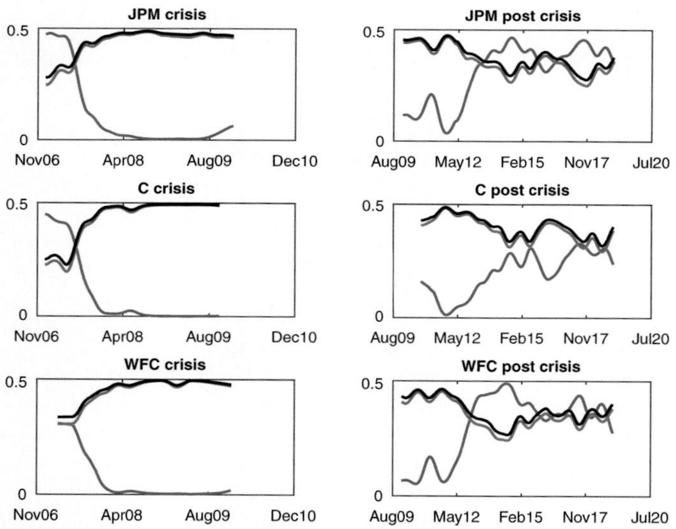

Figure 16.7 Equity capital in blue. Regulatory capital based on VAR in red and CVAR in black. The assets are JPM in the top panel, C in the middle panel, and WFC in the bottom panel.

Figures 16.7 and 16.8 present the three capital levels through the crisis period of 2007 through 2009 and between 2010 and the end of 2018 as the post-crisis period for the US assets.

In all cases the equity capital is insufficient in the crisis period and recovers with a dip in 2011 to being just adequate near the end of the period. Figure 16.9 presents the results for RBS and ING.

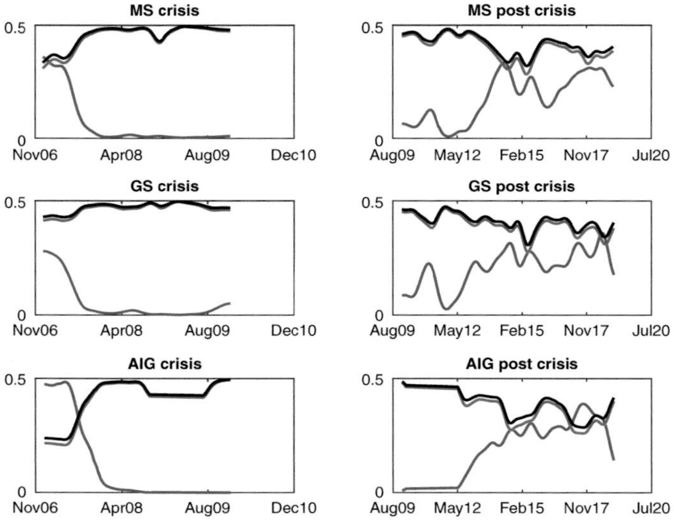

Figure 16.8 Equity capital in blue. Regulatory capital based on VAR in red and CVAR in black. The assets are MS in the top panel, GS in the middle panel, and AIG in the bottom panel.

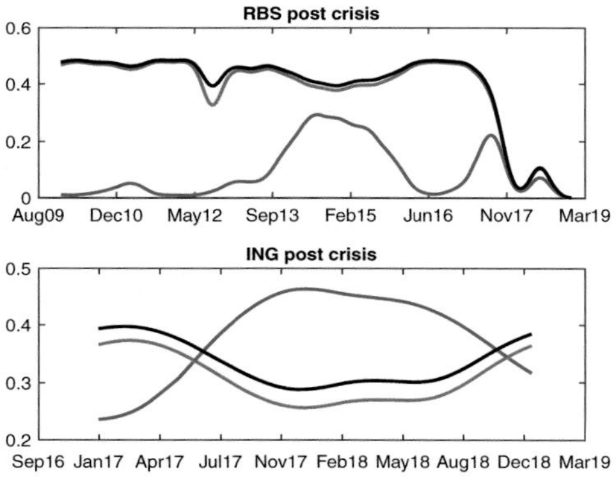

Figure 16.9 Equity capital in blue. Regulatory capital based on VAR in red and CVAR in black. The top panel is for RBS. The bottom panel is for ING.

We observe that for RBS the regulatory capital has fallen sharply to meet the equity capital while for ING the equity capital rises to be more than adequate and then meets the regulatory level.

17

Trading Markovian Models

The return from a position in a financial asset, be it long or short, over a short period has an outcome recognized to be hopefully positive, but it could be negative. The set of possible outcomes is therefore at least two with generally many more possibilities. The outcome may then be viewed as random. However, the fact that outcomes are random does not imply that they are random walks with independent and identically distributed return distributions. If they were such random walks, then there would be little basis for developing what are called technical trading rules as studied in Brock et al. (1992) and Murphy (1999). Evidence of the possible absence of random walk considerations is presented in Lo and MacKinlay (1999) and Madan and Wang (2016b). Here we model a proper Markov process; for another such example we cite Kakade and Kearns (2005). In this context we employ measure-distorted valuations for the design of optimal trading policies. For the use of probability distortions in a discrete-time context we cite Madan and Schoutens (2016).

A dynamic trading policy or trading rule must decide what position is to be taken at each time point and how it will be unwound. For a start consider trading a single financial asset with all positions to be taken at market close that are unwound at the close of the following business day. For a position of a dollars with x being the log price relative between the two days the outcome is $a(e^x - 1)$ and the trading policy must choose a, the level of the position. The return x is a random outcome with some distribution.

A Markovian trading policy chooses the position a as a function of a state variable s that helps predict the distribution of the realized return x. The policy function could be stationary, and we may write the policy function as the deterministic function $a(s)$. Alternatively it could be nonstationary and then be written as $a(s, t)$. One could put time as part of the state definition, but then the state space is not compact and you can never be in the same state at two different time points. Learning how return distributions depend on the state then becomes nearly impossible. Though technically returns are modeled as real-valued random variables, they are actually quite compact and well within a 100% in absolute value.

Given the compactness of returns, the state variables influencing returns should likewise be compact with ample opportunities for visiting the same state many times so that we can learn how return distributions depend on the state. From this perspective, the stock or asset price itself is not a reasonable state variable as its range is not bounded and very similar return distributions are possible at widely differing levels for the asset price. Besides, stocks can be split with prices halved and returns unaffected. The return is a relative price and the state variable should also be a relative price. These considerations lead us to consider as a state variable the ratio of the asset price to an exponentially weighted average of past prices. Many technical rules exploit similar information.

For an asset price process $S(t)$ we then define the exponentially weighted average of past prices by

$$Y(t) = \theta \int_{-\infty}^{t} e^{-\theta(t-u)} S(u) du.$$

The state variable considered is

$$Z(t) = \frac{S(t)}{Y(t)}.$$

For $\theta = 0.1$ using daily data one observes that $Z(t)$ lies well within 0.85 and 1.15. One anticipates that in the tails for Z asset price drifts should be positive/negative for values of Z that are extremely low/high. This is the presence of Z tail reversion. Locally in the neighborhood of unity for Z one may have increases in Z associated with rising drifts or the presence of a local momentum or the other way around for local mean reversion.

For a compact state space and a stable dependence of return distributions on the state, there is no reason for the trading policy to depend on calendar time. This can happen when optimizing expected terminal utility of cumulated returns at some finite horizon. This is the nonstationarity induced by calendar time getting closer to the horizon. For personalized life-cycle problems such horizons may be a critical concern, but for designing trading policies managing the investment activities of enterprises it may be more appropriate to consider infinite horizons with discounting, thereby removing the dependence on calendar time.

Another source of time dependence can arise in expected utility maximization problems as risk aversions may vary with wealth accumulation and the accumulated wealth to date is taken as a state variable. Here one is modeling the effects of increasing dissatisfaction with wealth reflected in the diminishing marginal utility of money as one gets richer. These are not concerns of relevance to the design of trading policies for corporate enterprises. Such issues are avoided by ensuring that objectives scale with wealth and the marginal utility of wealth is always unity.

When the objective is taken to be the market value of the enterprise and market values are modeled as scaling with size, then wealth to date is not a state variable. It is also not contained in a compact set and trading policies do not take account of such entities. The trading policy will be just a function of asset price relative to its exponentially weighted average. In generalizations one may consider other higher-dimensional compact sets as state variables. The methods developed here may be applied to construct trading strategies using machine learning methods and reinforcement learning where state transitions are explored by data analysis conducted simultaneously with the development of the strategy (Carapuço et al., 2018).

Here we analyze the time series of state variables and daily forward returns to estimate the dependence of return distributions on the current state. In this context we formulate the problem of maximizing the measure-distorted value of discounted cumulated returns to infinity. The policy is a deterministic function $a(s)$ of the state variable that is solved for on time reversing a nonlinear partial integro-differential equation. Explicit solutions are illustrated for a set of 10 assets that are among the larger capitalizations of the S&P 500 index.

17.1 Return Dependence on States

We take the return distribution to be in the four parameter BG class of distributions with parameters b_p, c_p, b_n, c_n that are now modeled as arbitrary functions of Z the ratio of price to an exponentially weighted average of past prices. Given data S_t on an asset price, we construct the theoretical average of past prices by

$$Y_t = (1-\lambda)\sum_{k=1}^{\infty}\lambda^k S_{t-k}.$$

In application the series is truncated and the normalization factor appropriately adjusted. We then obtain observations each day on

$$Z_t = \frac{S_t}{Y_t},$$

$$r_t = \log\left(\frac{S_{t+1}}{S_t}\right).$$

We observe, as already noted, that Z_t is well within 0.85 and 1.15 for $\lambda = 0.9$. We wish to employ data on the pairs (Z_t, r_t) to model the dependence of the return distribution on Z. The density of r given Z is of the form

$$f\left(r; b_p(Z), c_p(Z), b_n(Z), c_n(Z)\right),$$

and one may construct the conditional log likelihood of r_t given Z_t as

$$\sum_t \log\left(f\left(r_t; b_p(Z_t), c_p(Z_t), b_n(Z_t), c_n(Z_t)\right)\right),$$

once we have parameterized the functions $b_p(Z), c_p(Z), b_n(Z), c_n(Z)$. The parameters in these functions may then be estimated by maximizing the conditional log likelihood.

We do not, however, wish to select a particular functional form for the dependence of parameters on the level of Z. Consider a grid of the Z interval from 0.85 to 1.15 in steps of 0.025 with a total of 13 grid points. Denote these points by Z_k. Suppose we observe a function with noise taking the values y_k at the points Z_k, then a smooth kernel estimator for a function of Z is given by

$$g(Z) = \frac{\sum_{k=1}^{13} k(Z, Z_k) y_k}{\sum_{k=1}^{13} k(Z, Z_k)}.$$

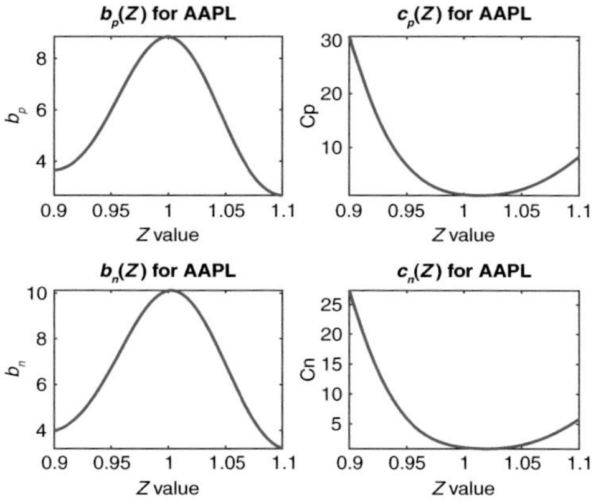

Figure 17.1 Return dependence of Apple stock

The function g can take a variety of shapes depending on the values of y_k. We employ a Gaussian kernel for k with a bandwidth of 0.0025. Each of the four parameters is modeled using this kernel smoother and there are 13 y values for each of the four parameters for a total of $52 = 13 \times 4$ parameters to be estimated in modeling the dependence.

Given these 52 sets of y values one may evaluate $b_p(Z_t), c_p(Z_t), b_n(Z_t), c_n(Z_t)$ at each data point and then evaluate by Fourier inversion the density

$$f(r_t; b_p(Z_t), c_p(Z_t), b_n(Z_t), c_n(Z_t)).$$

For 1,000 data points this would require 1,000 Fourier inversions to compute the likelihood once. We take 99 percentile points for the data on Z and denote these points Z_j, for $j = 1, \ldots, 99$ and compute the density at just these points. For an arbitrary value Z_t we use the density at the closest Z_j. The number of Fourier inversions in a log likelihood computation is then just 100.

The dependence was estimated for 10 stocks by such a maximum likelihood procedure using data for 1,000 days ending December 31, 2018. There is not much point in observing the estimates for y_k, but rather the functional dependence given the four functions $g(Z)$,one each for each parameter. There are then 10 sets of four graphs for each of the 10 stocks. We present in Figures 17.1 to 17.3 the graphs for three stocks: Apple (AAPL), Microsoft (MSFT), and Amazon (AMZN).

For each asset one may compute the asset drift rate as a function of the state Z by

$$\mu(Z) = \left(\frac{1}{1 - b_p(Z)}\right)^{c_p(Z)} \left(\frac{1}{1 + b_n(Z)}\right)^{c_n(Z)}.$$

We present in Figure 17.4 the dependence of drifts on the state Z for all 10 stocks.

We observe that generally the drifts rise to positive values in the left Z tail and fall to sometimes negative values in the right Z tail. MSFT is a clear case of local mean reversion

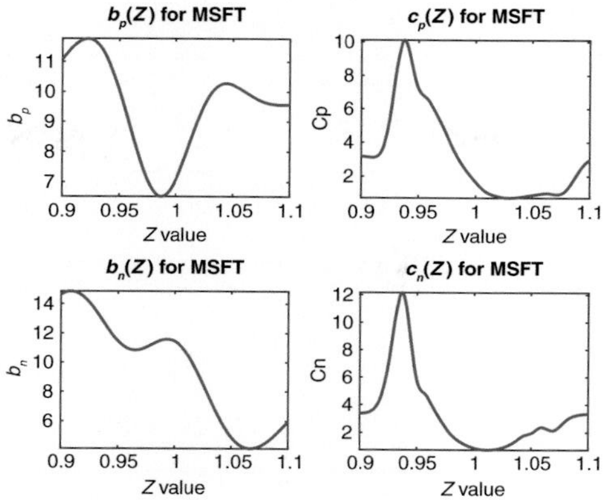

Figure 17.2 Return dependence of Microsoft stock

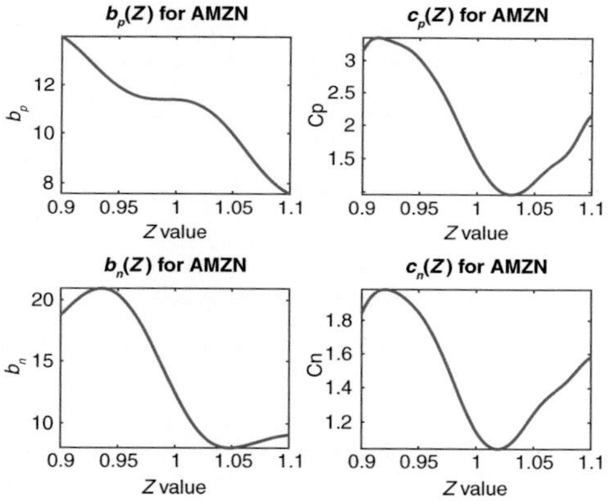

Figure 17.3 Return dependence of Amazon stock

while AAPL shows some local momentum. The drifts are presented in basis points and are quite low all across for T.

17.2 Markovian State Dynamics

The dynamics of Z when the logarithm of the asset price is the BG process is as follows:

$$\begin{aligned}
dZ &= Z(t)\frac{dS}{S} - Z(t)\frac{dY}{Y} \\
&= Z(t)\left((e^x - 1) * \mu(dx, du)\right) - \theta Z(t)\left(Z(t) - 1\right) dt \\
&= Z(t)\left((e^x - 1) * \widetilde{\mu}(dx, du)\right) \\
&\quad - \left(\theta Z(t)\left(Z(t) - 1\right) - Z(t)\int_{-\infty}^{\infty} (e^x - 1)\nu(Z(t), dx)\right) dt.
\end{aligned}$$

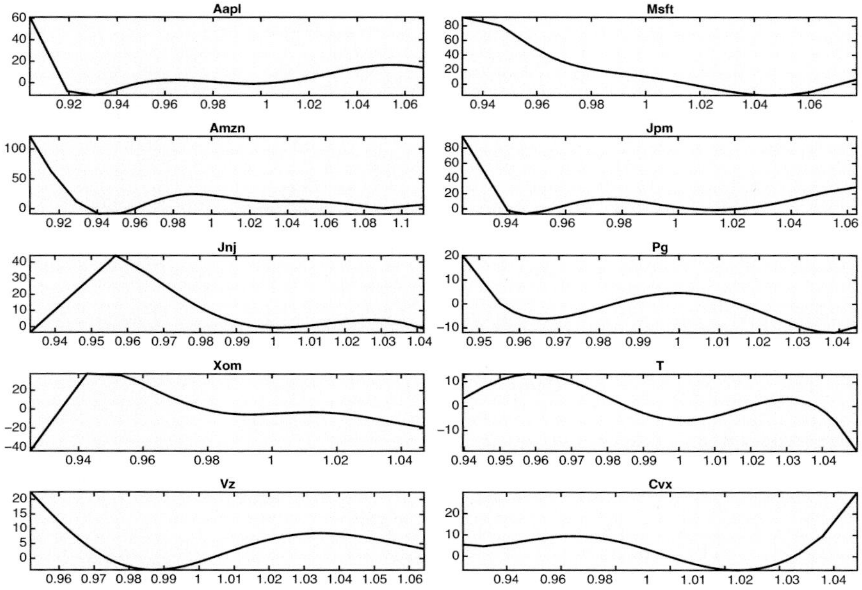

Figure 17.4 Asset drifts as functions of Z for each of 10 stocks

Here $\mu(dx, du)$ is the integer-valued random measure associated with the jumps of the BG process $X(t)$ and $\widetilde{\mu}(dx, du)$ is the compensated measure. The compensator for $X(t)$ is

$$\nu(Z, dx)\, dt = \left(\frac{c_p(Z)}{x} \exp\left(-\frac{x}{b_p(Z)} \right) \mathbf{1}_{x>0} + \frac{c_n(Z)}{|x|} \exp\left(-\frac{|x|}{b_n(Z)} \right) \right) dxdt.$$

17.3 Formulation and Solution of Market Value Maximization

We seek a stationary policy $a(Z)$ that maximizes the market value of the discounted cumulated returns taken to infinity. One may then first consider the discounted terminal payout defined by

$$\chi = \int_t^\infty \int_{-\infty}^\infty e^{-\eta(u-t)} a(u)\,(e^x - 1)\, \mu(dx, du),$$

for an adapted process a of dollar positions in the asset with the dynamics provided. The discounting is critical for the finiteness of the integral over time. The discount rate to be used is a choice to be made on the basis of how far in the future one wishes to go to consider liquidation or evaluation of realized outcomes.

A nonlinear valuation of this terminal value may be defined as a solution of a nonlinear backward stochastic partial integro-differential equation. Let the discounted value be $e^{-\eta t} B_t$ for a driver $g(s, H_s)$ defining the risk charges along the way. We then have that

$$e^{-\eta t} B_t + \int_t^\infty g\left(s, H_s\right) ds + \int_{(t,\infty]\times\mathbb{R}\setminus\{0\}} H_s(x)\widetilde{\mu}(ds \times dx) = \chi.$$

In a Markovian case we may write

$$B_t = V(t, Z(t)).$$

The function V satisfies the semilinear partial integro-differential equation

$$\dot{V} + \mathcal{K}V(t, Z) - g(t, \mathcal{D}V_{t,Z}) = \eta V,$$

$$\mathcal{K}V(t, Z) = d_t^{\mathrm{T}} \nabla V + \int_{\mathbb{R}\setminus\{0\}} (a(Z)Z(e^x - 1) + \mathcal{D}V_{t,Z} - \nabla V(t, Z)^{\mathrm{T}} Z\left(e^x - 1\right))\nu(Z, x)dx,$$

$$\mathcal{D}V_{t,Z} = V(t, Ze^x) - V(t, Z),$$

$$d_t = \int_{\mathbb{R}\setminus\{0\}} Z\left(e^x - 1\right) \nu(Z, t, dx) - \theta Z(Z - 1).$$

Recognizing that $\mathcal{K}V - g$ is our measure distortion and the integral in the drift term cancels out for finite variation processes, one may write

$$\begin{aligned}
&\eta V(Z) + \theta V_Z Z(Z - 1) - \dot{V} \\
&= -\int_0^\infty G^+\left(K\left([a(Z)(e^x - 1) + (V(Ze^x) - V(Z))]^- > w\right)\right) dw \\
&\quad + \int_0^\infty G^-\left(K\left([a(Z)(e^x - 1) + (V(Ze^x) - V(Z))]^+ > w\right)\right) dw.
\end{aligned} \tag{17.1}$$

One may approximate the derivative on the left-hand side by

$$\eta V(Z) + \theta Z(Z - 1)\frac{V(Z_u) - V(Z_d)}{Z_u - Z_d},$$

for Z in the interior and Z_u, Z_d adjacent Z points above and below the current grid point. Near the upper boundary we write

$$\eta V(Z) + \theta Z(Z - 1)\frac{V(Z) - V(Z_d)}{Z - Z_d},$$

and at the lower boundary we write

$$\eta V(Z) + \theta Z(Z - 1)\frac{V(Z_u) - W(Z)}{Z_u - Z}.$$

This gives us the value function vectorized on a grid by the equation

$$\begin{aligned}
HV - \dot{V} &= q(Z, a), \\
q(Z, a) &= -\int_0^\infty G^+\left(K\left([a(e^x - 1) + (V(Ze^x) - V(Z))]^- > w\right)\right) dw \\
&\quad + \int_0^\infty G^-\left(K\left([a(Z)(e^x - 1) + (V(Ze^x) - V(Z))]^+ > w\right)\right) dw.
\end{aligned}$$

The matrix H is fixed across time steps. For the first row we have

$$\eta - \frac{\theta Z(Z-1)}{(Z_u - Z)}, \frac{\theta Z(Z-1)}{(Z_u - Z)},$$

and for the last row we have

$$-\frac{\theta Z(Z-1)}{(Z - Z_d)}, \eta + \frac{\theta Z(Z-1)}{(Z - Z_d)},$$

and in the middle we have

$$-\frac{\theta Z(Z-1)}{(Z_u - Z_d)}, \eta, \frac{\theta Z(Z-1)}{Z_u - Z_d}.$$

We now introduce time reversal by

$$W(Z,u) = V(Z, -\ln u)$$

whereby

$$\dot{W} = -\frac{1}{u}\dot{V}$$

and hence write

$$HW + u\dot{W} = q.$$

On discretization of $\dot{W}$ we write

$$HW + u\frac{W' - W}{h} = q,$$

or

$$W' = W + \left(\frac{h}{u}\right)(q - HW). \tag{17.2}$$

This equation is solved in the unit interval for $0 \leq u \leq 1$ with $W(Z,0) = 0$. At each level of u, Z we maximize q over the choices for a to determine $a(Z,u)$. The value for a was constrained to be below unity in absolute value. The choice for $a = 0$ was not allowed and two optimization problems were solved for a between 0.01 and 1, and the other for a between −1 and −0.01. The maximum of the two solutions defined the position for the particular level of u, Z. Our interest is just in $a(Z,1)$.

17.4 Results on Policy Functions for 10 Stocks

The recursive forward solution (17.2) of the time-reversed equation (17.1) was conducted with the null boundary condition at the origin. At the start with a zero boundary condition and a zero investment there is no exposure and no value function is defined. Hence as noted, we excluded the zero solution by solving for both a positive solution and a negative solution and then selecting the greater of the two. The value of θ was 0.1 and that of η was 0.2.

We present in Figure 17.5 the 10 policy functions for each of the 10 stocks for which the dependence was estimated and reported on.

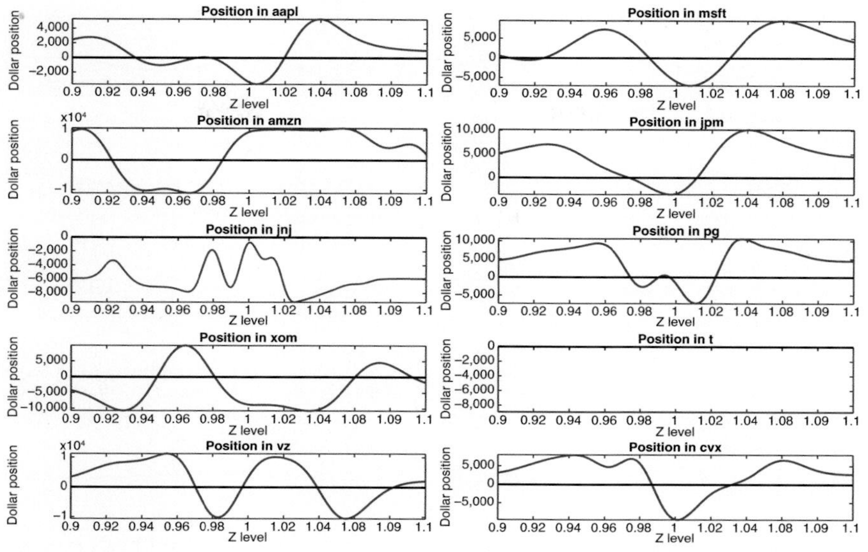

Figure 17.5 Policy functions for 10 stocks based on bilateral gamma motion with all parameters depending on state variable Z. This is the ratio of price to an exponentially weighted average of past prices.

Some remarks on the results are as follows. First we note that the drift was small across the board for T and this stock was shorted at all Z levels. Amazon is a clear case of a momentum stock for which some mean reversion occurs only deep in the two tails. Apple show momentum on the up side and mean reversion on the down side with positions rising as the average rises above unity. But as the average falls below unity, positions rise in anticipation of mean reversion. A similar picture holds for Microsoft. Local mean reversion is observed for Exxon on both sides though it is much stronger on the down side. Verizon displays momentum locally on both sides but not as pronounced as in Amazon.

17.5 Results for Sector ETFs and SPY

A similar analysis was conducted for the nine sector ETFs of the US economy and the market index SPY. Figure 17.6 presents the dependence of assets drifts on Z, the ratio of the index level to its exponentially weighted average of past prices. All the indices display strong levels of mean reversion in the extreme left tail with mixed results for the right tail. For example, the financial sector displays momentum in the extreme right tail. There is a consistent mean reversion all across for XLP.

The Z contingent positions for the these 10 underlying assets are presented in Figure 17.7.

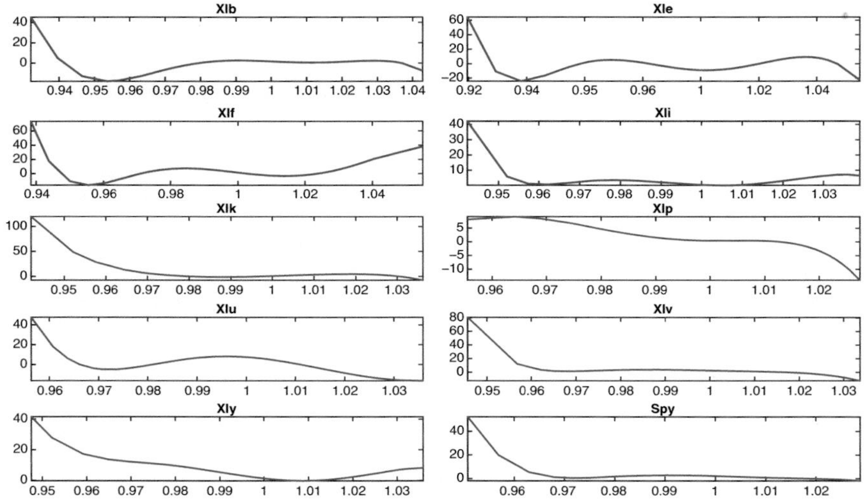

Figure 17.6 The dependence of assets drifts on Z, the ratio of the index level to its exponentially weighted average of past prices

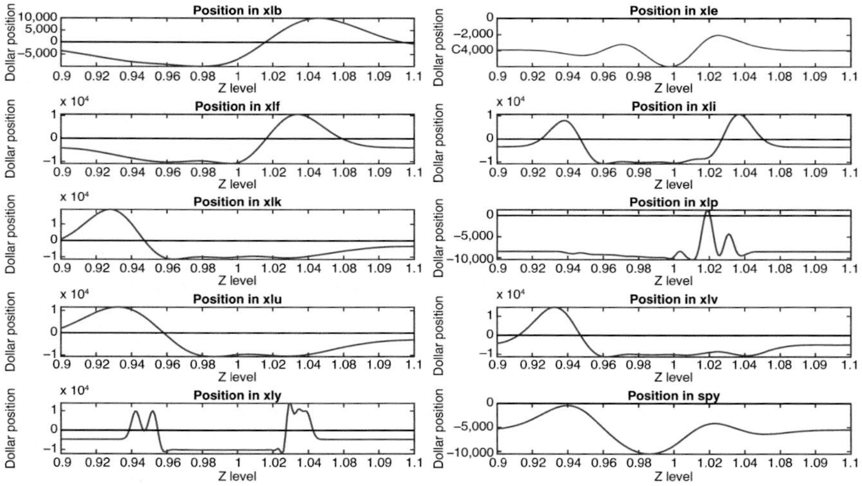

Figure 17.7 The Z contingent positions for the these 10 underlying assets

18

Market-Implied Measure-Distortion Parameters

Prices of option or other contracts that promise a contingent or fixed future payoff may be seen as valuations of this specified payout to be received or delivered. Prices of stocks or equity indices have a different status as there is no promised specific payout other than the value of the stock in the future but at no specific time. On division of future prices by the current price, attention may be focused on the return to a dollar and for all practical purposes all prices are then unity and there are no prices to be explained or understood. What is at issue is the structure of the possible random returns. They have a distribution and one may ask what distributions are possible. As already noted on numerous occasions, the distributions arise only at a horizon and the horizon is a questionable entity as no specific horizon is of any particular relevance. Working with self-decomposable limit laws, the risk is represented by arrival rates of moves and the question turns on the market possibilities for these entities. In the interests of balancing tractability and sufficient generality, we restrict the question here to the class of BG arrival rates that are characterized by their four parameters. Alternative representations may also be considered in generalizations.

The basic principle driving market possibilities is the absence of not just arbitrages, but more generally of highly attractive outcomes by taking long or short positions (Cochrane and Saá-Requejo, 2000; Cherny and Madan, 2009). From the perspective of exposures, these are outcomes that have high levels for their measure-distorted valuations evaluating the structure of arrival rates for gains and losses. Critical to ascertaining market attractiveness is an evaluation of the parameters of measure distortions, however parametrized, from observed market realities. We have explored the path of calibrating or estimating these parameters from data on option prices at a point of time. Here we explore and implement a strategy for doing the same from the time series data on equity prices. This typically includes information on market open, high, low, and close prices for each day.

18.1 Designing the Time Series Estimation of Measure-Distortion Parameters

Markets give access to random returns to an invested dollar with outcomes $e^x - 1$ where x is the log price relative. Theoretically there are lower, L_t and upper values, U_t for this exposure, that are solutions to backward stochastic partial integro-differential equations. It is then conjectured that

$$
\begin{aligned}
L_t &+ \int_t^{\infty}\int_0^{\infty} \Gamma^{+}\left(\nu_s\left([e^x-1]^{-}>w\right)\right) dw ds \\
&+ \int_t^{\infty}\int_0^{\infty} \Gamma^{-}\left(\nu_s\left([e^x-1]^{+}>w\right)\right) dw ds \\
&+ \int_t^{\infty}\int_{-\infty}^{\infty} (e^x-1)\widetilde{\mu}(dx, ds) \\
&= \chi,
\end{aligned}
$$

where $\widetilde{\mu}$ is the compensated jump measure for the jumps in the log price relative and

$$
\begin{aligned}
\Gamma^{+}(x) &= a\left(1-e^{-cx}\right)^{\frac{1}{1+\gamma}}, \\
\Gamma^{-}(x) &= \frac{b}{c}\left(1-e^{-cx}\right).
\end{aligned}
$$

Given estimates of ν as a BG compensator from the time series data on returns we wish to to estimate a, b, c, and γ. It follows on taking conditional expectations that

$$
\begin{aligned}
\widetilde{U_t} = U_t - E_t\Big[&\int_t^{\infty}\int_0^{\infty} \Gamma^{-}\left(\nu_s\left([e^x-1]^{-}>w\right)\right) dw ds \\
&+ \int_t^{\infty}\int_0^{\infty} \Gamma^{+}\left(\nu_s\left([e^x-1]^{+}>w\right)\right) dw ds\Big] \\
= E_t\left[\chi\right]&
\end{aligned}
$$

is a martingale as is

$$
\begin{aligned}
\widetilde{L_t} = L_t + E_t\Big[&\int_t^{\infty}\int_0^{\infty} \Gamma^{+}\left(\nu_s\left([e^x-1]^{-}>w\right)\right) dw ds \\
&+ \int_t^{\infty}\int_0^{\infty} \Gamma^{-}\left(\nu_s\left([e^x-1]^{+}>w\right)\right) dw ds\Big].
\end{aligned}
$$

The martingale conditions assert that

$$
E\left[d\widetilde{U_t}|\mathcal{F}_t\right] = 0, \qquad E\left[d\widetilde{L_t}|\mathcal{F}_t\right] = 0
$$

and they give access to a large number of orthogonalities. For example, let Z_t be $\mathcal{F}_t$ measurable, then

$$
E\left[Z_t d\widetilde{U_t}\right] = 0, \qquad E\left[Z_t d\widetilde{L_t}\right] = 0,
$$

and these orthogonalities may be employed to build estimates of a, b, c, and γ using generalized methods of moments estimation as developed in Hansen (1982).

Implementation would, however, require data on U_t, L_t and the computation of the differential for the expectation of the remaining risk charges. We proceed, at this early stage, with a more myopic construction on a daily basis. Let U_t, L_t be upper and lower valuations for the investment of a dollar for a single day. The next day, positions are liquidated with upper and lower valuations being equal and equal to one dollar. In this case the differential in the upper price is $U_t - 1$ and that in the lower price is $L_t - 1$. The former is the current risk charge for upper valuations while the latter is the negative of the current risk charge for lower valuations. More precisely we write

$$U_t - 1 = \int_0^\infty \Gamma^- \left(\nu_t \left([e^x - 1]^- > w\right)\right) dw + \int_0^\infty \Gamma^+ \left(\nu_t \left([e^x - 1]^+ > w\right)\right) dw \tag{18.1}$$

and

$$1 - L_t = \int_0^\infty \Gamma^+ \left(\nu_t \left([e^x - 1]^- > w\right)\right) dw + \int_0^\infty \Gamma^- \left(\nu_t \left([e^x - 1]^+ > w\right)\right) dw. \tag{18.2}$$

For candidate data on the upper prices for a dollar investment we take the ratio of maximum of daily highs over the last five days to the average of the close prices for last five days. For the candidate lower price we take the ratio of the minimum of daily lows over the last five days to the average close. The data may be partitioned into nonoverlapping five-day intervals to construct candidate data for the left-hand side of equations (18.1) and (18.2).

For the right-hand side we use estimates of BG model parameters constructed daily using 252 days of immediately prior returns as log price relatives. The closed form for the integrals of bilateral gamma Lévy tails provides access to the measures $\nu_t \left([e^x - 1]^- > w\right)$, $\nu_t \left([e^x - 1]^+ > w\right)$ for a grid of values on w. These are precomputed for each of 201 stocks for each of 2,768 days from January 3, 2008, to December 31, 2018.

The right-hand side of equations (18.1) and (18.2) is then just a numerical integral using this tail measure data and the four parameters a, b, c, and γ. The right-hand side was computed every day and five-day tail measure averages were used for the input into the numerical integrations involved.

The parameters were estimated by minimizing over a, b, c, and γ the sum across all 201 stock underliers of squared deviations between the left- and right-hand sides for 553 nonoverlapping five-day intervals. The result was a time series of estimates for the measure-distortion parameters.

18.2 Estimation Results

After deleting outliers for each parameter in the two tails we constructed a smoothed kernel estimate of the time path of the four measure-distortion parameters. Figure 18.1 presents the time paths for the four parameters.

In addition we present eight quantized points along the proportion of points represented by each of these in Table 18.1.

We observe that the maximum risk charge on the down side given by b/c is typically smaller than the maximum on the upside given by a. The large values for c focus attention on the smaller tail counts that are lifted substantially by the large values for γ.

18.3 Distribution of Measure-Distorted Valuations for Equity Underliers

Markets do not allow arbitrages nor do they permit very attractive return distributions or arrival rate structures to emerge. From the perspective of arrival rates, one may evaluate

Table 18.1 *Representative measure-distortion parameters*

a	b	c	γ	Proportion
0.2049	0.3206	26.7899	49.2341	0.0438
0.2888	0.5000	2.4887	44.9047	0.0471
0.3248	0.6243	29.5470	28.6940	0.0572
0.3794	0.6601	2.5819	23.3759	0.2357
0.2716	0.4232	17.4786	37.1619	0.0606
0.3122	0.5968	3.8022	31.8134	0.1347
0.3691	0.4731	14.2941	23.2993	0.1785
0.5200	0.7588	2.7775	17.4809	0.2424

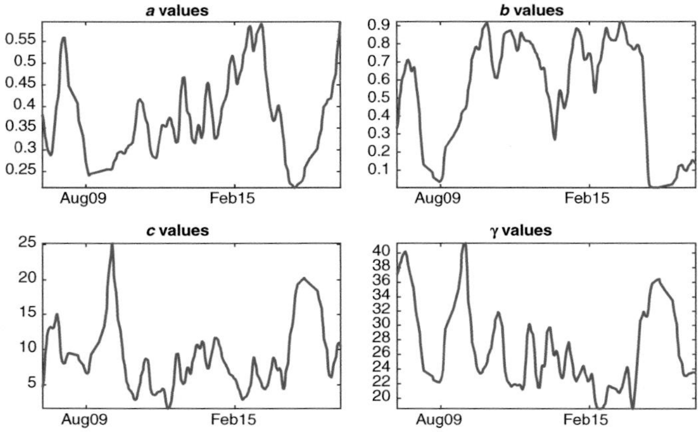

Figure 18.1 Smoothed time paths for the measure distortion parameters as estimated from time series data on high, low, and close prices for 201 underliers across 2,768 days partioned into 553 nonoverlapping five-day intervals

the attractiveness of positions by their measure-distorted valuations. To assess the access to measure-distorted valuations permitted in markets we take the fourth row of Table 18.1 for a stylized set of measure-distortion parameters. For these distortion parameters we evaluate the measure-distorted valuation for each of 201 equity underliers for each of 2,768 days from the estimated BG arrival rate structure. The arrival rate structure was estimated using 252 immediately prior returns. Figure 18.2 presents the quartiles for the measure-distorted valuations each day across the 201 underliers for the day. The valuations are for the larger of those for going short and going long the asset. We observe the deep dive taken by the valuations for the period of the financial crisis. In all cases the measure-distorted valuations are negative, reaching a negative of two percentage points at their highest point. Toward the end of the period the valuations lie between negative three and six percentage points for the upper and lower quartiles respectively with a median near five percentage points. The exact values are of course dependent on the choice of distortion parameters.

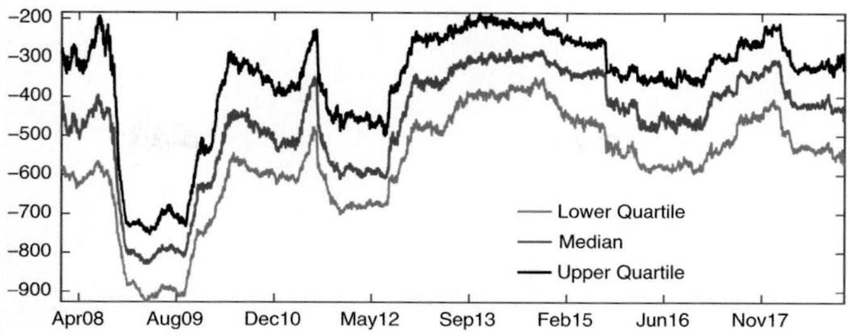

Figure 18.2 Quartiles across 201 underliers for each day of measure-distorted valuations. The valuations are the greater of the two for long and short positions.

18.4 Structure of Measure-Distorted Valuation-Level Curves

The measure-distorted value for either long or short positions is a function of the four BG parameters. By construction it is a concave function of the random variables, but the interest here is how it is structured as a function of the BG parameters. It is clear that increases in b_p, c_p raise value while increases in b_n, c_n lower value for the long position and the reverse for the short position. A value maximizer is interested in the structure of the greater than sets or sets where value exceeds a given level. It is not possible to view these sets in the four dimensions of the four parameters. We consider instead interesting two-dimensional subsections. These are (b_p, b_n), (c_p, c_n), (b_p, c_p), and (b_n, c_n) with the other parameters held at fixed levels. We report on the measure-distortion parameters from the fourth row of Table 18.1 and a typical BG parameter set as provided by those for SPY on December 31, 2018. These parameters were as follows:

$$b_p = 0.0056, \quad c_p = 1.2981, \quad b_n = 0.0104, \quad c_n = 0.6484.$$

Such a typical set reflects the property of smaller and more frequent up moves than the corresponding down moves. We construct two level curves, for valuation levels of negative 4% in blue and negative 4.5% in black for all four subsections.

Figure 18.3 presents the structure of level sets in the (b_p, b_n) subsection. We observe that the greater than sets in this subsection are not convex, but instead it is the less than sets that are convex. Hence convex combinations of acceptable points may be unacceptable at given valuation levels.

Figure 18.4 presents the level curves in the c_p, c_n subsection. Here too the less than sets are mildly convex. However, the level curves are almost linear and the departures from combinations of acceptable points becoming unacceptable are small.

Figure 18.5 presents the subsection (b_n, c_n) with the less than sets being convex.

Figure 18.6 presents the level curves in the b_p, c_p subsection and here the greater than set is convex.

Given the existence of subsections for which the greater than sets are not convex, the general structure of better than sets in the four dimensions of the BG parameters is difficult to describe or characterize.

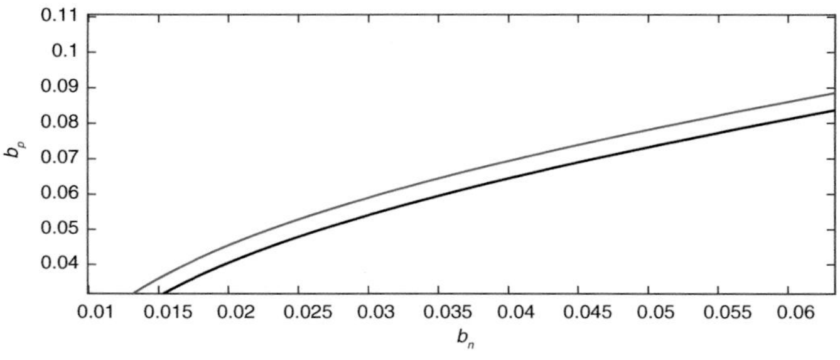

Figure 18.3 Valuation-level curves in the b_p, b_n subsection

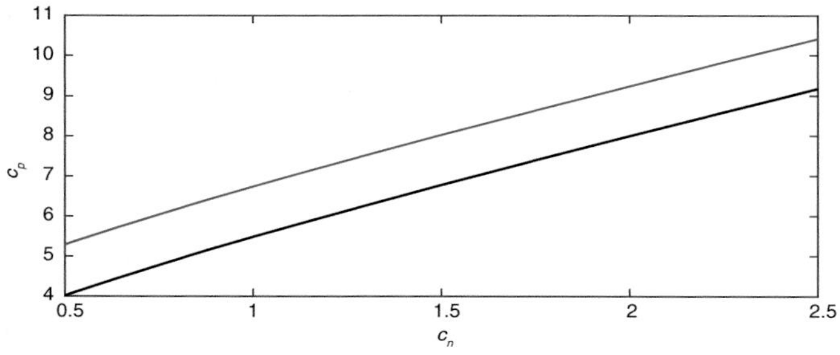

Figure 18.4 Valuation-level curves in the c_p, c_n subsection

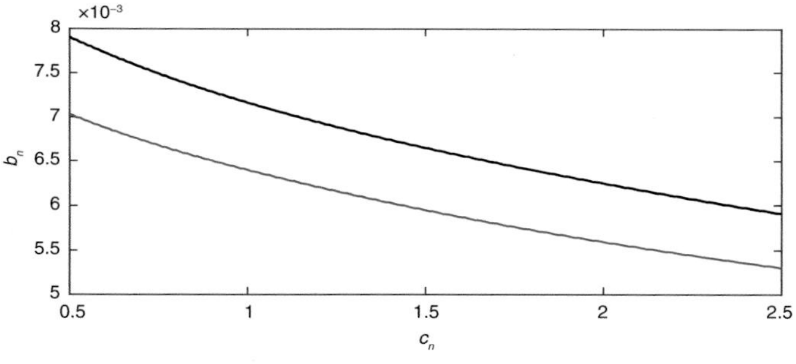

Figure 18.5 Valuation-level curves in the b_n, c_n subsection

18.5 Valuation Frontiers

With a view to exploring the frontier for BG parameters associated with a given level of the measure-distorted valuation for going long in the asset, we generated the closest point attaining the required measure-distortion level. For BG parameters of SPY over the entire period we extracted all points with a measure-distortion level below negative four percentage points. There were 1,558 such points out of a total of 2,768 points. For each

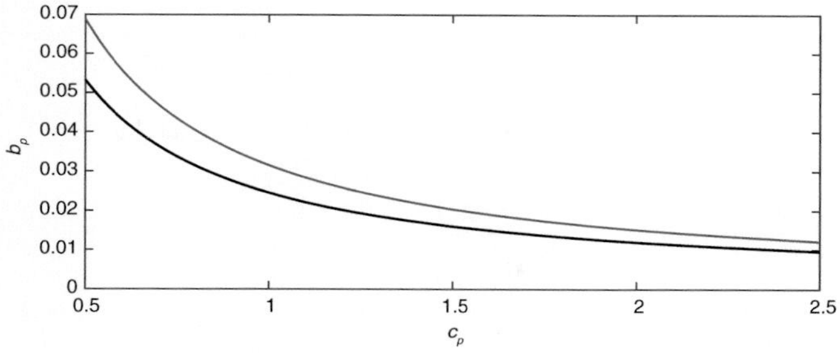

Figure 18.6 Valuation-level curves in the b_p, c_p subsection

such point we minimized a weighted Hellinger-type distance between the point and a point attaining the desired measure-distortion level −4%.

Given two Lévy measures $k(x)$, $k'(x)$, for $x > 0$ one may define a weighted Hellinger-type distance by

$$d(k, k') = \int_0^\infty \left(\sqrt{k(x)} - \sqrt{k'(x)}\right)^2 x^\eta dx.$$

When both k, k' are in the BG class, on the positive side this distance is explicitly given by

$$d_\eta(k, k') = \int_0^\infty c_p x^{\eta-1} e^{-x/b_p} + c'_p x^{\eta-1} e^{-x/b'_p} - 2\sqrt{c_p c'_p} x^{\eta-1} e^{-\left(\frac{1}{2b_p} + \frac{1}{2b'_p}\right)x} dx$$
$$= \Gamma(\eta)\left(c_p b_p^\eta + c'_p b_p'^\eta - 2^{1+\eta}\sqrt{c_p c'_p}\right)\left(\frac{b_p b'_p}{b_p + b'_p}\right)^\eta.$$

A similar expression holds for the negative side and the total distance is the sum of distances for the positive and negative sides.

Figure 18.7 presents the prior and posterior values of all four parameters that just enter the acceptability level of −4% and are closest to the original point using the distance $d_{.5}$. A similar graph was constructed for the distance d_1 for which the movements in b_n were much larger.

We observe that one may increase b_p, c_p, and even c_n while substantially reducing b_n to enter the associated set of acceptability for the measure-distorted value.

18.6 Acceptability Indices

The measure-distorted value is a conservative valuation of the risk exposure. One may seek to assess the quality of a risk by evaluating the highest level of stress the risk can bear. For this one needs to create an ordering of distortions. For distributions at a horizon the use of single-parameter distortions like minmaxvar permits an ordering of distortions with higher concavity of distortions representing greater stress levels. One may then evaluate the acceptability index of BG exposures. This was done for our set of BG estimates on 2, 768

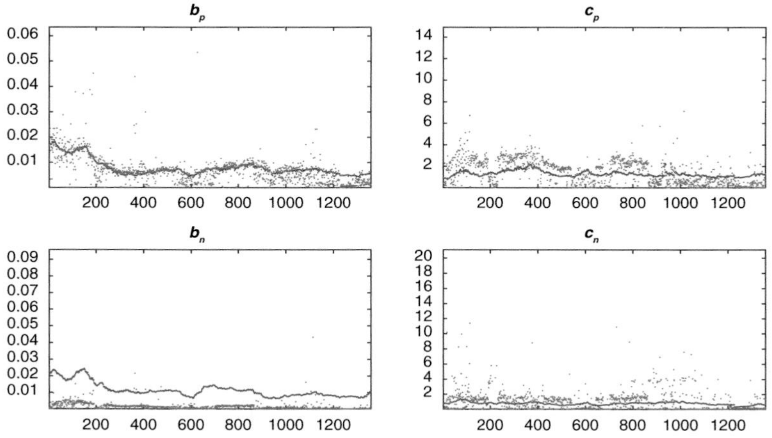

Figure 18.7 Movements in parameters required to just attain a −4% measure-distorted valuation that minimizes the distance d_1 between the Lévy measures

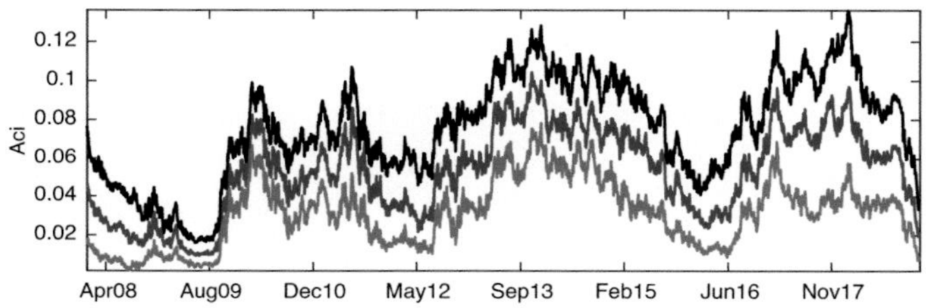

Figure 18.8 Quartiles for the acceptability indices based on bilateral gamma estimates on 201 underliers and 2, 768 days covering the period 2008 to 2018

days for 201 underliers. Figure 18.8 presents the quartiles across the 201 underliers for each day. The acceptability index is the larger of the two for going long or short the asset. Except for the crisis period it is mainly the long position that has the higher acceptability level.

The positions are short or long based on the sign of the drift. Figure 18.9 presents the drift quartiles for the same underliers over the same period. We observe that excepting the crisis these drifts were generally positive with the positions being long the asset.

18.7 Acceptability-Level Curves

Comparable to the analysis of measure-distorted values one may fix the asset speeds in the upward and downward directions and graph the curves of equal levels of acceptability attained by the upward and downward scales. Figure 18.10 presents the relationship between the two scales for a variety of settings for the speed parameters. The acceptability level was 0.05 and the speed parameters were those for SPY on December 31, 2018, and variations thereof.

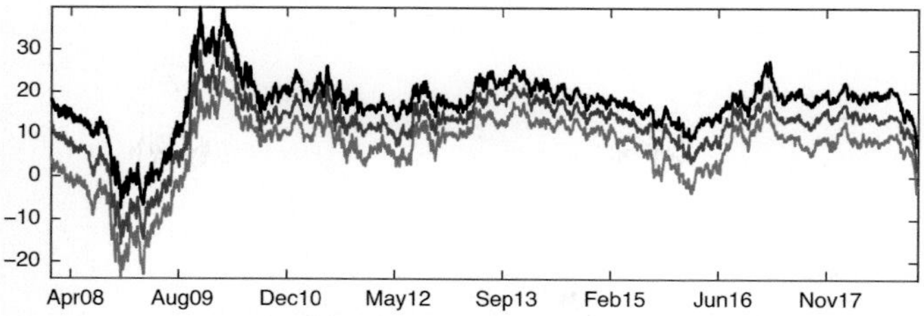

Figure 18.9 Quartiles for asset drifts in basis points based on bilateral gamma estimates for 201 underliers and 2, 768 days covering the period 2008 to 2018

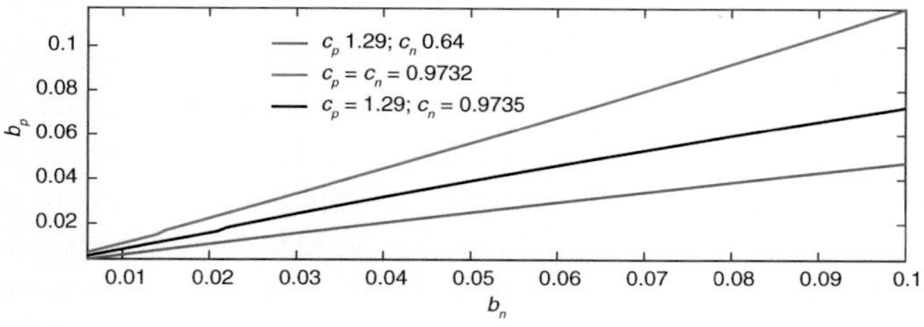

Figure 18.10 Upward scale as a function of the down scale for acceptability using minmaxvar at 0.05 for a variety of speed settings

In general we observe up speed generally bigger than down speed. For any setting of these parameters the scales must move in tandem to maintain a given level of risk acceptability. For the general space of BG distributions the maintenance of a given level of acceptability or a measure-distorted valuation places the four parameters on a three-dimensional manifold in four-dimensional space. A considerable variation is possible in the distributional structure even when acceptability is constrained. The following table provides two sets for acceptability 0.0215:

b_p	c_p	b_n	c_n
0.0073	1.4645	0.0144	0.7146
0.0091	1.5580	0.0101	1.3547.

18.8 Equilibrium Return Distributions

A single measure of performance, be it a level of acceptability or a level for the measure-distorted value, places the possible BG parameters on a three-dimensional manifold in four-dimensional space. Equilibrium return distributions identifying all four parameters require the simultaneous satisfaction of four independent equations. By way of illustration, consider fixing the speed parameters at, say, $c_p = 1.29$ and $c_n = 0.65$, the levels observed for SPY on December 31, 2018.

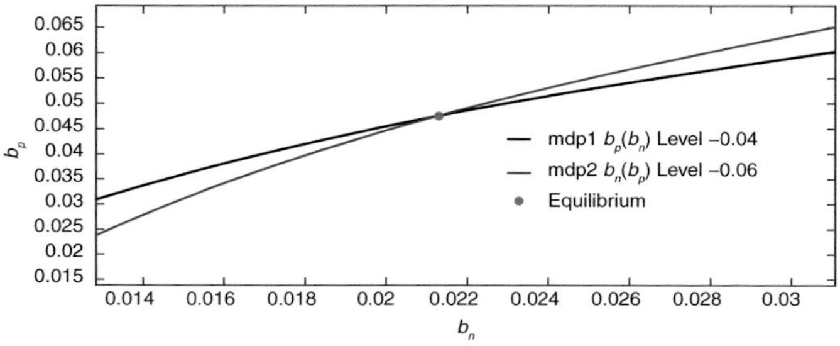

Figure 18.11 Equilibrium determination of scale parameters given speed parameters

Consider now two measure distortions given by rows 4 and 8 of Table 18.1. Suppose the upward and downward scale parameters are monitored by different groups of individuals. Further suppose the upward scale is monitored using measure distortions with the row 4 parameters for a valuation of −4%. One may then determine b_p as a function of b_n such that the measure-distorted value is −4%. Similarly we consider monitoring the downward scale parameter using row 8 parameters and a target valuation of −6%. This condition determines b_n as a function of b_p to meet this condition. Figure 18.11 displays equilibrium scales up and down, determined to meet both requirements.

For a full equilibrium, determining all four parameters simultaneously would require four monitoring equations. Madan and Schoutens (2019c) argued for speed parameters being monitored by market buy and sell order participants for the up and down motions. Limit order participants may be viewed as monitoring scale parameters. Madan and Schoutens empirically estimated parameter response functions for each parameter as a function of the other three with the equilibrium computed as a fixed point of this mapping. A more detailed study of such equilibria is here left as a topic for future research.

18.9 Empirical Construction of Return Distribution Equilibria and Their Properties

The four parameters of the BG model are anticipated in markets to be related to each other in a nonlinear way. A linear relation may be anticipated between the positive and negative, linear and quadratic variations. For 201 stocks over the 2, 768-day period from January 3, 2008, to December 31, 2018, we estimated each day for each stock the bilateral gamma model one 252 days of immediately prior returns. These were transformed to linear and quadratic variations for the positive and negative moves. The linear variation on the positive side is scale b_p times speed c_p while the square root of the quadratic variation is the scale times the square root of speed. These were recorded in basis points for both sides.

We then performed a distorted least-squares projection of each variation across the 2, 768 days on the other three variations for each stock to construct a lower and upper expectation for each variation given the other three. The stress level used in the distorted projection was 0.25. Given these projections, we extracted the days on which each variation was between the lower and upper expectation given the other three. These were taken as the equilibrium

Table 18.2 *Amazon*

b_p	c_p	b_n	c_n	prop.
100.82	2.6290	95.39	2.6279	20.34
107.49	2.2901	101.56	2.2886	21.57
252.36	1.6370	178.92	2.2550	1.47
114.62	1.8734	105.08	1.8732	15.69
48.30	5.7977	44.62	5.8052	5.15
99.64	2.3481	92.91	2.3431	20.34
272.89	1.4241	192.95	1.9562	4.17
114.74	1.7167	103.39	1.7168	11.27

Table 18.3 *GE*

b_p	c_p	b_n	c_n	prop.
186.81	2.0837	315.48	1.3618	7.83
62.24	4.0381	59.85	4.0391	18.28
227.88	1.4863	334.82	1.1414	3.13
70.61	1.9941	66.13	1.9925	42.04
201.80	1.8410	314.74	1.3161	11.75
71.13	2.8699	67.72	2.8673	6.01
258.96	1.1787	363.71	0.9571	1.57
71.99	1.3367	65.26	1.3392	9.40

Table 18.4 *WFC*

b_p	c_p	b_n	c_n	prop.
64.13	1.9138	60.85	1.8396	27.74
102.16	2.5365	100.46	2.5370	15.07
647.30	0.7059	544.86	0.8691	4.45
90.65	1.5191	91.16	1.4383	12.84
596.43	0.9382	558.30	1.0754	8.05
111.14	1.8647	108.00	1.8745	20.03
470.96	0.7623	441.96	0.8509	1.88
68.42	1.5309	61.05	1.5155	9.93

sets of variations for each stock. The number of equilibrium points varied between 12 and 1,017. The quartiles as percentages of the 2,768 days were 5.09, 8.89, and 14.74. There were 80 stocks with more than 300 equilibrium points. We pooled the equilibrium variations for these 80 stocks to get 25,173 sets of equilibrium variations across all stocks and days. A principal components analysis of these equilibrium variations revealed two components that explained more than 99.5% of the variation. The two components essentially linked the positive and negative linear variations to their quadratic counterparts with

$$\mu_p = a + 3.1394\sigma_p, \qquad \mu_n = b + 7.0014\sigma_n.$$

The equilibrium return distribution variations may therefore be seen to lie on a two-dimensional submanifold of four-dimensional space.

We quantized the equilibrium variation sets into eight groups for each of the 80 stocks that had more than 300 equilibrium points. The group variations were then transformed back to parameters. We present the eight equilibrium bilateral parameters for Amazon, GE, and WFC along with the proportion of equilibrium points they represent in Tables 18.2, 18.3, and 18.4.

References

Aase, Knut K. (1984). Optimum portfolio diversification in a general continuous-time model. *Stochastic Processes and Their Applications*, **18**, 81–98.

Acerbi, C., and D. Tasche (2002). On the coherence of expected shortfall. *Journal of Banking and Finance*, **26**, 1487–1503.

Anderson, T. W., and D. A. Darling (1952). Asymptotic theory of certain "goodness of fit" criteria based on stochastic processes. *Annals of Mathematical Statistics*, **23**, 193–212.

Angelos, B. (2013). "The Hunt Variance Gamma Process with Applications to Option Pricing". PhD Dissertation, University of Maryland.

Archakov, I., and P. R. Hansen (2018). A new parameterization of correlation matrices. Working Paper, University of Vienna.

Arrow, K. J., and G. Debreu (1954). Existence of an equilibrium for a competitive economy. *Econometrica*, **22**, 265–290.

Arrow, K. J., and F. H. Hahn (1971). *General Competitive Analysis*. Holden-Day.

Artzner, P., F. Delbaen, M. Eber, and D. Heath (1999). Coherent measures of risk. *Mathematical Finance*, **9**, 203–228.

Atlan, M., H. Geman, D. Madan, and M. Yor (2007). Correlation and the pricing of risks. *Annals of Finance*, **3**, 411–453.

Bakshi, G., F. Chabi-Yo, and X. Gao (2018). A recovery we can trust? Deducing and testing the restrictions of the recovery theorem. *Review of Financial Studies*, **31**, 532–555.

Bakshi, G., D. B. Madan and G. Panayotov (2010). Returns of claims on the upside and the viability of U-shaped pricing kernels. *Journal of Financial Economics*, **97**, 130–154.

Barkhagen, M., J. Blomvall, and E. Platen (2016). Recovering the real-world density and liquidity premia from option data. *Quantitative Finance*, **16**, 1147–1164.

Barles, G., R. Buckdahn, and E. Pardoux (1997). Backward stochastic differential equations and integral–partial differential equations. *Stochastics and Stochastic Reports*, **60**, 57–83.

Barndorff-Nielsen, O. E. (1998). Processes of normal inverse Gaussian type. *Finance and Stochastics*, **2**, 21–68.

Basak, S., and G. Chabakauri (2010). Dynamic mean-variance asset allocation. *Review of Financial Studies*, **23**, 2970–3016.

Basak, S., and G. Chabakauri (2012). Dynamic hedging in incomplete markets: A simple solution. *Review of Financial Studies*, **25**, 1845–1896.

Basel Committee on Banking Supervision (2010). Basel III: A global regulatory framework for more resilient banks and banking systems. *Bank for International Settlements Communications*, Basel, Switzerland.

Bass, R. F. (1988). Uniqueness in law for pure jump Markov processes. *Probability Theory and Related Fields*, **79**, 271–287.

Bertsimas, D., L. Kogan, and A. Lo (2001). Hedging derivative securities and incomplete markets: An ε arbitrage approach. *Operations Research*, **49**, 372–397.

Black, F., and M. Scholes (1973). The pricing of corporate liabilities. *Journal of Political Economy*, **81**, 637–654.

Borovička, J., L. P. Hansen, and J. A. Scheinkman (2016). Misspecified recovery. *Journal of Finance*, **71**, 2493–2544.

Boyarchenko S., and S. Levendorskii, (2000). Option pricing for truncated Lévy processes. *International Journal of Theoretical and Applied Finance*, **3**, 549–552.

Brennan, M. J., and R. Solanki (1981). The pricing of contingent claims in discrete time models. *Journal of Finance*, **34**, 53–68.

Brock, A., J. Lakonishok, and B. Lebaron (1992). Simple trading rules and the stochastic properties of stock returns. *Journal of Finance*, **47**, 1731–1764.

Buchmann, B., D. B. Madan, and K. Lu (2019). Weak subordination of multivariate Lévy processes and variance generalized gamma convolutions. *Bernoulli*, **25**, 742–770.

Campbell, J., A. Lo, and C. MacKinlay (1997). *The Econometrics of Financial Markets*, Princeton University Press.

Carapuço, J., R. Neves, and N. Horta (2018). Reinforcement learning applied to Forex trading. *Applied Soft Computing*, **73**, 783–794.

Carr P., H. Geman, D. Madan, and M. Yor (2002). The fine structure of asset returns: An empirical investigation. *Journal of Business*, **75**, 305–332.

Carr, P., H. Geman, D. Madan, and M. Yor (2007). Self-decomposability and option prices. *Mathematical Finance*, **17**, 31–57.

Carr, P., and D. B. Madan (1998). Towards a theory of volatility trading. In *Volatility*, R. Jarrow (ed). Risk Publications, pp. 417–427.

Carr, P., and D. B. Madan (2001). Optimal positioning in derivative securities. *Quantitative Finance*, **1**, 19–37.

Carr, P., and D. B. Madan (2005). A note of sufficient conditions for no arbitrage. *Finance Research Letters*, **2**, 125–130.

Carr, P., and D. B. Madan, (2014). Joint modeling of VIX and SPX options at a single and common maturity with risk management applications. *IIE Transactions*, **46**, 1125–1131.

Carr, P., D. B. Madan, M. Melamed, and W. Schoutens (2016). Hedging insurance books. *Insurance, Mathematics and Economics*, **70**, 364–372.

Chen, S., C. Lee, and K. Shrestha (2001). On the mean-generalized semivariance approach to determining the hedge ratio. *Journal of Futures Markets*, **21**, 581–598.

Cherny, A., and D. B. Madan (2009). New measures for performance evaluation. *Review of Financial Studies*, **22**, 2571–2606.

Choquet, G. (1953). Theory of capacities. *Annales de l'Institut Fourier*, **5**, 131–295.

Christoffersen, P., R. Elkhami, B. Feunou, and K. Jacobs (2010). Option valuation with conditional heteroskedasticity and non-normality. *Review of Financial Studies*, **23**, 2139–2183.

Cochrane, J. H., and J. Saá-Requejo (2000). Beyond arbitrage: "Good deal" asset price bounds in incomplete markets. *Journal of Political Economy*, **108**, 79–119.

Cohen, S., and R. J. Elliott (2010). A general theory of backward finite state difference equations. *Stochastic Processes and Their Applications*, **120**, 442–466.

Cont, R., and P. Tankov (2004). *Financial Modelling with Jump Processes*. Chapman and Hall.

Copeland, T., T. Koller, and J. Murrin (1994). *Valuation: Measuring and Managing the Value of Companies*. Second edition. Wiley.

Cox, J. C., S. A. Ross, and M. Rubinstein (1979). Option pricing: A simplified approach. *Journal of Financial Economics*, **7**, 229–263.

Cuesdeanu, H., and J. C. Jackwerth (2018). The pricing kernel puzzle: Survey and outlook. *Annals of Finance*, **14**, 289–329.

Cvitanić, J. (2000). Minimizing expected loss of hedging in incomplete and constrained markets. *SIAM Journal of Control and Optimization*, **38**, 1050–1066.

Cvitanić, J., and F. Zapatero (2004). *Introduction to the Economics and Mathematics of Financial Markets*. MIT Press.

Davis, M., and D. Hobson (2007). The range of traded option prices. *Mathematical Finance*, **17**, 1–14.

Davis, M. H. A., and S. Lleo (2015). *Risk-Sensitive Investment Management*. World Scientific.

Debreu, G. (1959). *Theory of Value*. Wiley.

Delbaen, F., and W. Schachermayer (1994). A general version of the fundamental theorem of asset pricing. *Mathematische Annalen*, **300**, 463–520.

Delbaen, F., and W. Schachermayer (1998). The fundamental theorem of asset pricing for unbounded processes. *Mathematische Annalen*, **312**, 215–250.

Duffie, D., V. Fleming, M. Soner, and T. Zariphopoulou (1997). Hedging in incomplete markets with HARA utility. *Journal of Economic Dynamics and Control*, **21**, 753–782.

Duffie, D., and H. R. Richardson (1991). Mean-variance hedging in continuous time. *Annals of Applied Probability*, **1**, 1–15.

Eberlein, E. (2001). Application of generalized hyperbolic Lévy motions to finance. In *Lévy Processes: Theory and Applications*, O. E. Barndorff-Nielsen, T. Mikosch, and S. Resnick (eds). Birkhäuser pp. 319–336.

Eberlein, E. and Keller U. (1995). Hyperbolic distributions in finance. *Bernoulli*, **1**, 281–299.

Eberlein, E., and D. B. Madan (2010). The distribution of returns at longer horizons. In *Recent Advances in Financial Engineering: Proceedings of the KIER–TMU Workshop*, M. Kijima, C. Hara, Y. Muromachi, H. Nakaoka and K. Nishide (eds). World Scientific, pp. 1–18.

Eberlein, E., D. B. Madan, M. Pistorius, W. Schoutens, and M. Yor (2014b). Two-price economies in continuous time. *Annals of Finance*, **10**, 71–100.

Eberlein, E., D. B. Madan, M. Pistorius, and M. Yor (2014a). Bid and ask prices as non-linear continuous time $\mathcal{G}$-expectations based on distortions. *Mathematics and Financial Economics*, **8**, 265–289.

Edwards, E., and P. Bell (1961). *The Theory and Measurement of Business Income*. University of California Press.

Elliot, R. J. (1982). *Stochastic Calculus and Applications*. Springer.

Elliot, R. J., D. B. Madan, and T. K. Siu (2020). Lower and upper pricing of financial assets. Working Paper, Robert H. Smith School of Business, University of Maryland.

El Karoui, N., and S. J. Huang (1997). A general result of existence and uniqueness of backward stochastic differential equations. In *Backward Stochastic Differential Equations*, N. El Karoui and L. Mazliak (eds.). Longman, pp. 27–36.

Fernholz, E. R. (2001). Equity portfolios generated by functions of ranked market weights. *Finance and Stochastics*, **5**, 469–486.

Fernholz, E. R. (2002). *Stochastic Portfolio Theory*. Springer.

Fernholz, E. R., and I. Karatzas (2005). Relative arbitrage in volatility-stabilized markets. *Annals of Finance*, **1**, 149–177.

Feuerverger, A., and P. McDunnough (1981). On the efficiency of empirical characteristic function procedures. *Journal of the Royal Statistical Society, Series B (Methodological)*, **43**, 20–27.

Flannery, M. J., and A. A. Protopapadakis (2002). *Review of Financial Studies*, **15**, 751–782.

Föllmer, H., and M. Schweizer (1991). Hedging of contingent claims under incomplete information. In *Applied Stochastic Analysis, Stochastics Monographs*, Vol 5. M. H. A. Davis and R. J. Elliott (eds.). Gordon and Breach, pp. 389–414.

Föllmer H., and D. Sondermann (1986). Hedging of non-redundant claims. In *Contributions to Mathematical Economics*, W. Hilberbrand and A. Mas-Colell (eds.). North Holland, pp. 205–223.

Gordy, M. B. (2003). A risk-factor model foundation for ratings-based bank capital rules. *Journal of Financial Intermediation*, **12**, 199–232.

Grothe, O., and S. Nicklas (2013). Vine construction of Lévy copulas. *Journal of Multivariate Analysis*, **119**, 1–15.

Hansen, L. P. (1982). Large sample properties of generalized method of moments estimators. *Econometrica*, **50**, 1029–1054.

Harrison, J., and D. Kreps (1979). Martingales and arbitrage in multiperiod securities markets. *Journal of Economic Theory*, **20**, 381–408.

Haugen, R. (2001). *Modern Investment Theory*. Prentice Hall.

Heath, D., E. Platen, and M. Schweizer (2001). A comparison of two quadratic approaches to hedging in incomplete markets. *Mathematical Finance*, **11**, 385–413.

Henderson, V., and D. Hobson (2009). Utility indifference pricing: An overview. In *Indifference Pricing: Theory and Applications*, R. Carmona (ed.). Princeton University Press.

Huang, C. F., and R. H. Litzenberger (1988). *Foundations for Financial Economics*. Prentice Hall.

Jacod, J., and A. N. Shiryaev (2013). *Limit Theorems for Stochastic Processes*, second edition. Springer.

Jaynes, E. T. (1957). Information theory and statistical mechanics. *Physical Review*, **108**, 171–190.

Jensen, M. C. (2002). Value maximization, stakeholder theory, and the corporate objective function. *Business Ethics Quarterly*, **12**, 235–256.

Joe, H. (1997), *Multivariate Models and Dependence Concepts*. Chapman and Hall.

Jorion, P. (2006). *Value at Risk: The New Benchmark for Managing Financial Risk*, third edition. McGraw Hill.

Kakade, S. M., and M. Kearns (2005). Trading in Markovian price models. In *Learning Theory (COLT)*, P. Auer and R. Meir (eds.). Lecture Notes in Computer Science, vol. 3559. Springer, pp. 606–620.

Kallsen, J., and P. Tankov (2006). Characterization of dependence of multidimensional Lévy processes using Lévy copulas. *Journal of Multivariate Analysis*, **97**, 1551–1572.

Karatzas, I., and S. E. Shreve (1988). *Brownian Motion and Stochastic Calculus*. Springer.

Keen, S. (1995). Finance and economic breakdown: Modeling Minsky's "financial instability" hypothesis. *Journal of Post Keynesian Economics*, **17**, 607–635.

Keynes, J. M. (1936). *The General Theory of Employment, Interest and Money*. Macmillan.

Keynes, J. M. (1937). The general theory of employment. *Quarterly Journal of Economics*, 209–223.

Khintchine, A. Y. (1938). *Limit Theorems for Independent Random Variables* (in Russian). ONTI.

Kotz, S., and S. Nadarajah (2004). *Multivariate T Distributions and Their Applications*. Cambridge University Press.

Küchler, U., and S. Tappe (2008). Bilateral gamma distributions and processes in financial mathematics. *Stochastic Processes and Their Applications*, **118**, 261–283.

Kusuoka, S. (2001). On law-invariant coherent risk measures. *Advances in Mathematical Economics*, **3**, 83–95.

Leland, H. (1980). Who should buy portfolio insurance? *Journal of Finance*, **35**, 581–594.

Lévy, P. (1937). *Théorie de l'Addition des Variables Aléatoires*. Gauthier-Villars.

Lintner, J. (1965). The valuation of risky assets and the selection of risky investments in stock portfolios and capital budgets. *Review of Economics and Statistics*, **47**, 13–37.

Lo, A., and C. MacKinlay (1999). *A Non-random Walk down Wall Street*. Princeton University Press.

Madan, D. B. (2015a). Estimating parametric models of probability distributions. *Methodology and Computing in Applied Probability*, **17**, 823–831.

Madan, D. B. (2015b). Asset pricing theory for two-price economies. *Annals of Finance*, **11**, 1–35.

Madan, D. B. (2016a). Risk premia in options markets. *Annals of Finance*, **12**, 71–94.

Madan, D. B. (2016b). Conic portfolio theory. *International Journal of Theoretical and Applied Finance*, **19**. https://doi.org/10.1142/S0219024916500199

Madan, D. B. (2017a). Measure-distorted arrival rate risks and their rewards. *Probability, Uncertainty, and Quantitative Risk*, **2**, 1–21.

Madan, D. B. (2017b). Efficient estimation of expected stock price returns. *Finance Research Letters*, **23**, 31–38.

Madan, D. B. (2018a). Instantaneous portfolio theory. *Quantitative Finance*, **18**, 1345–1364.

Madan, D. B. (2018b). Financial equilibrium with non-linear valuations. *Annals of Finance*, **14**, 211–221.

Madan, D. B. (2019). Nonlinear equity valuation using conic finance and its regulatory implications. *Mathematics and Financial Economics*, **13**, 31–65.

Madan, D. B. (2020). Multivariate distributions for financial returns. *International Journal of Theoretical and Applied Finance*, **23**, 2050041. https://doi.org/10.1142/S0219024920500417

Madan D., P. Carr, and E. Chang (1998). The variance gamma process and option pricing. *Review of Finance*, **2**, 79–105.

Madan, D. B., M. Pistorius, and M. Stadje (2017a). On dynamic spectral risk measures, a limit theorem and optimal portfolio allocation. *Finance and Stochastics*, **21**, 1073–1102.

Madan, D. B., and W. Schoutens (2012). Tenor specific pricing. *International Journal of Theoretical and Applied Finance*, **15**(6). https://doi.org/10.1142/S0219024912500434.

Madan, D. B., and W. Schoutens (2016). *Applied Conic Finance*. Cambridge University Press.

Madan, D. B., and W. Schoutens (2017). Conic option pricing. *Journal of Derivatives*, **25**, 10–36.

Madan, D. B., and W. Schoutens (2018b). Zero covariation returns. *Probability, Uncertainty and Quantitative Risk*, **3**(5). https://doi.org/10.1186/s41546-018-0031-1

Madan, D. B., and W. Schoutens (2019a). Nonlinear equity valuation using conic finance and its regulatory implications. *Mathematics and Financial Economics*, **13**, 31–65.

Madan, D. B., and W. Schoutens (2019b). Conic asset pricing and the costs of price fluctuations. *Annals of Finance*, **15**, 29–58.

Madan, D. B., and W. Schoutens (2019c). Equilibrium asset returns in financial markets. *International Journal of Theoretical and Applied Finance*, **22**. https://doi.org/10.1142/S0219024918500632

Madan, D. B., and W. Schoutens (2020). Self-similarity in long horizon asset returns. *Mathematical Finance*, **30**, 1368–1391.

Madan, D. B., W. Schoutens, and K. Wang (2017b). Measuring and monitoring the efficiency of markets. *International Journal of Theoretical and Applied Finance*, **20**, 1–32. https://doi.org/10.1142/S0219024917500510

Madan, D. B., W. Schoutens, and K. Wang (2018). Bilateral multiple gamma returns: Their risks and rewards. SSRN Abstract No. 3230196.

Madan, D. B., and E. Seneta (1987). Simulation of estimates using the empirical characteristic function. *International Statistical Review*, **55**(2), 153–161.

Madan, D. B., and Seneta, E. (1990). The variance gamma (VG) model for share market returns. *Journal of Business*, **63**, 511–524.

Madan, D. B., and K. Wang (2016a). Acceptability bounds for forward starting options using disciplined convex programming. *Journal of Risk*, **19**, 1–13.

Madan, D. B., and K. Wang (2016b). Non-random price movements. *Finance Research Letters*, **17**, 103–109.

Madan, D. B., and K. Wang (2017). Asymmetries in financial returns. *International Journal of Financial Engineering*, **4**, 1750045. https://doi.org/10.1142/S2424786317500451

Madan, D. B., and K. Wang (2020a). Signed infinitely divisible signed probability models in finance. SSRN paper no. 3489946.

Madan, D. B., and K. Wang (2020b). Correlated squared returns. SSRN paper no. 3560226.

Madan, D. B., and K. Wang (2020c). Multivariate bilateral gamma, copulas, co-skews and co-kurtosis. SSRN paper no. 3611853.

Madan D. B., and M. Yor, (2008). Representing the CGMY and Meixner–Lévy processes as time-changed Brownian motions. *Journal of Computational Finance*, **12**(1), 27–47.

Markowitz, H. M. (1991). Foundations of portfolio theory. *Journal of Finance*, **46**, 469–477.

McNeil, A., R. Frey, and P. Embrechts (2005), *Quantitative Risk Management: Concepts, Techniques and Tools*. Princeton University Press.

Merton, R. C. (1969). Lifetime portfolio selection under uncertainty. *Review of Economics and Statistics*, **51**, 247–257.

Merton, R. C. (1971). Optimum consumption and portfolio rules in a continuous time model. *Journal of Economic Theory*, **3**, 373–413.

Merton, R. C. (1973). Theory of rational option pricing. *Bell Journal of Economics and Management*, **4**, 141–183.

Milne, F. (1981). The firm's objective function as a collective choice problem. *Public Choice*, **37**, 473–486.

Minsky, H. M. (1982). *Inflation, Recession and Economic Policy*. Wheatsheaf.

Minsky, H. M. (1986). *Stablizing an Unstable Economy*. Yale University Press.

Modigliani, F., and M. Miller (1958). The cost of capital, corporate finance and the theory of investment. *American Economic Review*, **48**, 261–297.

Mossin, J. (1966). Equilibrium in a capital asset market. *Econometrica*, **34**, 768–783.

Murphy, J. (1999). *Technical Analysis of Financial Markets*. New York Institute of Finance.

Nakagawa, K., T. Uchida, and T. Aoshima (2018). Deep factor model: Explaining deep learning decisions for forecasting stock returns with layer-wise relevance propogation. ArXiv: 1810.01278v1[q-fin.ST]

Neufeld, A., and M. Nutz (2017). Nonlinear Lévy processes and their characteristics. *Transactions of the American Mathematical Society*, **369**, 69–95.

Ohlson, J. (1995). Earnings, book values and dividends in equity valuation. *Contemporary Accounting Research*, **11**(2), 661–687.

Peng, S. (1992). A generalized dynamic programming principle and Hamilton–Jacobi–Bellman equation. *Stoch. Stoch. Reports*, **38**, 119–134.

Peng, S. (2007). G-expectation, G-Brownian motion and related stochastic calculus of Itô type. In *Stochastic Analysis and Applications*, F. E. Benth, G. DiNunno, T. Lindstrom, B. Øksendal, and T. Zhang (eds.). Abel Symposia, vol. 2. Springer.

Peng, S. (2019). *Nonlinear Expectations and Stochastic Calculus under Uncertainty with Robust CLT and G-Brownian Motion*. Springer.

Philips, T. K. (1999). Why do valuation ratios forecast long-run equity? *Journal of Portfolio Management*, **25**, 39–44.

Pitman J., and M. Yor (2003). Infinitely divisible laws associated with hyperbolic functions. *Canadian Journal of Mathematics*, **55**, 292–330.

Pratt, J. W. (1992). Risk aversion in the small and in the large. In *Foundations of Insurance Economics*. J. W. Pratt and G. Dionne (eds.). Springer, pp. 83–98.

Preinreich, G. (1938). Annual survey of economic theory: The theory of depreciation. *Econometrica*, **6**, 219–241.

Protter, P. (2005), *Stochastic Integration and Differential Equations*, second edition, version 2.1. Springer.

Rasmussen, C. E., and C. K. I. Williams (2006). *Gaussian Processes for Machine Learning*. MIT Press.

Rockafellar, R. T., and S. Uryasev (2002). Conditional value-at-risk for general loss distributions. *Journal of Banking and Finance*, **26**, 1443–1471.

Rosazza, Gianin E. (2006). Risk measures via g-expectations. *Insurance, Mathematics and Economics*, **39**, 19–34.

Ross, S. A. (1976). The arbitrage theory of capital asset pricing. *Journal of Economic Theory*, **13**, 341–360.

Ross, S. A. (1978). A simple approach to the valuation of risky streams. *Journal of Business*, **51**, 453–475.

Ross, S. A. (2015). The recovery theorem. *Journal of Finance*, **70**, 615–648.

Rothschild, R., and J. Stiglitz (1990). Equilibrium in competitive insurance markets: An essay on the economics of imperfect information. *Quarterly Journal of Economics*, **90**, 629–649.

Royden, H. L., and F. M. Fitzpatrick (2010). *Real Analysis*. Prentice Hall.

Royer, M. (2006). Backward stochastic differential equations with jumps and related nonlinear expectations. *Stoch. Proc. Appl.*, **116**, 1358–1376.

Sato, K. (1991). Self-similar processes with independent increments. *Probability Theory and Related Fields*, **89**, 285–300.

Sato, K. (1999). *Lévy Processes and Infinitely Divisible Distributions*. Cambridge University Press.

Schoutens, W. (2003). *Lévy Processes in Finance: Pricing Financial Derivatives*. Wiley.

Schoutens W., and J. L. Teugels (1998). Lévy processes, polynomials and martingales. *Communications in Statistics: Stochastic Models*, **14**, 335–349.

Schroder, M., and C. Skiadas (1999). Optimal consumption and portfolio selection with stochastic differential utility. *Journal of Economic Theory*, **89**, 68–126.

Schweizer, M. (1995). Variance-optimal hedging in discrete time. *Operations Research*, **20**, 1–32.

Schweizer, M. (2005). A guided tour through quadratic hedging approaches. In *Handbooks in Mathematical Finance: Option Pricing, Interest Rates and Risk Management*, E. Jouini, J. Cvitanić, and M. Musiela (eds.). Cambridge University Press.

Sharpe, W. F. (1964). Capital asset prices: A theory of market equilibrium under conditions of risk. *Journal of Finance*, **19**, 423–442.

Skiadas, C. (2008). Dynamic portfolio selection and risk aversion. In *Handbooks in Operations Research and Management Science*, **15**, J. R. Birge and V. Linetsky (eds.). Elsevier, pp. 789–843.

Spiegeleer, J. D., D. B. Madan, S. Reyners, and W. Schoutens (2018). Machine learning for quantitative finance: Fast derivative pricing, hedging and fitting. *Quantitative Finance*, **18**, 1635–1643. https://doi.org/10.1080/14697688.2018.1495335

Staum, J. (2008). Incomplete markets. In *Handbook in Operations Research and Management Science*, **15**, J. R. Birge and V. Linetsky (eds.). Elsevier, pp. 511–563.

Stroock, D. W. (1975). Diffusion processes asociated with Lévy generators. *Probability Theory and Related Fields*, **32**, 209–214.

Stroock, D. W. (2003). *Markov Processes from K. Itô's Perspective*. Princeton University Press.

Stroock, D. W., and S. R. S. Varadhan (1979). *Multidimensional Diffusion Processes*. Springer.

Szegő, G. P. (2002). Measures of risk. *Journal of Banking and Finance*, **26**, 1253–1272.

Wang, S. S. (2000). A class of distortion operators for pricing financial and insurance risks. *Journal of Risk and Insurance*, **67**, 15–36.

Williams, J. (1938). *The Theory of Investment Value*. Harvard University Press.

Zhang, X. J. (2000). Conservative accounting and equity valuation. *Journal of Accounting and Economics*, **29**, 125–149.

Index

acceptable risk, 63
ADAM optimizer, 142
additive process, 102
αVG process, 49
alpha generation, 152
Anderson–Darling weighting, 32
arbitrage, absence of, 59
arrival rate function, $k_t(x)$, 5
asset pricing equations, 157
asset pricing model, 164

backward stochastic differential equation, 171, 204
backward stochastic partial integro-differential equation (BSPIDE), 174, 240
bandwidth, 238
bang bang control, 129
BGSSD model, 112
BGSSD risk-neutral parameters, 116
bilateral double gamma (BGG), 11
bilateral gamma (BG), 10
book of options, 173
Brownian motion, 99
  multidimensional, 49
Brownian motion, filtration, 99
bull spread, 136

certainty equivalents, 100
CGMY model, 14
characteristic function, 6
  Fourier inversion, 30
Choquet-type integral, 23
comonotone, 71
comonotone additive, 71
comonotone additivity, 203
compensated jump measure, 174
compensating measure, 150
complementary convex distortion, 68
complete market, 110
complex exponential variation, 16, 23, 30
  multivariate, 23
compound Poisson process, 154
concave distribution, 67
conditional distorted expectation, 111
  conservative, 115
conditional log likelihood, 237
conic hedging, 111
conjugate dual of a distortion, 79
conservative capital valuation, 196
conservative conditional distorted expectation, 115
conservative valuation of assets, 60, 65
conservative valuation, lower, 151
continuously compounded return, x_H, 5
copula, 29, 48
  Lévy, 48
  vine, 48
covariance functional, 102
cumulated wealth, 194

data centering, 33
deep learning functions, 143
delta hedge, 98
delta hedging, 90
delta neutral, 90
demeaning data, *see* data centering
depreferencing, 43
digital moment, 30
digital moment estimation, 31
distance measure, 225
distance quantiles, 192
distorted expectation, 61, 114
  conservative conditional, 115
distorted expectation, upper, 115
distorted least squares, 113
distorted least-squares hedging, 136
distorted valuation optimization, 100
distortion
  conjugate dual, 79
  maxminvar, 71
  maxvar, 71
  minmaxvar, 71
  minmaxvar2, 71
  minvar, 71

Wang, 71
distortion parameters, 88
distortion, complementary convex, 68
distortion, probability of, 67
distortions, 61
dollar vega level, 180
double gamma tilted bilateral double gamma, 43
double gamma tilted bilateral gamma, 43
driver function, 174
dynamic portfolio theory, 184

ε-insensitive optimization, 112
economic acceptability, 197
efficient frontier portfolio, 157
efficient frontiers, 152
eigen analysis of responses, 35
empirical cumulative distribution function, 31, 45
equilibrium return distribution, 256
equity capital, 196
expectation operator, dynamically consistent translation-invariant nonlinear, 204
expectations, distorted, 61
expected shortfall, 209
expected shortfall, CVAR, 210
expected utility, 39, 59
exponential kernel, 102
exponentially weighted average, 236
exposure design, 127

feedforward neural network, NNET, 185
filtered probability space, 173
finite variation process, 19
finite variation, process of, 7
finite-variation jump compensator, 173
Fourier inversion, 30
Fourier transform, 30

$\mathcal{G}$-expectation, 172
gamma density, 12, 57
gamma-distributed elliptical radius, 57
gamma-distributed elliptical radius distribution, 138
Gauss–Laguerre quadrature approximation, 41
Gaussian kernel, 238
Gaussian kernel smoother, 179
Gaussian process regression (GPR), 44, 102
generalized hyperbolic, GH, model, 14
generalized methods of moments, 246

Hellinger distance, 225
Hellinger-type distance, weighted, 251
Herfindahl concentration index, 192

implied volatility, 99
indifference pricing, 100
infinite variation process, 19
infinitely divisible, 19
integer-valued random measure, 173
integral of Choquet type, 23
iteration, 191

joint distributions, 150
joint Lévy measure, 144
joint Markov process, 178
jump compensator, finite variation, 173
jump measure, compensated, 174
jump process, pure, 7

kernel estimator, 237
kernel interpolator, 148
Kolmogorov–Smirnov distance, 225
Kullback–Leibler distance, 225
kurtosis, 110

law invariance, 203
least-squares hedge, 110
leveraged parameter, 221
Lévy compensator, 150
Lévy copula, 48
Lévy measure, 14
  generalized, 14
  joint, 144
  multivariate, 54, 138
  multivariate standard BG, m_{sbg}, 55
  risk-neutral, 88
Lévy process, 18
liability management, 135
life-cycle problem, 236
limit law, 19, 197
long-term solution, 185
lower conservative valuation, 151

machine learning, 236
marginal utilities of wealth, 184
market efficiency, 60
market valuation, nonlinear, 60
market value maximization, 59
market, complete, 110
Markov martingale process, 101
Markov process, joint, 178
Markov process, pure jump, 173
Markovian trading policy, 235
martingale model, 102
martingale representation theorem, 99
maximum entropy principle, 8
maximum likelihood estimation, 30
mean absolute deviation, 110, 112
mean variance, 100
mean variance hedging, 100
measure change, 40
measure-distorted value-maximizing hedging, 136
measure distortion, 73, 78
  risk charge, 86
measure-distorted valuation, 61, 127
  dynamic, 224

measure-distortion calibrations, 90
Meixner process, 15
moneyness, 39
Monte Carlo simulation, 20
multidimensional Brownian motion, 49
multidimensional exponential variation, 17
multidimensional tail probability, 167
multidimensional variations, 57
multivariate bilateral gamma, 52
multivariate complex exponential variation, 23
multivariate Lévy measure, 138
multivariate standard BG Lévy measure, $m_{\mathrm{sbg}}(x)$, 55
multivariate tail arrival function, 23
multivariate variance gamma component, 138
multivariate variance gamma, mvg, 49
myopic problem, 191
myopic solution, 185

negative required returns, 158
nonlinear integro-differential equation, 197
nonlinear partial integro-differential equation, 206
normal inverse Gaussian model, NIG, 13

optimal frontier portfolio, 157
optimal trading policy, 235
optimal trading, Markovian, 235
option pricing, 39

partial integro-differential equation, 189
pattern search, 140
penned forward, 107
policy function, 191
positive covariations, 144
power variation, 27
preference, 39
preference parameters, 44
premium, 85
probability density, $p_{t,t+H}(x)$, 5
probability distortions, 67
probability measures, supporting, 64
processes of independent increments, 102
prudential capital valuation, *see* conservative capital valuation
prudential valuation function, lower, 206
prudential valuation function, upper, 206
pure jump Markov process, 173
pure jump process, 7, 19

quantization, 140, 190
quantization algorithm, 35

random measure
  integer-valued, 173
rank correlation, 216
recovery theorem, 40
regulatory capital requirements, 224
reinforcement learning, 236
required regulatory capital, 196
required returns, 157
risk acceptability, 63
risk aversion
  degrees of, 88
risk charge gradient, 158
  zero, 157
risk charge, 66, 86
risk-neutral arrival rate, 133
risk-neutral distribution, 110
risk-neutral Lévy measure, 88
risky streams,valuation of, 59

Sato process, 112, 137
self-decomposable, 19
  infinitely divisible, 19
semi-variance, 110
semilinear partial integro-differential equation, 174
short gamma, 90
simultaneous perturbation, 146
skewness, 110
spatially inhomogeneity, 173
spot slide, 126
static arbitrage, 101
stationary policy, 188
stochastic exponential, 174
stochastic integral, 99
stochastic logarithm, 197
  signed, 197
stochastic portfolio theory, 186
stochastically domination, second-order, 141
strangles, delta, 98
structured products, 98
support vector machine regression (SVMR), 185
supporting probability measures, 64
symmetric variance gamma, svg, 9

tail measure, 16
tail probability, 30
  digital moment, 30
  multidimensional, 167
target exposure function, 116, 136
target quadratic variations, 144
term structure, 89
Tikonov regularization, 118
time inconsistency, 110
time reversal, 242
trading policy, optimal, 235
two-dimensional manifold, 256
two-price economy, 60

unleveraged parameter, 221
upper distorted expectation, 115
utility
  expected, 39, 59
  function, 39

valuation of assets, conservative, 60, 65
valuation operators, upper and lower, 65
valuation
  linearity of, 59
  lower, 65
  lower conservative, 151
  measure-distorted, 61
variance gamma
  multivariate, mvg, 49
  symmetric, 9
  vg, 10
variation measure, 25
  conditional, 25
variation outcome, 22
variation space, 22
vine copula, 48
viscosity solution, 172

Wasserstein distance, 225

zero risk charge gradient, 157
zero-theta exposures, 128

Printed in the United States
by Baker & Taylor Publisher Services